MEDWAY PUBLIC LIBRARY
3 1852 00105 2270

P9-DMO-961

Seinfeld

Also by Jerry Oppenheimer

State of a Union: Inside the Complex Marriage of Bill and Hillary Clinton

Just Desserts: Martha Stewart, The Unauthorized Biography

The Other Mrs. Kennedy: Ethel Skakel Kennedy, An American Drama of Power, Privilege, and Politics

Barbara Walters: An Unauthorized Biography

Idol: Rock Hudson, The True Story of an American Film Hero

Seinfeld

The Making of an American Icon

Jerry Oppenheimer

HarperCollins*Publishers*

HarperCollins books may be purchased for educational, business, or sales promotional use. For information, please write: Special Markets Department, HarperCollins Publishers Inc., 10 East 53rd Street, New York, NY 10022.

FIRST EDITION

Designed by Sarah Gubkin

Photo editor: Vincent Virga

Printed on acid-free paper

Library of Congress Cataloging-in-Publication Data

Oppenheimer, Jerry.
Seinfeld / Jerry Oppenheimer.
p. cm.
Includes index.
ISBN 0-06-018872-3
1. Seinfeld, Jerry. 2. Comedians—United States—Biography. 3. Television actors and actresses—United States—Biography. I. Title

PN2287.S345 O65 2002
791.45'028'092—dc21
[B] 2002068723

02 03 04 05 06 ❖/RRD 10 9 8 7 6 5 4 3 2 1

For Caroline and Cuco

Contents

"Stars can succeed by concealing who they are.
Comedians can't."

—Jerry Seinfeld

Prologue

His city—the New York City he introduced to middle America; the city whose foibles and characters and neurosis and buzzwords have become part of the national lexicon and consciousness; the city everyone thought they knew from watching his show; his city of movie theater lines and Chinese restaurant waiting lists and gymnasium saunas and fascist soup stand proprietors and inane coffee shop booth chatter and parking space battles and oddball neighbors and eccentric bosses and *yada, yada, yada* about nothing; his city, which was as much a character of his show as the players themselves—was burning on that sparkling Tuesday, September 11, 2001.

Jerry had awakened early, but it was still dark where he was, which was some fifteen hundred miles west of, and two time zones earlier than, what would become known as Ground Zero. Like so much of his adult life, Jerry was on the road, in a hotel room that horrific morning, in a suite at the Charter at Beaver Creek, a luxurious Rocky Mountains getaway, in Colorado.

It was shortly after 6:30 A.M., mountain daylight time, and he had requested an early wake-up call because he was scheduled to take the wheel of one of his $300,000 classic '50s Porsche Speedsters and participate with eighty-five other high-rollers in the Colorado Grand Vintage Car Rally, a charity event. In the past, most of the money had been raised for HIV patients; this year it was for research and treatment of bipolar manic depression. At a pre-rally brunch, the watch company Girard-Perregaux auctioned off for $20,000 one of its vintage chronographs—the kind of timepiece Jerry likes to sport on his wrist—for the cause.

After showering, washing his hair, flossing and brushing, Jerry automatically reached for the remote, tuning to his old network, NBC, and the *Today* show, to find out what, if anything, was happening that might impact his very narrow and focused world. Stunned and shocked, he watched when the twin towers were hit, sat paralyzed when they came tumbling down.

"That," he later intoned, "was the end of the old world."

Unlike the years when he suffered from fear of commitment and had no one but himself to think about, Jerry now had familial responsibilities: nearing fifty, a bit paunchy in the face, his hair shorter, but boyish looking, like an aging Jerry Lewis, he had a young wife—who would turn thirty the next day, September 12—and a ten-month-old daughter, waiting for him in New York. Indeed, the sneak attack had a very direct, profound, and frightening impact on America's most beloved comic.

He immediately tried to call home, but the lines were already fried by millions of other concerned callers; he ordered the front desk to continue trying, as he attempted unsuccessfully to get through on his cell phone, but all service was disrupted by the unprecedented assault and national panic.

Some thirty minutes north by limo from the World Trade Center was Jerry's nineteenth-floor palatial duplex, purchased from the violinist Isaac Stern in the tony Beresford, an elegant aerie overlooking Central Park.

Jerry had spent tens of thousands to renovate his new home beyond the $4 million and change he had paid for it, and thousands more in fines for exceeding the year-long renovation period tenants are allowed, all of which was chump change to the one-time poor boy from blue-collar Long Island who had become very wealthy from being very funny. Just weeks before, the Seinfelds had finally moved in to their sleek home, having previously stayed in a backup apartment nearby, or at his $32 million East Hampton estate. Not far from his new apartment was a $1.39 million state-of-the-art garage undergoing renovation for twenty of his valuable Porsches.

For many years, Jerry had lived in predominantly gay West Hollywood, not that there's anything wrong with that, and had considered living in Greenwich Village, another gay bastion, before he moved back to New York after *Seinfeld* ended its run and he had met his future wife. He chose the Upper West Side—the setting for his show—a community of young families and singles, after asking himself, "Which is more annoying to me, babies or homosexuals?"

Because he was in Colorado, Jerry apparently had no plans to be home to celebrate the milestone birthday of his wife, Jessica, whose relationship and marriage caused a scandal of seismic scale when she dumped her first wealthy husband within days of their summer 1998 honeymoon for the wealthier and more famous comedian. Now, with the Seinfelds' marriage going on two years, they had a show-business marriage: though a loving father and devoted husband, Jerry often traveled—his lifelong mistress was stand-up, and now that *Seinfeld* was over he was back on the road, hoping to fulfill a dream, being the best stand-up ever, and doing his shtick on the stage of the Palladium in London when he reached a hundred.

Luckily for both of them, the former Jessica Sklar Nederlander was able to balance her husband's dual lives—the professional and the domestic. Even if he was just playing with one of his cars in Colorado on her birthday accompanied by a boyhood pal, she thrived on being Mrs. Jerry Seinfeld, on seeing her name in the columns. If he ever were to get married, Jerry promised himself years earlier, it would be to a woman who was Jewish and who was willing to happily share him with his first love. Jessica filled the role perfectly.

With a nanny for the baby—the servant had her own $2,000-a-month apartment paid for by Jerry—Mrs. Seinfeld started a charity for needy children, and was free to hang out with her girlfriends, and to shop. If anything, the once-poor girl from upstate Vermont had developed into a world-class shopper, buying only the best for herself and baby Sascha.

When word of the attack reached the hotel in Colorado, a number of New Yorkers in the rally immediately drove their collector cars east; they had no choice because all planes were grounded for the duration of the emergency. Jerry didn't return to New York until the end of the week, when he was able to charter a private plane. On arrival, he felt as though he had walked into a room "where somebody had just died," he said later. "The quality of the air had changed. There was a density to it."

Back home, Jerry followed in the footsteps of other great funnymen, childhood comedic idols—the Bob Hopes, the Abbott and Costellos—

who rallied for their nation during previous times of war: he went on the road to entertain a shaken citizenry. In Charlotte, North Carolina, he did two shows for five thousand of the anguished who responded positively, needing the laughter, forgetting for an hour or so the horror, to deal with what Jerry saw as the new world.

"In a way it identifies us as Americans," Jerry said, after feeling the audiences' reaction, ". . . what I would call classic American behavior, an act of defiance and a way of striking back. Ironic humor . . . people coming back with a vengeance, returning to our lives with great glee."

About a month after the attack, he was at the forefront of a comedy fund-raiser at Carnegie Hall, with the proceeds going to the Twin Towers Fund, the relief charity established by Mayor Giuliani for the families of the police and firemen who had perished, raising $1,859,400. Among the comics he enlisted was Bill Cosby, whose brilliant comedy albums inspired Jerry as a kid in the 60s to become a stand-up. He also brought on board his longtime friend, George Wallace, a black stand-up comic, who had been Jerry's roommate, confidant, and the best man at his wedding.

Jerry was overjoyed with the help he was able to give his fellow Americans. However, the state of the world and its impact on the future of his daughter plagued him. Thinking about his own personal struggles and triumphs—much kept hidden from scrutiny by this most private of public figures, he said somewhat cryptically, "The world that I grew up in had some problems, too."

Seinfeld

—1—

Family Roots

If Jerry inherited from anyone his remarkable sense of humor, his singular drive, his uncommon focus, his tenacious entrepreneurial spirit, his behemoth ambition, his unbridled optimism, his unflinching confidence, and even the luck of being in the right place at the right time—the ingredients that combined to make him so successful—it was from his father, Kalmen Seinfeld.

A diminutive, sprightly door-to-door peddler who started huckstering fake holy water from Lourdes and ended hustling commercial signs, Kal Seinfeld was a street-smart salesman who used shtick to make friends, to influence people, and to weather adversity.

Indeed, it was tragedy and misfortune that greeted Kal when he came into the world just as the Great War in Europe was ending. His mother, Celia, died while giving birth to him, her one and only son, on October 20, 1918, in the family's tenement apartment in the Borough Park section of Brooklyn. Jerry's paternal grandmother had been weakened by the virulent influenza that was killing millions worldwide; she was one of forty thousand Americans who died that week, at the height of the epidemic in the United States.

Despite Celia Seinfeld's sad and untimely end, the baby, Kal, though tiny and scrawny, was healthy and kicking, with a shock of blond hair, eyes that were clear and blue, and with fair skin, more Irish than Jewish in appearance. He was the third of the Seinfelds' children; two daughters, Jennie and Jetti (called Yetta), had preceded him by a few years.

For Kal's father, Samuel Seinfeld, Jerry's paternal grandfather, the birth of a son and the death of a wife were both a blessing and a curse.

Sam Seinfeld had come to America as a boy, from Poland, during the great wave of Jewish immigration in the late 1800s to escape pogroms and repression. But he had no reason to be happy with his life in the new land.

From morning until night, except for the Sabbath when he went to shul to pray for better times, Seinfeld schlepped a pushcart through the teeming streets of Brooklyn and the Lower East Side, hawking halibut and flounder for a more business-minded brother who had a stand in the Fulton Fish Market.

Now, with the sudden death of his wife, life became even more difficult and complex for Seinfeld the fishmonger. As the family breadwinner, he faced a formidable dilemma: who would care for his infant son and toddler daughters, and for him, too?

Enter a new woman, Lena Bernstein, a stocky brunette immigrant from Riga, Latvia, a much contested region in Eastern Europe, whom Seinfeld married about a year after Celia's death.

Moving to another apartment in predominately Jewish Borough Park, Seinfeld started a new family with Lena, who was handed the burden of raising his first three—Kal, Jennie, and Yetta. Along with that formidable task came two tragic births: Lena's firstborn son with Seinfeld, Edwin, was crippled by polio while still a baby, and Lena's first daughter, Frieda, died in infancy of a gastrointestinal illness. However, two other healthy children followed—a daughter, Toby, born in 1923, and a son, Harold, in 1926, the last of Sam Seinfeld's second brood.

Years later Toby Seinfeld Rosenfeld, Jerry's half-aunt, expressed the feeling that her father had probably married her mother so that he wouldn't have to raise his first three children alone, and for no other reason—certainly not love. "He married my mother and she took care of his kids," she stated bluntly. "It was stupid of her. I say marry a man with little babies and you're a single person all the time, I mean that's my way of thinking."

Like Wall Street, the Seinfeld marriage eventually laid an egg. Around the time of the stock market crash and the beginning of the Depression, Seinfeld—remembered as being a cold, stern, and abusive man, emotionally and physically—flew the coop. By that time, his daughters were in their teens, and Kal was coming into adolescence. Lena had done the job for which Sam had married her: she'd gotten his kids through their difficult childhood years, and now, it was apparent, she was expendable.

"After he left we grew up *very* poor," Toby Rosenfeld recalled ruefully. "We were forced to go on welfare. Today they would have schlepped him in [for abandonment]. He got away with it in those days."

Ever since he became a performer in the late 1970s, Jerry has remained extremely private about his family roots—along with his own personal life. But he did acknowledge what he described as the "severe environment" in which his father grew up. "My grandfather," he stated firmly, "was not a good man."

If Sam Seinfeld had been a character in his grandson's TV series—and members of the Seinfeld clan did become models for a few zanies on the show—Jerry might have dubbed him the "Fish Nazi" because of his cruel persona and difficult nature.

After the Seinfelds' marriage ended, Lena, all but penniless, had another brief, unhappy marriage with a man named Deitz, a stormy union that also ended in divorce. For a time Sam Seinfeld, who eventually took a third wife, came to visit the children in the Bronx where Lena also had moved from Brooklyn. "He used to come over on Jewish holidays once in a while and put his hand on my head and say a prayer," Rosenfeld remembered. "I felt uncomfortable with him. I didn't know what type of person he was."

The parental split also sparked an ugly rift between both sets of Seinfeld children—Kal, and his sisters on the one side; Toby and Harold on the other. Their crippled brother, Edwin, was financially cared for, in part, by his father, but often spent time with other families who had summer homes at the ocean in Rockaway, where the air was better for the sickly boy.

For the Seinfeld children there would be a lifetime of icy silence interspersed with bouts of bickering and jealousy.

"Everyone was kind of broken up," Harold Seinfeld acknowledged years later. "We never got together too much. We lived apart."

His sister, Toby, agreed. "I tried to get close to them, but it didn't work out. They seemed to be like in a clique. They were *very, very* distant. We were like strangers. There was a lot of sadness in the Seinfeld family."

Even by the late 1990s bad blood still existed. When asked what she thought about the oceanfront palace in East Hampton that her nephew Jerry had bought from the singer Billy Joel with some of the riches from his hit television series, the widowed Toby, a mother and grandmother who had lived for years in a Long Island subdivision, voiced her indignation: "He can keep it! I have no interest in them. They were all cold to me, so I don't care. Jennie [Seinfeld Bloom], my half-sister, started talking to me. But all those years, trying to get warm to her, trying to get to be like one of the family, it never materialized."

Despite their differences, however, she believed sincerely that her half-brother, Kal, was a mensch. "He was a very nice person, and was *always* like that—very funny, easy, happy-go-lucky. If he could have done it, he would have made out just like Jerry. He was a comedian, and that's where Jerry got it from."

Worse than the chasm between the two sets of Seinfeld children was the schism that exploded between Sam Seinfeld and his first set of children. Family observers attributed the problems to the father's abusiveness. By the time Kal and his sisters were in their mid-to-late teens, they had fled from their father's grip to live on their own.

"All the kids left," Jerry has said of his father's family. "They lived by themselves. They would work, sell stuff on the street."

Jerry's half-uncle, Harold Seinfeld, explained that they had moved away from home "because of father problems, problems I don't want to talk about. Kal and his two older sisters, Jennie and Yetta, they lived together because of this disagreement with our father. There was problems with the family. That's about it."

Having led a tumultuous, trying life, Lena Bernstein Seinfeld Deitz died in 1939, a few years before Sam Seinfeld passed away following a massive stroke.

Looked after by his sisters, virtual surrogate mothers, scrawny Kal Seinfeld did what any resourceful boy in his position would do in those days—he took to the streets of Brooklyn working odd jobs to help with the siblings' self-imposed ragamuffin existence. By the time he was thirteen, always fascinated by things mechanical, Kal was driving and had learned to fix car engines to bring in much-needed money.

"They were all very responsible, looking out for each other," Harold Seinfeld observed.

Despite their hardships, the offspring of Sam Seinfeld managed to survive—a tough little band with an overabundance of grit, spirit, and independence. The women—Jennie, Yetta, and Toby—found hardworking, middle-class husbands, the most colorful and controversial of whom was said to be Yetta's. She also had the most education, became a teacher, and lived during the late 1930s and early 1940s in Havana, Cuba. Her husband, Melech, is remembered as a writer for Jewish publications, a Marxist, and a nudist of sorts. "He worked for the *Forward*," said Toby Rosenfeld. "Her husband was a communist, that's what I heard." And in what could be a scene ready-made for an episode of *Seinfeld*, Rosenfeld clearly recalled a visit many years ago to her half-sister's home in Florida, where "I saw her husband sitting in the kitchen, by the window, reading the newspaper. He was naked—*stark naked*!"

Except for Kal, the Seinfeld boys led rather mundane, lonely lives. The polio-stricken Edwin, forced to walk with braces on both legs and with the aid of a cane, never married and lived his entire adult life in a small apartment in lower Manhattan, where he worked as a bookbinder. "When he got older," his sister, Toby, recounted, "he tripped on a crack in the sidewalk and that, along with arthritis, really put him down, and he couldn't work no more and had to go on pension."

The other Seinfeld scion, Harold, also remained a lifelong bachelor, living alone in Brooklyn. "For most of Harold's life," his sister, Toby, noted, "he worked for the post office as a carrier. He used to get on the subway train different places, stop off, and deliver mail."

Harold's only claim to fame was as the apparent model for Jerry's rotund, stolid *Seinfeld* neighbor and nemesis, the postman whom Jerry greeted disdainfully with the words "Hel-lo, *New*man!"

—2—

Tojo, Television, and Holy Water

Perhaps due to his coming from a broken home, Kal Seinfeld's formal education was a disaster, on a par with the revolving doors at Macy's. Constantly in and out of schools, he finally and permanently dropped out of Washington Irving High School in Manhattan in June 1940, after just two years; at the age of twenty-one, he conceivably held the record as the oldest sophomore in the history of the New York City public school system. For Seinfeld, though, scoring money, not grades, was primary; he could read and write and that was enough. After dropping out, he bounced through a series of menial jobs—typist, auto mechanic, truck driver.

In his early twenties, living on his own in public housing in the shadow of the Williamsburg Bridge on the Lower East Side, he finally secured a position with some promise—an $18-a-week job, the most he had ever earned—as a route man for Ben Shinkman's Lincoln Towel Company on West Fortieth Street in midtown Manhattan. The job involved delivering towels to hotels and businesses in a panel truck. The boss liked the young man's sense of humor, spunk, his *sechel*, and before long he gave him a shot as a salesman.

Kal had finally found his calling, successfully hustling new business and earning commissions for his hard work. But his career was swiftly cut short by the flames of World War II.

On March 4, 1941, with Rommel's Afrika Korps in Tripoli, with the Nazis set to invade Yugoslavia, and with Premier Tojo secretly planning Japan's December 7 strike on Pearl Harbor, the twenty-two-year-old rookie salesman was called to arms, reporting to the induction center at Fort Dix, New Jersey, and becoming U.S. Army Private Seinfeld, Serial Number 32-010-413.

Standing just 5'5½" in his stocking feet, and weighing a bantam 117 pounds, Kal Seinfeld was no John Wayne. But he was as patriotic and gung-ho about winning the war as the millions of other boys who went off to serve their country.

After basic training, where he qualified as a marksman with the M-1 rifle, the lowest ranking on the range despite twenty-twenty vision, Kal went to Fort Jackson, in South Carolina, for eight weeks of radio-telegraph school. But on December 16, 1941, nine days after Pearl Harbor, America's entry into the war, Kal shipped out to the bloody South Pacific. His first duty station was the Japanese-devastated Schofield Barracks, in Hawaii. After that, he served in Hollandia, New Guinea; Leyte, Philippines; and Rock Hampton, Australia.

During his first two months in the Asiatic Pacific Theater, Seinfeld was an ammunition bearer, carrying bullets and shells and keeping records of the supply and consumption of ammo. But for more than three years his military occupation was that of a mail orderly, safely ensconced behind the front lines, classifying and separating mail for the three hundred men of his unit—Company D, 34th Infantry, 24th Division.

In his four years, five months, and nine days of service, Seinfeld never received a promotion above the rank of Private First Class, nor had he ever received a wound in the war, although he did suffer from a variety of ailments: jungle rot; a first-degree stomach and thigh burn caused by sunbathing aboard a ship; constipation; prickly heat; bronchitis (he smoked two packs a day); and the loss of four teeth and fillings due to the mineral effect of the water he drank while stationed in Australia.

Seinfeld arrived back on U.S. soil on February 1, 1945, and became a civilian once again on August 12, 1945, the same week President Truman ordered the atomic bomb dropped on Hiroshima and Nagasaki. Thankfully, he had come home in one piece, honorably discharged, and proudly wearing the Asiatic Pacific Service Medal, the Good Conduct Medal, the Philippines Liberation Ribbon, and the Combat Infantry Badge.

The immediate postwar years were seemingly as dismal for Kal Seinfeld as those that preceded his military service. What with all the other returning servicemen, he was unable to find a decent job, and had difficulty coping with civilian life. He was forced to live on Veteran's Administration unemployment benefits and subsistence-and-readjustment allowances.

A new technology, however, sparked the floundering GI's interest, one from which he hoped he could make a steady buck. It was called television. In the postwar recovery years, TV was on the verge of a major boom. The next decade, the Fifties, only a few years away, would become known as the golden age of television, when movies, radio, Monopoly, and virtually every other form of family entertainment took a back seat to the tube—Uncle Miltie, Red Skelton, Jack Benny, George Burns and Gracie Allen, were now seen and not just heard, albeit on a small, fuzzy, black-and-white screen. A visionary of sorts, Seinfeld believed correctly that TV would be the next big thing.

But despite his comedic flair, he didn't see himself as a performer. Nor did Seinfeld have an interest in the complexities of television production. Rather, he saw himself as the repairman of the small-screen Crosleys and Admirals that young families were buying on credit in ever-increasing numbers as they moved from Brooklyn, the Bronx, and Queens to the tract houses that were sprouting on potato fields on Long Island and elsewhere around the country.

Thus, with a whopping $65-a-month in VA benefits in hand, Seinfeld enrolled at the RCA Institutes, Inc., in Manhattan, to take one of the first courses ever offered in television servicing.

But three months later he abandoned his dream of becoming the proprietor of a business he had tentatively called "Seinfeld TV Repair—You Watch 'Em, I Fix 'Em." Bored with classroom studies about the intricacies of replacing tubes and fuses, he immediately quit the class when a friend of a friend offered him a commission sales job peddling shingles and siding door-to-door, à la Danny DeVito in *Tin Men*.

A born huckster and kibitzer, Seinfeld jumped at the opportunity. Jovial, friendly, with a wide grin, he loved meeting new people, adored schmoozing, delighted in being on the road and not confined to a classroom or an office. Housewives and blue-collar workers, from Queens to as far away as Montgomery, Alabama, where door-to-door sales jobs took him, fell for his Brooklynese fast-talk, lighthearted banter, corny puns, and semihilarious knock-knock-who's-there-style jokes as he sweet-talked and conned them into placing an order.

He felt he could sell anything—and he did in the early years. From shingles and siding, he went on to pitch encyclopedias, pots and pans, Fuller brushes, storm doors and windows. And he sold items that weren't entirely kosher.

"Kal was basically *very* entrepreneurial," recalled a longtime friend, Mike Wichter, an electrical contractor, who got a kick out of Seinfeld's salesmanship and still chuckled about it years later. "Kal got the idea of selling miniature grottoes with water that he said was from Lourdes, where people come to pray to the Virgin Mary. Now whether this was just plain tap water or not, people bought them and hung them on their walls. There was a little pump inside to make the water trickle and flow, and the customers thought this was very pretty and religious. Kal was doing great right up until the time the things developed a leak. And that was the end of that. He went on to something else.

"Kal knew exactly how to do business. He taught me how to be a good businessman. His biggest tip was, 'Go out and sell as much as you can, and then stay with the one that's going to make you the most money.' Now if that isn't good advice, I don't know what is. His son took it, too. Whatever Jerry Seinfeld is today he got from his father."

—3—

And Betty Makes Two

For years Kal avoided any romantic entanglements and was still a bachelor at thirty-two when, around 1950, he met the woman he would marry. Though his joke-a-minute style was impishly charming, Kal was by all accounts simple and unsophisticated, not a ladies' man in looks, demeanor, or style. Nor did he have the time and money to chase women, let alone become involved with one. Since boyhood, he had focused solely on survival, pursuing whatever get-rich-quick scheme came down the pike. On the road as a salesman, he had no roots, no home base, no complications, no commitments, and that was the kind of minimalist life he preferred. He felt secure with his independence, with not having to be responsible for, or beholden to, anyone.

Years later his one and only son would follow the same path.

But in 1950 Kal fell for Betty Hesney, a petite, buxom, olive-complexioned, brunette bookkeeper four years his senior, and his attitude toward settling down did a one-eighty. Like Kal, Betty had only basic schooling, an easygoing nature, liked to laugh, was shrewd, street-smart, and Jewish. They were a perfect match, their bond made more intense by the fact that Betty had also had an extremely difficult childhood, in some respects even more severe and touching than Kal's.

The story of Betty's early years are sealed in social service records, but what has surfaced is that she was born December 12, 1914, four months after the start of World War I, to Syrian parents who had immigrated to the United States from Damascus and Aleppo, Jewry's oldest diaspora communities. For years Jews in Syria lived in fear, their quarter under surveillance by secret police. Relations with foreigners were watched closely, and travel was limited. Syrian Jews were stereotyped as clannish, suspicious, secretive, frugal, and covetous. But the Hesneys, according to family lore, fled Syria for reasons other than governmental persecution or anti-Semitism.

Betty's father is said to have been banished from his family, but the facts behind that story are now long-buried. Jerry once gave scant details, asserting only that his maternal grandfather "was ostracized from the family in Syria, so they came to America." In fact, Syrian Jews are usually ostracized from their community by the rabbinate, which is tantamount to a social death sentence, and is more likely why Hesney fled his homeland.

Whatever the reason—familial conflict or rabbinical banishment—the Hesneys settled into the densely populated, highly observant, well-established Syrian Jewish community in the Bensonhurst section of Brooklyn—just blocks from Borough Park where Kal grew up and where he returned after the war. Sometime after the Hesneys arrived in America, tragedy struck. Betty's mother died suddenly—the cause of her death is unknown today—and her husband fell apart. Without means to support Betty, an older sister, and a brother, Jerry's grandfather was forced to give up custody of Betty, and she was raised in an orphanage and a succession of foster homes.

Seinfeld family friend Ruth Goldman, in whom Betty had confided details of her difficult childhood, noted, "There was no one there to help her, and she had to grow up on her own, make her own decisions, and keep everything within herself, because these people were not her parents."

Later, Betty moved in with her sister and brother-in-law who lived in Bensonhurst, where Kal met her.

Though Betty's Middle Eastern background was far more exotic than Kal's Eastern European roots, she readily adjusted to American customs and assimilated easily. However, some of her Syrian Jewish convictions were ingrained, and at least one would impact her son: from the time he was a boy, Jerry was warned that he should not marry a gentile woman, or even a convert to Judaism, an edict that negatively affected his relationships with women his mother referred to as shikses.

Because of the hardships his parents endured, Jerry once observed, they had "developed a lot of spine growing up . . . they hit it off because they realized they had this common background."

Kal and Betty courted for about a year. In 1951, they tied the knot in a simple ceremony, and she moved into his tiny walk-up apartment at 1874 Fifty-second Street in Borough Park, near the Chinese take-outs, bagel shops, and kosher butchers on Ditmas Avenue and Ocean Parkway—a noisy, crowded section that stood in the shadow of the dreary elevated tracks of the Long Island Railroad, with the constant rumble of the IND subway line below.

Carrying a sample case, Kal escaped most days to go on the road, sometimes hopping the Montauk branch of the LIRR to another world—green, sweet-smelling, developing suburban towns on Long Island, with lyrical names—Valley Stream, Lynbrook, Merrick, Seaford, Massapequa. Out there, a short ride from Brooklyn, he found a thriving market among the first-generation, blue-collar Irish- and Italian-American postwar refugees from the boroughs of New York—the cops, the carpenters, the electricians—for his faux holy water, crucifixes, bibles, statues of the Virgin Mary, portraits of Jesus, and other religious tchotchkes that he bought from a wholesaler in the city—adding a hundred percent or more markup.

With his blondish hair, mustache, and twinkly blue eyes, he often got away with using an Irish brogue or saying his mama was Italian and his father Irish, in order to appear convincing to his goyish customers and make a sale. Who, he figured rightly, would buy a crucifix from a crafty little Brooklyn Jew? Kal was a lovable con artist, which Jerry has acknowledged. "He was very supportive of the scamming type of lifestyle that a comedian would have, because he was a salesman."

—4—

Near the Mall

Betty Seinfeld's biological clock was running down when she married Kal. But coming from a broken family, and having been bounced from orphanages to foster homes, she desperately wanted children—a stable family to call her own.

After the Seinfelds married she became pregnant for the first time in her late thirties, and after a difficult pregnancy, Carolyn Seinfeld was born in 1952.

Soon after, Betty became pregnant with her second and last child, and on April 29, 1954, after another difficult pregnancy, Jerome Seinfeld came into the world, in Brooklyn, happy and healthy.

Jerry was three weeks shy of fourteen months when, on a steamy August day in 1955, the Seinfelds moved from their crowded apartment in Brooklyn into a modest, newly constructed split-level at 50 Eastgate Road, in a subdivision of cookie-cutter homes called Hamilton Park in the village of Massapequa on Long Island, a half-hour by train but a world away from the teeming city.

The village was named "great water land" by its first inhabitants, the Marsapaque Indians. In 1653, troops led by a British army captain named John Underhill, who was working for the Dutch, battled with the peaceful native Americans, and history books say more than a hundred were killed. An enormous statue of an Indian chief is one of the town's landmarks.

From the turn of the century until the Roaring Twenties, Massapequa was considered a resort because the best part of the town overlooks sparkling, blue South Oyster Bay, a short trip across the water to famed Jones Beach on the Atlantic. Celebrities often visited or vacationed in Massapequa—from movie cowgirl Annie Oakley to the dance team of Irene and Vernon Castle to the Duke of Windsor who played polo there. Many of the wealthy WASPs who summered in Massapequa restricted Jews and other minorities from their exclusive enclaves.

But in the postwar years when the developers arrived and slapped up tacky housing and shopping centers, Massapequa became part of Long Island's suburban sprawl, attracting working-class Italians and Jews like the Seinfelds, who had come for their piece of the American dream, and Massapequa was dubbed "matzoh-pizza" by the new arrivals. Jerry, who injected autobiographical observations into his stand-up act, famously quipped that Massapequa was Indian for "near the mall."

A conservative, Republican stronghold, the hardworking, taxpaying, churchgoing citizens are proud of the kind words President Ronald Reagan bestowed upon them during his tenure in the White House. He eloquently described Massapequa—home to such tabloid icons as Long Island Lolita's paramour Joey Buttafuoco; televangelist temptress Jessica Hahn, history-making male-to-female sex-change diva Christine Jorgenson, and "Boss of all Bosses" Carlo Gambino—as "quintessential suburban America."

Struggling financially, Kal and Betty became home owners with a no-money down, thirty-year $10,700 VA loan. With little money or taste, the Seinfelds furnished their place à la Sears.

The house where Jerry spent the next seven years of his life was advertised by the contractor, Eastgate Builders, Inc., as a "front-to-back split," presumably because the three small bedrooms and one bath were six steps up and in the back, and the recreation room was six steps down from the living room and dinette kitchen in the front.

About six months after the Seinfelds moved in, the Katzes—Norman, his wife, Frances, and their three-and-a-half-year-old daughter—bought the new split-level next door. "Most of us were struggling Jewish families," Frances Katz observed. "It was difficult making ends meet, but we enjoyed it because we were young and because it was brand new, and it was a nice neighborhood."

While many of the newcomers to 'Pequa, as the town was also known, felt isolated in their new suburban environment, Kal was lucky to have family just two doors away. His sister, Jennie, long a comforting shoulder to lean on during rough times, bought a similar split-level at the same time with her husband, Manny Bloom, a stockbroker. "I used to kid around we were the sandwich in the middle between their two houses," Norman Katz said. The Seinfelds and the Blooms were a tight-knit, supportive, extended family; Jerry and his sister, Carolyn, and the Blooms' daughters were more like siblings than first cousins.

Like his father, Jerry felt enormous affection for his aunt Jennie who was very outdoorsy, a Birkenstock kind of woman who was into organic food and drove a Saab. "He just thought she was the beginning and the end of the world in a lot of ways," recalled Caryn Trager, Jerry's first serious girlfriend. "She was a tough, strong woman."

The Seinfelds settled easily into suburban life—playing poker, canasta, and mahjong with the neighbors, gardening, doing little do-it-yourself projects on their postage-stamp property. It was the mid-Fifties, the Wonder Bread-and-mayonnaise Eisenhower years—a time of Cold War, McCarthyism, and the emergence of a new music form called rock 'n' roll. It was an era when Massapequans, like the rest of America, ate just-introduced Swanson frozen TV dinners and watched Luci and Desi, Uncle Miltie, Klem Kadiddlehopper, and what NBC billed as a "sofa-and-desk" program starring a riotous hipster named Steve Allen. A quarter-century and three hosts later, the *Tonight Show* launched Jerry Seinfeld's career.

The Seinfelds joined a new conservative synagogue, Temple Beth Shalom, in nearby Amityville. Betty was inducted into the sisterhood for social reasons, and Kal found an audience for his jokes. But congregants noted that the Seinfelds didn't usually attend religious services as a family.

"Kal was an outgoing guy, so he liked to go to the temple and schmooze," Norman Katz explained. "When one of the guys who used to blow the shofar quit, Kal picked it up and started fooling around with it, and the next thing you know he's blowing the shofar on the holidays."

Rabbi Leon Spielman, who joined the temple at its founding in 1955 and was still active almost a half-century later, remembered Kal as a major-league ham. "He thought he was a comedian—and sometimes he was," the rabbi stated fondly. "I used to have the system of offering the officers—Kal became a vice president—the opportunity to make announcements on each of the high holy days. One Kol Nidre night, it was Kal's opportunity to make a straightforward announcement—'services tomorrow'—nothing funny. At that moment a truck passed on the street in front of our temple and backfired, and Kal said, 'If he were Jewish he'd be in here.' Someone who was put off by Kal's attempt at humor came running down the aisle and yelled, 'If we want funny stories, we'll invite Milton Berle!'

"When Eisenhower was running for president they had a button that said, 'I Like Ike.' If they had button for Kal Seinfeld, it would have said, 'I Like a Mike.' "

"Some people got annoyed," added the rabbi's wife. "If the rabbi gave Kal the opportunity to speak, he'd never get finished. He'd stand there saying one thing after another. He loved the mike. He'd come to a meeting and people would argue, get excited, but not Kal. If someone stood up and said 'the temple owes money,' he'd say, 'Oh, really? Hmmm. Is that so?' It would almost agitate you; you'd be so upset and he'd be saying calmly, 'Oh, really? Is that so?' Like, 'It's not my problem.' He always had a light touch and was a wonderful stabilizing factor in the congregation."

Besides having Jerry Seinfeld in its Hebrew school in the 1960s, the temple had one other claim to fame—its close proximity to the real-life murder house that became the basis for a hokey 1970s best-selling book and hit movie called *The Amityville Horror*, which dealt with the supernatural.

With little Jerry in tow, Betty shopped with the other young mothers at the Foodtown in one of the shopping centers, or at a market appropriately named Two Guys From Massapequa. When the kids were playing in the Seinfelds' small backyard—Jerry's forte was a pail and shovel in a sandbox—she and the neighborhood girls would get together in the kitchen to gossip over coffee and cake while watching the kids. "Betty was pretty much like the rest of us, the way we were in the Fifties," Frances Katz observed. "We played mahjong together, we went on weekend vacations as a group, leaving the kids at home with someone, and go up to the country, the mountains, and spend a weekend there."

Older than the other fathers in the neighborhood, Kal joined in their poker games—he was a card shark of sorts, having picked up all the tricks during his stint in the army, and in the streets of Brooklyn. They were small-stakes games, nickel-and-dime stuff, played in the Seinfelds', or Katzes', or Blooms' rec rooms over beer and salami sandwiches, while the women gabbed upstairs. On those carefree card-playing evenings the Katzes and the Seinfelds wired up a little walkie-talkie system so they could hear Jerry and the other kids in the unoccupied house to make sure they were okay.

"Sometimes even Rabbi Spielman would play with us, too," Norman Katz said. "One night we were over at Kal's house and it must have been about a quarter to ten and Kal says, 'I'll be back soon,' and we played until about quarter of eleven and he never showed up the rest of the night. We never knew where he went but he just left us there. We finished the game and went home. He just disappeared, a real comedian."

Betty helped with the family's limited budget by taking in sewing. She was an expert seamstress, a trade she picked up in the orphanage. Her customers, whom she got by word of mouth, came from the more affluent section of town called Harbour Green, near the water, a tony area she envied and where she vowed to live one day.

Meanwhile, Kal, the family breadwinner, was still peddling dreck door-to-door, everything but *The Watchtower* and Girl Scout cookies, and if he could have gotten a franchise, he would have.

"He was out on the road most of the time," Katz recalled. "He'd go away for a while selling all kinds of stuff, like religious articles—crosses, bibles, knickknacks, icons, that sort of thing—then he'd come back, and then he'd be off again."

But the pint-sized peddler was barely making ends meet, and Betty was furious with him for being on the road so much and away from the family. Both the tight finances and Kal's career choice caused friction and often sparked fights that could be heard all over Eastgate Road on summer evenings when the windows were open. Kal was looking for that one niche product that he hoped would put him on easy street, and after about three or four years he came up with a possible answer. Among the do-dads he bought to peddle from a wholesaler in New York was a selection of small wooden signs, which could be personalized for a family.

"There was a sign on my parents' house that said 'The Egans,' and it was cut out of something like Masonite and had a couple of kids and a dog painted on it, and my father told me Kal sold him that thing, walking around, door-to-door," remembered John Egan, who years later worked for Kal for almost a decade as a sign painter. "So actually what happened was people started asking Kal for signs, and he decided that was the business for him. He started with nameplates and addresses for houses, and then people started asking for signs for their businesses. But Kal never picked up a brush in his life. He was a salesman. He hired sign painters; he made connections. He'd get an order, then he'd have some guy in Queens make the sign, and Kal would mark it up and sell it for a good profit. At first he was a middleman."

As Norman Katz remembered, "Kal started reading up on sign painting, and started fooling around with them. He was an entrepreneur. He would go to a doctor and he'd say, 'Give me a check-up and I'll make you a sign.' He'd go to a dentist and say, 'Clean my teeth and I'll make you a sign.' He got word of mouth going. He got a lot of business out of the temple, too. A lot of them were in business for themselves and they used him as their sign man. He was personable and very friendly and people went for it."

Kal initially set up shop in the small attached garage closest to his other next-door neighbor, an ex-Marine and Nassau County police detective, who didn't appreciate the late-night clammer caused by Kal's hammering, drilling, and cutting as he experimented with signs and letters. "I heard there was a little tension there because of the noise; there's not that much space between the houses," said Kathy Greaney who, along with her husband, Brian, became the fourth owner of the Seinfeld house in 1985 and learned the Seinfelds' history from the Katzes and others. In a place of honor in the Greaney house is a framed memento from Jerry arranged for by a family member with connections.

". . . let this letter stand as official recognition of the fact that along with the rest of my family I lived at 50 Eastgate Road," the comic wrote to the Greaneys on December 1, 1992. "We moved from Brooklyn in 1955 when I was one year old. We then moved to another part of Massapequa in 1962 when I was 8 years old. I hope your family makes as happy memories in this house as we did. Sincerely, Jerry Seinfeld."

Publicly, the very private Jerry never mentioned the family's first, simple starter home in Massapequa.

By the time he was in his terrible twos, Jerry became fascinated with monkeys, having a tantrum if Betty or Kal didn't take him for his weekly visit to nearby Monkey Mountain, which was within walking distance of the Seinfeld home. Monkey Mountain was part of a Jungle Camp amusement park established some years earlier by Bring-'Em-Back-Alive animal hunter Frank Buck. One year Massapequa was said to have been overrun by hundreds of monkeys when a park worker accidentally left a plank down over a moat.

In September 1959, Betty enrolled Jerry in kindergarten at Eastlake Elementary School, a six-block walk from the house. He went through first and second grades without leaving a mark. Small and skinny, with glasses, was the way he was remembered—a quiet, shy, very neat and mannerly child.

"My children and Jerry all used to play together," said Frances Katz, "and I never thought of him as remarkable in any way. He was just an ordinary kid. All the kids were pretty much the same."

"He was," Rabbi Spielman observed, "what I call a late bloomer."

But in the privacy of the Seinfeld home Jerry was outspoken, demanding, driven. "There was this big chocolate cake, and Jerry, who spoke very clearly for his age, told me he wanted a piece," said Betty Seinfeld, who has often told this story to underscore her son's drive and ambition. "Well, he never wanted just a piece of cake; it had to be the *whole* cake—and he always waited until he got what he wanted. I sliced him a good portion, but he said he wanted a bigger piece. When I told him to take what I was giving him, he turned on his heels and walked away. I remember saying to myself, 'That boy will always get what he wants and never settle for less.'"

Jerry was demanding and acquisitive and even then wanted his closet filled with sneakers, Keds in particular. "I've always liked sneakers—that was something that I responded to even at six years old," he says. "I drove my mother crazy about getting me sneakers, the dark blue kind that you could only get in the city. On Long Island, they only had black. I've got a picture of me wearing them in my first-grade class. Every other kid in the picture has regular shoes. I'm in high tops."

His sister, Carolyn, with whom Jerry has always had a very close and caring bond, said her brother was "very driven" as a youngster. "If he wanted a toy, he'd sit at the table crying or arguing or carrying on. He'd obsess about things like that. He still does, even now that he's gotten to where he is."

—5—

Moving On Up

By 1962, when Jerry was seven, his father's sign-painting business had started to take off. Kal moved out of his small, cluttered Eastgate Road garage, and set up shop in a small, cluttered ground-floor space at 1065 North Broadway in North Massapequa's business district; joined the Massapequa Chamber of Commerce; bought his first van; and hired a veteran sign painter named Abe Rosenbloom to do the actual painting while he hustled up the customers.

Because of the coincidental combination of name and product, Kal changed the spelling of his moniker (for business purposes only) to *SIGN*feld, and called his start-up Kal Signfeld Signs. His first display ad in the Yellow Pages reads:

> Kal Signfeld . . . 'SIGNS' is our middle name—Show Cards, Real Estate, Doctors-Lawyers. PLASTIC LETTERS OUR SPECIALTY, 4 inches to 4 feet.

Later ads included a cartoon drawing of Kal—a dapper little figure wearing a big smile, a baggy-pants suit, and carrying a cane—more like a burlesque character, which people compared him to, than a sign painter. Later he used the slogan "Kal Signfeld Signs With Advertising Punch."

By the fall of 1962, with his business growing, Betty convinced Jerry's father that it was time to sell their house and move to a residence befitting a businessman and his family on their way up. She had had her eye on Harbour Green, a community of new houses that overlooked South Oyster Bay, home to a number of clients for whom she sewed and to the mobster who was the prototype for the Godfather trilogy.

The Seinfelds put their house on the market, advertising it privately so as to save on the cost of the broker's commission, and almost doubled their money when they sold it to Joel Halle, another Brooklyn native. After he closed, Halle discovered that the patio was faulty and the outside underground oil tank was environmentally questionable.

Just after Thanksgiving Day 1962, the Seinfelds became the proud owners of the last home to be built in Harbour Green—a $29,990 red-brick, high ranch-style with white columns, at 311 Riviera Drive South, on a peninsula bordered by two canals, Massapequa Cove and South Oyster Bay.

Florence Jacks, who lived next door, was in her dining room, which faced the front of the Seinfelds' house, when she spotted a little mustachioed man wearing a fedora, standing outside, gazing up at his new house. "It was Kal and I guess they had just closed and I remember him looking at it with such great pride, and I thought, 'How nice, he got what he wanted.' "

Jerry transferred to Birch Lane Elementary School, where he started to be noticed for the first time, as evidenced by a comment his third-grade teacher, Miss Brown, wrote on his report card. "Jerry does a little too much fooling around," she observed, "and not enough constructive activities."

Years later, after he got involved in some constructive activities, Jerry took Barbara Walters on a walking tour of Riviera Drive South for one of her celebrity specials, and he asserted that he had had a lonely, unexceptional childhood in Massapequa.

"For quite a few years I really didn't have any friends around this neighborhood," he said, "and that's when I started to be funny. That's what makes you funny. You have to use your own mind to keep yourself entertained."

Life at the Seinfelds was much like the television programs Jerry watched in the early Sixties, a vast wasteland, as Newton Minnow once described it. The house was devoid of books, except for pop paperbacks, *Reader's Digest*, romance and women's magazines, *Life*, *TV Guide*, and a daily subscription to *Newsday*.

"There was a simplicity about Jerry's parents," observed David Elkin, who was Jerry's close friend beginning in the seventh grade, spent considerable time in the Seinfeld home, and who, along with Jerry and one or two other boys, was shuttled every Saturday to Temple Beth Shalom by Kal for their Hebrew lessons and bar mitzvah preparation. "They weren't overly sophisticated, educated people. They were not intellectual. Jerry was not on the academic fast-track. School was less emphasized in his home. We didn't go home and study. We came home and played and had dinner and watched TV."

There was no art, no music beyond pop tunes, because Kal liked to jitterbug and do the Lindy. The furnishings and decorations, such as the silver wallpaper in the bathroom and the blue plaid wallpaper in Jerry's room—blue is his favorite color—were indiscriminate, uncoordinated, and inexpensive.

"*Everything* was a schlocky, ugly design," a critically observant close family friend said. "No piece had anything to do with any other piece. It was like—you need a couch, here's a couch. There was a sofa that was off-white, goldish, faux antiquish. But it all looked secondhand. It was very, very sparse. Jerry's room was known as 'the blue room' because of the blue plaid wallpaper. It had very little in it—except some posters."

One of the few interesting pieces, a grandfather's clock, even had the name "Kal Seinfeld" carved into it with a penknife by the man of the house.

The Seinfelds' taste, or lack of it, was also reflected on the exterior. Above the front door was the Seinfeld name—in foot-high brass letters—installed by Kal Signfeld Signs almost immediately after they moved in, and the marquee rubbed some neighbors, such as Florence Jacks, the wrong way.

"That huge sign over the doorway," she said, "drove me a little crazy. Kal put it up because he was proud of his house, proud of his name, proud to be living where he was living, proud of his children—just a happy man—a proud man in a good sense. But frankly I thought the sign was in poor taste, and I tried to encourage them to take that down, but it wasn't going to happen."

In terms of the family's values since the move up from Eastgate Road nothing had changed much except location. Making money was still the motivating factor for Kal and Betty, and Jerry and his sister were left to mostly fend for themselves. As Betty Seinfeld has acknowledged, "Kal and I had no time to take our kids skating or to ballparks." But Jerry didn't mind the independence the family's lifestyle provided. "I didn't care," he says. "They could do whatever they wanted because I was busy watching TV. We were an independent family and went our own way . . . it wasn't that kind of cloying, got-to-talk-every-day thing. There was plenty of breathing room. We were all just kind of roommates."

Much of the time, Kal was on the road selling signs and overseeing installations, and his free time was spent with synagogue business. At the same time, Betty continued to bring in extra money from her tailoring business. "Betty wasn't a typical Jewish mother," Ruth Goldman observed. "She gave Jerry a tremendous amount of freedom, with no guilt trips, because she had her own things to do."

Jerry boasted in later years to Caryn Trager that he always thought his mother was "so cool." He also said his mother never believed him when he told her he was sick, and was matter-of-fact about such things. "When he was a kid he walked around either on a broken foot, or with a broken arm, for two days before she finally took him to the hospital," Trager said. "She wasn't cruel, but everything with her was always fine—'You aren't sick, everything's fine.' And that's kind of how Jerry was—everything's always fine, everything's okay."

As Jerry himself candidly put it: "There was a very healthy benign neglect that we were raised with. It was [good] for me because I didn't want to be bothered. I felt like I was rooming in my house growing up. My personality is similar to my parents'. I'm fiercely independent, very self-sufficient."

—6—

Tough Times

At the age of eight, Jerry showed signs of an inventive sense of humor, giving an early stand-up performance with a schoolyard pal named Lawrence McCue for a class fair at Birch Lane Elementary.

Jerry conceived a comedy skit that poked fun at President Kennedy. The idea came to him after watching the first appearance of comic Vaughn Meader on the *Ed Sullivan Show*. Glued to the family's only TV, Jerry was struck by Meader's impersonation of Kennedy.

"Basically, I was Jerry's straight man," recalled McCue, who became a physician's assistant. "He was the creative genius. Jerry played President Kennedy and I played a reporter asking him questions. I would set him up and Jerry would deliver the line. For a week we sat out on the playground and rehearsed. We made jokes about Kennedy. It was quite a hit, very funny, and we were the only ones in the class who did a comedy skit."

Around the same time, Jerry got laughs from his playground audience with a routine about an obese female classmate, a shtick that supposedly was so hilarious it made a pal toss his milk and cookies. "I felt the milk and I saw it coming at me, and I said, 'I would like to do this professionally,'" was the way Jerry graphically remembered and humorously exaggerated the moment.

"He knew he wanted to be a comedian because at a very early age we were always playing in his room and he was always making jokes, always laughing," Scott Brod, a pal at the time, said. "His father was really the impetus for Jerry, a very funny person, and that just permeated to Jerry. His father could have been a stand-up comic himself."

Looking back to that time, Jerry believed that from about the age of eight ". . . it was in me to do [comedy]. I watched comedians on Sullivan—I watched [Jonathan] Winters and Skelton with a different look on my face than other people did; I was transfixed by them. It was always the thing I wanted to do, but I never thought I would. It was like watching people on the moon, and you think, 'I'd like to try that,' but you don't think you will. It was that deeply imbedded a fantasy."

Most of what Lawrence McCue remembered about Jerry involved laughter. But there were also tears.

Despite Massapequa's melting-pot reputation—its "matzoh-pizza" sobriquet is still brought up today by longtime locals to underscore the town's tolerance—anti-Semitism was part of the community's fabric, and it even reared its ugly head among the innocents at Birch Lane.

"I get tearful thinking about it and I've never discussed it with anybody," McCue said. "We were in the playground—me, Jerry, and another friend of ours. There was this kind of a tough kid at Birch Lane, a bully, and this kid was with his friends and came over to Jerry and made a loud comment about 'dirty Jews,' and I heard him repeat the comment to our pal. I'll never forget the hurt I saw in Jerry's face from that horrible comment. I saw him wiping his eyes. I have a Christian upbringing but I felt bad for him having to hear it."

Former U.S. Representative Angelo Dominick Roncallo, a longtime Massapequa Republican politico and a close friend of Kal's, said the town had a history of discrimination. "When I first moved here in 1947 they had restrictive covenants, and some of that continued into the Sixties. People of my Italian background and Kal's Jewish background were not exactly accepted with open arms," he added, chuckling bitterly. "Of course that changed when there was that great influx of people from the city—half Jewish, half Italian. We probably got along together for our own self-protection."

Years later, Barbara Walters interviewed Jerry on Riviera Drive South, with the ABC cameras rolling, and intoned, "This seems like such a good place to live. Did you like living here?" Jerry responded with a resounding "No."

One reason for Jerry's unhappiness had to do with his size and fragility, friends maintained. "Jerry was a tiny boy," observed a playmate, Gene Hurwin, a street kickball enthusiast who became an occupational therapist. "You knew, instinctively, that you didn't want him on your team, or you'd be dead. I remember, for fifteen minutes this giant mess of a game was going on, and Jerry wasn't playing. He was sitting on the sideline, just watching."

At his parents' behest, Jerry reluctantly joined the Cub Scouts; Kal and Betty had hoped that if he were with other boys and with leaders, the socialization and regimentation might bring him out of his shell, but that didn't happen.

"You never knew the boy was in my den," observed his den mother, Marlene Schuss, a family friend and neighbor. "He was a shy, quiet, unassuming, little, skinny kid with glasses who *never* spoke and *never* had a friend. In his house he must have gone crazy and been funny. They did say that in the house he would jump around, make jokes all the time, but I really had to pull him out. I was given an assignment to produce a play for the den and I decided to do *The Lion and the Mouse.* Of course I made my son, Spence, who was a year younger than Jerry, the lion. Guess who was the mouse? Jerry. And he was a *little* mouse. He was *so* quiet. It went over beautifully. I always say I was the first person to put Jerry on the stage."

Years later, she saw Jerry perform at a comedy club on Long Island. "We're sitting in this smoke-filled room and he comes out, bright lights in his eyes, first time he was there, and he does this little routine. He says, 'I remember how ugly I was. I wore glasses. I was in the Cub Scouts. When they put the kerchief around my neck I was . . . '—and he made a joke of how ugly he looked—and all of a sudden someone at my table yells out, 'And your den mother is here!' He says, 'What?' He puts his hands up to take away the glare from his eyes, and he says, 'Marlene?' And then he says, 'Oh, my God.' It was adorable."

In his book *SeinLanguage*, Jerry reminisced about his Cub Scout experience: "I remember I'd get the outfit all set up: blue pants, blue shirt, little yellow handkerchief, the giant metal thing to hold the handkerchief together. Then I'd go outside, get beat up, come back, put my regular clothes on."

Whatever emotional pain and embarrassment he felt, Jerry hid it by losing himself in television—the comedy shows of Lucy, Danny Thomas, Red Skelton, Jackie Gleason, Jack Benny, and the kid programming—*Rocky and Bullwinkle, Jonny Quest, Batman, Superman,* and *Flipper.* While Superman was his all-time hero, Jerry, at ten, was also taken by *Flipper,* which was the story of two brothers, Sandy and Buddy, who, like Jerry, lived near the shore, but unlike Jerry—who had a mutt he named Ralph after Ralph Kramden, and a parakeet—they had a pet dolphin who was both friend and helper in their weekly adventures on NBC.

"I was really into *Flipper,*" Jerry recalls. "I wanted to be an oceanographer. If I couldn't be Bud, I figured I could be Ranger Ricks [a widower who was the boys' father on the show], get a job for the Florida Everglades Protection Bureau. I wanted to communicate with dolphins."

When he ventured outside, he was constantly on his bike, usually alone. "I led the swinging Stingray life," he jokingly boasted, referring to the popular model of Schwinn bicycles he owned. Jerry was the first kid on Riviera Drive to own a Stingray, a metallic blue model. "I had so many different Stingrays—I had every one made—the three-speed, the five-speed, you name it.

"The bike also was a superior paperboy vehicle because of the way the bags stood on the handlebars," pointed out Jerry, who had a short stint delivering *Newsday* at a salary of eight to twelve dollars a week, which he felt was a staggering wage. "I couldn't believe I made that much."

One of his neighborhood pals, Mitchell Silverman, who became an endocrinologist, recalled, "We used to run around on our Stingrays and ram each other."

Jerry liked to play stunt driver and besides riding his bike into

Silverman's he once performed a test where he slammed head-on into a street-lamp pole to see what it felt like. Luckily, he escaped without injury.

Jerry saw his Stingray mainly as a means of escape. "I'd get on my bike in the morning and take off for parts unknown, like Tackapausha Preserve or Bethpage State Park. Now, that was an exciting adventure."

The reason why Jerry had so many nifty, expensive Stingrays, which the struggling Seinfelds couldn't afford in those days, was because Kal *handled* for every set of handlebars for his demanding son.

"Kal was always trying to make some kind of deal," recalled Dominic Totino, who owned Sunrise Cyclery, the Massapequa bicycle shop that supplied Jerry's stable of Schwinns. "He'd come in and want a bike in exchange for making a sign. He never liked to pay cash. It was always, 'If there's anything we can do, let's do it.' Naturally, I preferred to sell the bike for cash than do any bartering. When I say 'dealmaker,' that's basically what Kal was. 'Let's make a deal and not get any money involved.' I think he always felt he got the better deal because his cost on the sign was very little. He *hated* to part with cash. I remember delivering many a Stingray to his house for Jerry."

Despite the fact that Kal's business was growing and his signs were showing up all over town he was still just getting by financially.

"They couldn't afford the house" was the assessment of family friend and neighbor, Victor Elkin, a psychologist, who, along with his therapist wife, Sybil, a professor of social work at Columbia, and their two boys, David and Jay, had moved to Harbour Green three years before the Seinfelds. "It's an exaggeration to say the Seinfelds had *no* money, but they *needed* money. They moved in on a shoestring. The house was above their heads. I think the Seinfelds were the poorest on Riviera Drive."

Because of their limited funds Betty continued to take in sewing. Marlene Schuss, whose second husband's family was distantly related to Kal, said, "Betty became the neighborhood dressmaker. She did all the alterations for weddings, for bar mitzvahs. You name it, she did it."

Another method Jerry's mother dreamed up for bringing in much-needed money was to become a landlord, which caused a neighborhood furor of sorts.

"Betty had a knack of finding ways to make money. She was a very,

very entrepreneurial person," said Elkin. "She redid the lower level into an apartment and made a two-family house and rented the downstairs to somebody, which was illegal, but she did it. They put in a private entrance so that the renter didn't have to go through the front door.

"Betty was the kind of person who would do that—make an illegal apartment—because she was a go-getter. I believe her growing up as an orphan had a tremendous influence on her working hard at finding ways to make money."

In a small community like Harbour Green, where everyone seemed to know everybody else's business, word quickly spread about the Seinfelds' clandestine rental, raising hackles among some of the neighbors who had fled crowded apartments in the city just so they could enjoy single-family suburban living.

"We were furious about the apartment," declared neighbor Bill Elio, who had been in wholesale meats and provisions, and had purchased Signfeld signs over the years for his business. "Kal was a little bit of an operator; he wasn't a saint—let's put it that way. Because he had a high-ranch he could easily make an apartment down below. Naturally, the word spread, and when we found out we were pissed off. It wasn't right, because we came here as young families to get away from apartment living.

"We didn't know what to do about it. I don't think people wanted to get involved, or confront Kal and Betty, which was stupid, because we all paid high taxes and his taxes were being paid by the rental income. It wasn't right. We paid for the luxury of the area and here's someone bringing down the neighborhood."

Vic Elkin remembered discreet discussion in the community about the Seinfeld housing problem, but to the best of his knowledge no one called authorities. "There was talk, there were mixed feelings," he stated. "Some neighbors felt that if the Seinfelds couldn't afford the house, they shouldn't have bought it. My feeling was—I don't like the idea about the apartment, but I understand their need."

Another neighbor, Harriet Harris, a friend of Betty's, also became aware of the situation. "I used to walk my children to the bus stop in the morning, and someone standing at the bus stop would invariably nod and whisper, 'Oh, *that's* the Seinfelds' tenant.' They were making it a point to show that Betty and Kal had turned their house into an apartment."

One family that didn't voice opposition to the Seinfelds' illegal abode was the Gambinos, Carlo and Kathryn, who moved from a seemingly modest but elegantly furnished and expensively decorated row house on Ocean Parkway, in Brooklyn—Kal's old stomping grounds—into a discreetly guarded complex at 30 Club Drive, just across Grand Lagoon from the Seinfelds.

Unlike some of the more outspoken members, like Kal, of the Harbour Green Civic Association, Gambino kept a low profile, and for good reason: by the time he had moved into the community in March 1961, at the age of sixty-one, he was the boss of reputedly the largest and most influential crime "family" in the country—two dozen crews and 950 men, a family that made the fictional Sopranos and Corleones seem like Mouseketeers. Marlon Brando's Don Vito Corleone, head of the Corleone family, was modeled on Gambino. And unlike Carmella Soprano, Kathryn Gambino came from a mob family of her own; she was the sister of the boss of the Castellano organization.

As neighbors, though, Jerry's father and the "godfather" hit it off at social occasions at the Harbour Green Shore Club, a private beach and pool for the residents, who early on received from Gambino an offer they couldn't refuse.

"When we were building our beach club," Vic Elkin said, "we were running out of money and Carlo went to a meeting and said, in a broken-Italian manner, 'You need money?' And the members said yes. 'How much do you need?' They said *X* thousands of dollars. A half-hour later he showed up with *X* thousands of dollars in a briefcase and said, 'Here, you got the money.' The members asked whether they needed to guarantee the loan and he said, 'Don't worry, you don't have to sign paper.'"

As a result of those kinds of community dealings, Kal and Gambino forged an odd-couple kinship. The two were similar in many ways: both were cunning, street-smart businessmen who had learned the ways of the world on the streets of Brooklyn; physically they were small in stature, Gambino taller by just a hair, but often described as sparrowy; both had prominent noses, and both appeared gentle and amiable. Kal was mesmerized by the mobster's style and reputation, and Gambino, who spoke little—he'd been wiretapped so many times by the FBI that he knew to keep his trap shut—was a perfect audience for the sign peddler who regaled the godfather with interminable stories of his small-potatoes business scams over the years.

Moreover, Gambino had a fondness and an affinity for spunky little Jews like Kal; one of the mafioso's closest associates, Meyer Lansky, with whom Gambino worked in building the strength of the crime syndicate with legitimate business ventures over the years, was considered the most influential boss in the history of American organized crime. Hounded by the police and the FBI, Lansky returned to his Jewish roots by trying to become a citizen of Israel, a country to which Kal professed a strong allegiance.

Kal often played poker at the Gambinos, who had a "beautiful finished basement with a bar that ran the entire length of the house for entertaining guests," according to Seinfeld family friend, Ruth Hinden, whose home was located next to the Gambino place. She remembered times when a feared mobster known as "Three Fingers" Brown and his colleagues showed up at the house for card games, and times when she went bowling with the wife of Terry Zappi, Gambino's reputed lieutenant, who also lived next door, and knew the Seinfelds.

"I used to say to Kal after he played cards with Gambino, 'Did you win?' And he always said, 'Are you crazy? I *always* let *him* win,'" chuckled Ruth Goldman. "They played cards together and naturally, because of Gambino's reputation, Kal always let the godfather collect the pot. Some people could mingle with all walks of life, and that's what Kal could do. He could be with a gorilla and have fun. There were always telephone trucks, cleaning trucks, vans parked near the Gambinos'—naturally it was the FBI watching him," she continued, "so the neighborhood was very safe."

One of Gambino's favorite *bambinos* in the neighborhood was the Seinfeld boy. Kal boasted to friends that every time Carlo saw Jerry, "He'd pinch his cheek and say, 'You're being a good boy now, Jerry, aren't you?' and give him a new silver dollar." On Halloween, neighborhood kids came to the Gambino house and collected not just a bag of candy but a five-pound box of candy. But when a costumed Jerry arrived at the don's door he got the VIP treatment: besides the candy, one of the soldiers dropped a handful of silver dollars into Jerry's trick-or-treat bag, orders from the boss. And when Jerry was bar mitzvahed in the summer of 1967, he is said to have received an eye-popping check from the Gambinos.

At the offices of Signfeld Signs, meanwhile, Kal had found a bulletproof method to get deadbeats to pay their overdue bills. "I'd hear Kal on the phone talking to the owner of an Italian restaurant or pizza parlor, a guy who owed him some money," said John Egan. "So he'd say to the guy, 'You know I have a friend, Carlo, in Massapequa here,' letting this guy know he knows the 'godfather.' Or Kal'd say, 'You know I was playing cards with my friend Carlo Gambino the other night'—and then the guy would get scared and pay up. Kal used that connection, and it worked."

—7—

Becoming a Man

Hundreds of students reported for their first day of school at Berner Jr. and Sr. High in early September 1966. There were the jocks—looking forward to bashing heads on the gridiron; there were the nerds—with PaperMate pens neatly aligned in shirt-pocket vinyl holders; and there were the hoods—with packs of Marlboros seated firmly in the rolled-up cuffs of their T-shirts, also looking forward to bashing heads.

But whatever their orientation and interests, all of them seemed to tower over an anxious and mousy Jerry Seinfeld, a little *pisher* who, at 4'9" and weighing a scrawny hundred pounds, saw only "belt buckles and knee caps," according to his classmate Lonnie Seiden, Jerry's first gal pal, who noted, "We were both small for our age, both under five foot."

Having turned thirteen four months earlier, Jerry not surprisingly hadn't a real clue as to what he wanted to do with his life, no concept whatsoever of where he was going, nor did he fit neatly into the student factions that had quickly established themselves that first day of junior high.

"If anything," maintained David Elkin, who went through school with him, "Jerry was somewhere between the nerds and the jocks. A lot of the kids in our neighborhood were intermingled in those different factions. It was much easier and freer to go hang out with the jocks, sometimes play football with them, and then go play handball against the school wall with the hoods, play for a quarter a game."

Looking back at that awful adolescent time, an unusually forthright Jerry once observed, "I wasn't the happiest kid in the world. I was out of it. I wasn't connecting with girls, with friends, with sports, with school, with beer and cigarettes and everything everybody was doing. I couldn't get into the social stream, the scholastic path, the sports. I got my head in the door of a lot of things and never was really a part of anything.

"I never liked the idea of being a part of a big group," he continued. "I didn't like the scapegoating. All of a sudden the group would turn on one guy and just humiliate him. It terrified me, and I felt very uncomfortable living with that potential. That's why I avoided the groups. I was too sensitive. If they'd have turned on me, I couldn't have dealt with it. I'd always find some other disaffected kid and hang out with him."

Among the clearly disaffected at Berner was Lonnie Seiden, the first of a number of petite, buxom, long-haired, brunette, mostly Jewish girls, many of whom resembled Jerry's mother, who were in and out of his life through the years.

Following the divorce of her parents when she was twelve, Lonnie, who was born a month after Jerry, had moved from the flashy and affluent "Five Towns" area of Long Island to Massapequa's tony but sedate Nassau Shores community—home to the acting Baldwin brothers—with her mother and her new father, an Italian hairdresser named Lenny Grosso, who wore gold chains, was a karate fanatic, and a decade her mother's junior. Lonnie was the first child of divorce Jerry had ever known.

Each needing a friend and a confidant for emotional support, Jerry and Lonnie discovered each other in seventh grade homeroom under *Seinfeld* the show–type of circumstances.

"He was S-E-I-N and I was S-E-I-D—but beyond that Jerry's zipper was down," she said, laughing at the moment. "That's how we met. My friend Shelley said, 'Oh my God, Jerry's fly's open,' and I said 'You've got to tell him,' and she said 'I'm not telling him, you tell him.' So I told him, and oh, God, he was *so* embarrassed.

"But I thought he was cute," she continued, "even though he looked like a geek—small, braces, physically very clumsy, he still had to grow into his legs. But it was nice to find somebody my own size, and we started to hang out."

Jerry was more than intrigued with life at the Seidens' house—far different from the world of the Seinfelds—and spent as much time as he could there, mostly in her ultrafeminine bedroom, with its hot-pink carpet, pink accents on the twin beds, matching pink bedspreads, beanbag chairs—one with funky tiger stripes—and an enormous inflatable pack of Lifesavers, among other mid-Sixties teenage kitsch. But it wasn't the campy decorating that enthralled him; it was the cast of characters, especially Seiden's mother, who, from all descriptions, acted like Kramer in drag.

"My mother owned a beauty parlor and she and Lenny worked long hours and she'd come home from work and have a cocktail to relax," Seiden explained. "And a couple of times she got a little drunky-poo and made a fool out of herself and, of course, Jerry was there when she totally lost it.

"Once, we were sitting out on the porch and she was yelling and screaming and we didn't know what was going on, and we come running in and she's on the floor and she's smacking the couch and Jerry's looking at me and saying to her, 'Are you all right Mrs. Grosso? Are you okay?' And all she could do was point at the TV. She was watching a Don Knotts movie and it really made her laugh and Jerry was like, 'Oh, that is pretty funny.' And he was like 'Let me help you up,' and he put her on the couch and that was it. I said to Jerry, 'Oh, my God, I'm so sorry.'

"The next time he was over, again she was in her same position prone on the couch staring at the television and this time she starts screaming and we go running in there and I'm like 'Mom, what's the matter now?' And Jerry spots what she's screaming about. She used to buy these gigantic candy bars, the really big size, and sticking out of the candy bar was her front tooth. She was being a pig and eating this humongous thing, trying to get it into her mouth, and her tooth fell out. Jerry just gave her that stare and laughed. That's why he enjoyed being at my house—it was a lunatic asylum.

"Jerry adored my stepfather, Lenny; absolutely *adored* him. Lenny was a swinger, a fucking nut job. He used to run around the house pretending he was Bruce Lee, a wild man. When Jerry saw my mother for the first time in years a number of years later the first thing he said to her was 'How's Lenny?' We had three dogs, we had a pool, we lived right on the bay. It was like a wild house. Jerry came from a normal household compared to mine. His parents were together and seemed to love each other and they were older. His family was like *Marcus Welby, M.D.,* or *The Wonder Years.* Mine was more like *Soap.*"

Naturally, because they'd grown fond of one another, Jerry and Lonnie were up for experimentation. After all, it was the "swinging sixties," free love and free dope were in the air everywhere, but not so much in conservative and conventional Massapequa. The war was raging in Vietnam and the parades and protests on American campuses and even on the White House lawn were getting bigger, but not so much in Massapequa. Shocking acts of violence were occurring around the country—ex-convict Richard Speck murdering eight student nurses in Chicago; architectural honor student Charles Whitman killing twelve and injuring another thirty-three with a rifle from atop a tower at the University of Texas—but not so much in Massapequa. "Negroes" were rioting in cities like Chicago, New York, and Cleveland, burning the ghettos, shooting at the "pigs," demanding jobs, while leaders like Stokely Carmichael and other "Black Power" advocates called for revolution, but not at all in Massapequa—which was lily-white.

"Another friend of mine had turned me on to marijuana, but Jerry was *very* straight," Seiden said. "That part of my life did not come into my relationship with him. But there was an incident where Jerry was at my house one day and we went to the health food store and we bought this organic stuff that was like marijuana but it wasn't, and it was supposed to get you high. It was an herb or a vitamin or something and my friends and I were smoking it and Jerry was sitting at my desk watching. And my mother came in and started going ballistic, started throwing everybody out of the house. I kept saying, 'Mom, we're not smoking pot, we bought this at the health food store.' But she was totally out of control, yelling at Jerry, and I'm like, 'Why are you yelling at him, he's not even smoking it.'"

The sexual experimentation between Jerry and Lonnie was much the same. Though still a virgin, Lonnie had some experience with boys and was into it. Jerry, on the other hand, was neither experienced nor into it, Seiden recalled.

"We were in his room after school," she said, "and we started fooling around—a little touching, a little kissing, a little feeling, and I just remember his hands were clammy, and his breath smelled like peppers and eggs. I was pressed up against the wall, and the next thing I knew we were on the ground and did a lot of rolling around, but we just never got anywhere. He got like three of my buttons open, or my whole shirt open, but it really never got further than that. He was a bumbling teenage boy, and my memories are, 'Oh, God, I don't think I'm enjoying this.' I wasn't anxious to do that again with him. I remember Betty coming in the door, but we weren't anywhere near being naked, and we ran down to the den."

While Jerry remained a boy sexually, he became a man religiously the same year he entered junior high and became buds with Lonnie Seiden. During the summer of 1967 there were ten bar mitzvahs in quick succession at Temple Beth Shalom, and Jerry's was one of them.

Despite the fact that Kal Seinfeld was obsessively involved in the machinations and politics of the temple—he was a vice president, board chairman of the Hebrew school, a member of the Hebrew school's house committee, blower of the shofar during the high holidays, opener of the shul in the morning and locker of the doors at night, the always available tenth man for last-minute minions, builder of bookcases for the prayer books, donor of signs for special events, and a fund-raiser and recruiter who helped the congregation grow—his son, Jerry, despised religious services and detested attending classes for his bar mitzvah.

As Scott Brod, who prepared for his bar mitzvah alongside Jerry, observed, "Some people's hobby is golf, Kal's was the temple. But Jerry hated Hebrew school. *Hated* it. Ninety percent of his time there was spent saying, 'I can't wait to get this bar mitzvah over with.' He only did what he had to do until the day he was bar mitzvahed, and that was the end of it. Kal wasn't the type of person to show anger if he was bothered, but maybe it did hurt him deep down."

Cantor Max Klein, who prepared Jerry for his bar mitzvah, took a different, more diplomatic view. "I knew Jerry from the time he was ten years old and he *always* came prepared for his bar mitzvah lessons, *always* on time," he asserted. "Admittedly, he was not overly eager to come to extra Torah readings, but he did everything he was assigned. He even came to a special Sunday morning class where they all learned to daven a little, to put their tallis and tfillin on properly. He never took advantage, and I'm sure Kal and Betty would never let him take advantage. Jerry wasn't solemn, but he was serious—and I have a feeling much of it was because Jerry knew the synagogue was a very special place for his dad. At his bar mitzvah Jerry read from the Torah and did a very creditable job, and I went to his house that Saturday night and they had a modest reception."

If, as some contend, Jerry and his father were at odds over the rites, rituals, and meaning of their religion, it was one of the few areas upon which they disagreed. Jerry has always expressed great affection, admiration, and love for his father, both publicly and privately.

For instance, the first few lines of Jerry's book of comic bits, *SeinLanguage*, which was published in 1993, becoming a huge best-seller because of the popularity of his show, underscored his affection.

Wrote Jerry, in part:

> There has never been a professional comedian with better stage presence, attitude, timing, or delivery . . . a comic genius . . . The thing I remember most . . . is how often my father would say to me, "Sometimes I don't even care if I get the order, I just have to break that face." He hates to see those serious businessman faces. I guess that's why he, like me, never seemed to be able to hold down any kind of real job. Often when I'm on stage I'll catch myself imitating a little physical move or a certain kind of timing that he would do "To break that face."

Therefore, it came as a shock to longtime friends that Jerry, who became rich and famous from his inherited ability to break face, had refused to honor his father by lending financial support to the ailing temple to which Kal Seinfeld had dedicated so much of his time and energy.

"The whole town was very disappointed that Jerry never donated to the temple after Kal's death," declared Marlene Schuss. "It was presented right to his mother. She was *begged* for donations. It was—'Can't you do anything with Jerry to give help, because the temple is totally falling apart?' So they really have it in for him. Jerry has his own reasons not to support his father's love for this particular temple. I have no idea what they are and I'll probably never know. I don't personally hold it against him, but that's not what I hear about the congregation."

Rabbi Leon Spielman, who in early 2001 had been the temple's religious leader for almost a half-century, was clearly saddened and surprised by Jerry's lack of support.

"There are people who feel Jerry should do something to help the temple, which is sinking," the rabbi's wife said, expressing her husband's feelings. "But we cannot sit in judgment. We would have liked to make it the 'Seinfeld Synagogue,' or something like that, but for whatever reason Jerry wouldn't help. Perhaps it had something to do with his growing up, people go through changes in their lives. People always say the temple's so broke, why not turn to Jerry, and the rabbi always says, 'Please do.' I don't know whether Jerry's cutting his Massapequa roots out, or what."

Scott Brod added, "There should be something at the temple saying—'In memory of Kal Seinfeld, for his years of dedication.' Jerry's father gave his life and blood to that temple. I went to the president of the temple and said I'd become a full member if you'd put something up to honor Mr. Seinfeld, and he went into this whole thing—'Don't talk to me about Seinfeld!' And I said, 'Kal, not Jerry.' It's a real sore spot. Jerry should have done it. It's never too late."

—8—

High School Loner

Jerry entered the hallowed halls of Massapequa High School as a ninth grader just after Labor Day, 1968, with only two interests—cars and Bill Cosby—as the world exploded around him.

Around the time of his graduation from junior high, Dr. Martin Luther King Jr. and then Senator Robert F. Kennedy were assassinated, and just days before Jerry got his first high school roster of classes, helmeted, club-wielding Chicago police attacked demonstrators outside the Democratic National Convention—horrific events that sent shockwaves around the world.

But Jerry was oblivious. Lost in a fantasy world of his own, isolated in his blue plaid bedroom, he pored obsessively through well-thumbed issues of *Road & Track* and *Car & Driver*, vowing to his few friends that one day he would own his dream car, a Porsche. Posters of the car, not *Playboy* centerfolds, bedecked his room.

Meanwhile, back in the world, young Massapequans like Jerry were increasingly participating in protests against the war, the policies of lame duck President Lyndon Johnson, and other burning social issues of that turbulent time in American history, taking moral and political stands. But Jerry was a blank slate, lacking passion or commitment.

"There was a lot of social activism among students at the high school," said Jerry's friend, Spence Halperin, son of Marlene Schuss. "But I don't remember Jerry or his family being involved in any of that. There were some student strikes, the whole 'Dump Johnson' movement was happening among the liberal left of Massapequa, which was really the neighborhood Jerry and I lived in. Massapequa High School was extremely polarized politically. The administration was very conservative, and they would try to repress any kind of demonstrations done by students. They would suspend people if they walked out to demonstrate in the middle of the day. I remember there was a day of mourning after the Kent State killings. We tried to organize a student strike, and a large group walked out of school, went to the flagpole out front, and attempted to pull down the flag. Of course, they were stopped. But I never once saw Jerry at any of those things. He didn't care."

Jerry's only apparent concession to the radicalism of the time was that he wore jeans, always neatly laundered, and had long hair—"a Jewish Afro, a lot of hair piled high on his head"—as Arnold Herman, a Massapequa High School teacher who taught Jerry social studies, remembered him.

"Jerry was a listener and a watcher, not an activist," continued Herman, who taught at Massapequa for four decades and saw all kinds of kids come and go. "I never felt Jerry was involved—just a quiet kid who did his work, had a few friends, and a wonderful smile that just lit up. I ran the 'It's Academic' program; the model United Nations; I ran a recycling center—fifty kids would come down on a Saturday to help me—but Jerry wasn't involved in any of the activities. If he'd been one of the top kids intellectually, I would have kidnapped him into one of my academic teams, but he wasn't one of the kids I went after."

The Seinfelds themselves were Republicans, and Kal, a veteran, viewed himself as patriotic and therefore a supporter of United States' involvement in Vietnam, which was probably the reason why Jerry didn't participate in any of the protests. As Massapequa teacher Don Nobile, who also taught Jerry social studies, observed, "Most of the kids reflected their parents' views. And Massapequa was and is more of a conservative town." Along those lines, one of the celebrities who came out of Massapequa was Peggy Noonan, who became a prominent conservative Republican author and President Reagan's speechwriter who coined the famous phrase "a thousand points of light" for President George H. W. Bush.

According to longtime Signfeld Signs employee John Egan, Kal's political views were, in the main, business-related. "He bellied up to the Republicans because that was the party in control in Massapequa, and Kal felt it would help business, help him get work." The senior Seinfeld knew how politics worked—that one hand rubs the other. So it came as no surprise that he once hired the seventeen-year-old son of a powerful local Republican to work as a sign painter in his shop.

"Kal treated me like a nephew, better than some of my uncles," recalled Jim Roncallo, son of one-time Massapequa GOP power broker Angelo Roncallo. "Kal had done campaign work for my dad, signs for his elections, and he was a Republican because of my father."

A popular Massapequa attorney who did some legal work for Kal, Angelo Roncallo served as a Republican councilman and Nassau County comptroller from the mid-1960s to early 1970s; was a delegate to the 1972 Republican National Convention that nominated Richard Nixon for his second term, with the Watergate scandal looming; and was elected as a Republican to the Ninety-third Congress, where he served from 1973 to 1975. Many of his campaign signs were contributions from Signfeld Signs.

Moreover, Kal continued to be loyal friend and supporter of Roncallo's during the dark days when federal prosecutors charged the congressman with allegedly extorting a political contribution from a contractor while Roncallo was an official of Nassau County, where Massapequa is located. Roncallo won acquittal in 1974, and three years later he was elected as a justice to the New York State Supreme Court.

"My son was a hell-raiser, but he was a pretty good artist and a very good sign painter, and Kal did me a favor and took him under his wing," Roncallo said fondly, years later. "In local politics we always needed signs, and Kal was there, and we would turn to him and he was very gracious about making signs for us and assisted us in every way he could. I was on the town board in 1965, and then I was comptroller, and Kal was very helpful to me. He did whatever he could for me. I don't know what his registration was; I don't know what his voting pattern was—all I know is that every time I ran for office he was very helpful to me and very solicitous of what I was doing, and I was very happy about it."

As with his love of humor, Jerry's lifelong affair with cars also was inherited from his father, who began fiddling with engines as a kid on the streets of Brooklyn. Besides his signature signs, Kal became well known around town for his replica Model-T Ford that rode on bicycle tires and was powered by a lawnmower engine. On weekends and summer evenings he'd take youngsters on rides around Harbour Green, and whenever there was a Massapequa event or parade Kal was always on hand behind the wheel of the little car, which was either draped in Old Glory, or regaled with signs advertising you-know-who. When the director Oliver Stone came to town to shoot portions of the Tom Cruise movie *Born on the Fourth of July,* based on the critically acclaimed 1976 memoir of Ron Kovic, the crippled Vietnam War veteran and antiwar activist who grew up in Massapequa, Kal, the ham, managed to sneak the car into a street scene or two.

Unlike most of the Jewish guys at Massapequa High who were on the college track, and whose only interest in cars was getting their daddy's to cruise the All-American hamburger joint and the Carvel on Friday and Saturday nights, ogling girls, Jerry actually took a course in auto mechanics as an elective at his father's urging—he thought that someday maybe Jerry could make a few bucks as a grease monkey. But Jerry's interest in taking the course also underscored his very real interest in cars and how they worked.

"Jerry and I talked about cars all the time, so we took automotive shop and engine shop together," said David Elkin, who combed the auto magazines with his friend and classmate. "In many ways the class was quite humorous to Jerry because these guys who were in shop with us—guys who actually *rebuilt* cars—had trouble understanding the technical aspects, things like how to figure the displacement of a four-cylinder engine, the math involved."

As a result of that course, Jerry learned to do tune-ups and could virtually take an engine apart and put it back together. "I don't think there was anything before the comedy where he showed that kind of discipline," observed a longtime friend from childhood, Chris Misiano.

Many years later, after he'd won virtually every accolade possible and had become incredibly wealthy and could write his own ticket, Jerry still felt that the happiest day of his life was when he got his driver's license—a routine rite of passage that most grown men have long forgotten by the time they reached the age of forty—and "the greatest moment" was when his mother gave him the keys to her car—a 1965 red Nash Rambler—which was known as a "bed on wheels" because the seats folded down. "I had it repainted blue"—like his bedroom—"for nineteen dollars," he said, adding the quick quip, "The job was not nearly as good as Earl Scheib's—snowflakes would actually chip the paint." While the Rambler was not the Porsche Carrera or Speedster of his dreams, it got Jerry on the road and, like the Schwinns of his childhood, gave him the freedom he craved.

If Jerry wasn't tinkering with, or fantasizing about, cars, he sat mesmerized by Bill Cosby, whose albums he listened to rapturously, at 33⅓ rpm—memorizing the bits, analyzing the comic's superb timing, imitating Cosby's inventive characters—the Weird Harolds and Fat Alberts. Jerry was the only kid in Harbour Green who could boast of having a complete collection of Cosby's albums, and on the wall of his bedroom, along with the pinups of Porsches, was a life-size poster of the comic who, Betty Seinfeld has acknowledged, was her son's "idol."

Jerry's devotion to Cosby began in sixth grade, when he became a fan of the TV show *I Spy*, which, Jerry realized even at that age, was a departure from typical espionage shows—just as *Seinfeld* one day would stand out from the usual sitcom fare.

Jerry felt a strong connection to the characters played by the glib Cosby, who provided much of the subtle humor, and his partner, a droll Robert Culp, as American cloak-and-dagger agents who could see the humor in situations and took a casual approach to life and their job—which was always Jerry's, and his father's, style. Moreover, Cosby, who was the first black performer to have a starring role in a regular dramatic series on American television, and who won three Best Dramatic Actor Emmys three years in a row, fascinated Jerry—who had never known a black person until he got to college.

Years later, he acknowledged that Cosby had an enormous impact on him. "When I first listened to his album it transformed me," Jerry pointed out. "I can't say why, but it makes a difference for people to know that they can go see somebody or turn on some show and it's well observed, it's respectful of their intelligence, and you laugh. It's like sex to me."

The obsession with Cosby led Jerry to begin experimenting with comedy in the privacy of his room, or at Chris Misiano's house. "There would be times when we would sit in my den with a tape recorder and try to come up with material," said Misiano, whose family moved from Queens to a modest section of Massapequa in 1968—the reason being so the family could be closer to a relative who could oversee Chris because his parents worked long hours. At fourteen, both loners, Jerry and Chris, whose mother was a JC Penney's office supervisor, and father a foreman in a print shop, bonded when they discovered that each had a still-burgeoning interest in comedy and performing. "We did stuff about the supermarket—the cart with the bad wheel, the old woman with the cart that squeaks all the time—kid stuff." Like Jerry, Misiano would go on to bigger things, beginning as a lowly grip, and later becoming a cameraman and director for such quality shows as *Law & Order, ER,* and *The West Wing*, among others.

"Jerry would be at the house drinking milk out of the carton at our refrigerator," recalled Misiano's older brother, Vincent, who also became a highly respected cinematographer and director for television. "Jerry was somebody who was clearly always interested in comedy, in performing, and Chris actually wanted to be a performer and took acting classes, but then decided, after having gone into it, that he was happier behind the camera than in front of it—but had a real appreciation for people like Jerry who would go up on stage.

"They'd sit in the basement and make comedy tapes trying to duplicate the Mel Brooks records, or the José Jiminez records, which they used to listen to and say, 'This is really cool.' And he and Chris would try to duplicate it. They would record themselves. And I suspect that within the Seinfeld household being funny was something that was very, very highly regarded. Kids aim to please, and that was one way for Jerry to get over with Dad."

The Misianos' mother, Tula, said she usually had no idea what the boys were up to in her home because of the long hours she and her husband spent at work, commuting back and forth into the city. "I think what bonded Jerry and Chris was their interest in theater, which I was unaware of at the time," she said. "I'm very straight arrow. I never thought of this business—television and movies. It was totally out of my ken. I thought CPA would be good, a lawyer or something. But it was a joy to have Jerry here with Chris. I just reveled in the fact that they were friends and that they weren't getting into trouble. They were never rowdy, never wild. There was just a lot of laughing, making up skits and performing for each other. I just thought of what they were doing as a goof, writing little skits and laughing. I never really thought of star quality. You would think two boys being together so much, they would argue, that some jealousy could get in the middle, but I never heard one argument. They had a very special relationship."

Jerry always felt comfortable with just a few close friends, like Chris Misiano, who shared his blossoming interest in comedy and performing. "I always wanted to talk *about* things—to get into things in a different way than what people were doing at that age," he observed. "The social experience was never as interesting to me—although I certainly felt painfully left out—but I wanted to talk about other things, like Jonathan Winters. 'Isn't it great the way he does this bit?' So I'd make friends with the real outcasts. Because they were grateful for any kind of contact, they'd discuss anything with me—I felt more comfortable around what would be generically called 'nerds.'"

His friend David Elkin, who became an anesthesiologist, saw those traits in Jerry. "He had an intellectual curiosity beyond the comedy. He was a very questioning young man, questioning all those things that life is filled with, like the creation of the universe. Jerry and I would sit outside his house at night and he'd ask questions—'If you could only have one question answered before you die, what would that one question be?' He'd ask about the existence of God. Jerry didn't always look at life like everybody else did. He was not a conformist. He would look at the angle rather than the straight line. He would not think about doing things in the most logical, most routine way."

By his junior year, Jerry had grown somewhat more confident socially and started to emerge from his shell. His eighth-period social studies teacher, Joseph McPartlin, couldn't help but notice the wit and jocularity of the short, skinny boy, who sat in the last seat of the first row.

"Jerry was a gentle cut-up, you couldn't get angry at him and so I gave him an award, which I called the 'Brattinella Award'—just a made-up name—for being the biggest cut-up in the class," McPartlin recalled. "The kids used to love these awards. I'd give them these big, twenty-four-hour lollipops, and the biggest one was the 'Brattinella.' I gave it once a year in June, and they actually vied to get that award. When Jerry won, he got up and wanted to make a speech, and started thanking everybody in the class for voting for him. He was hands-down the funniest in that class. I'd be locked into the War of 1812, and he'd be making funny comments in the back—but with a light touch. There were kids you could really explode at, but not Jerry. He was a nice kid, obviously bright, but he looked like he was floating through things—very casual, always meticulous, even in jeans. He had these jeans with a rip, but precisely where the rip was supposed to be. He took care of himself that way."

Jerry was in McPartlin's last class of the day and often, after the bell, he would wander up to the front of the class as the others raced out, to kill time chatting with his teacher as he marked papers. "He wasn't getting the bus anywhere, so he'd sit and talk—it could be sports—the Mets—and we'd do this a couple of times a week. He had a kind of quiet confidence, kind of knew himself, always saw the light side of things. But I didn't know that he had any kind of artistic bent."

Longtime driver's and phys ed teacher Al Bevilacqua, who dubbed himself "the coach of the stars" because he taught such Massapequa stalwarts as Jerry, Joey Buttafuoco, Jessica Hahn, Ron Kovic, and the Baldwin brothers, to drive or climb a rope in gym, was never impressed with Jerry's sense of humor, and, like his other colleagues, was shocked by Jerry's later success. "I'm a comedian too," said Bevilacqua, whose name and hard-bitten persona was once featured in an episode of *Seinfeld*, "so I can remember cracking jokes, trying to leave the driver's ed class smiling. I must have cracked a joke about some of the crazy things that happen in a car and Jerry said afterwards, 'You know, you're really funny, coach, you belong on the stage.' I said, 'Yeah?' And Jerry said, 'Yeah, there's one leaving in the morning,' and with that he walked out. Maybe that was his first joke, and I laughed, but he wasn't a funny guy—he was quiet, in the woodwork."

For one whole semester Bevilacqua looked at Jerry through a rearview mirror—and it was a face he never forgot. "Out of every million people there's one person who should not be on the road. We had this one girl in Jerry's driver's ed class and it was fifteen minutes of terror every time she took the wheel. And Jerry would be in the back seat and he'd have that expression, that same smirk that I noticed on his show. And then I'd tell the three kids in the car, 'Look, we gotta hang in with this girl,' and I'd see Jerry in the mirror, like shrieking with his teeth—the same exact expression of horror, anger, and exasperation that he had on the show with that fat guy Newman."

Bevilacqua obviously made an impression on Jerry. Not only did he use his character on the show, but he also commented in *SeinLanguage* on the "strange occupational choice" of a phys ed teacher. "What is this job? You're walking around in shorts with a whistle all day long. You've got an office next to a shower, you're torturing and humiliating young boys all day. It's weird. Always walking by the shower, it's like a porno movie."

—9—

Romeo and Juliet

Julius Caesar was one of the literature assignments in Norma Rike's tenth-grade English class, and when it came time to assign parts Jerry asked her if he could play the role of Caesar. "He was a good student, he enjoyed literature, so I said fine," Rike recalled. "Well, he was so disappointed to find out that he got knocked off in the first act. He said, 'Is that all?' So we just went over the death scene again and again so Jerry could emote a little bit. He was very good."

By eleventh grade, Jerry decided to take a shot at performing publicly on stage for the first time by joining Massapequa High School's drama club. Still very shy, extremely nervous about being in, or a part of, an organized group where he'd have to mix, he needed a bit of prodding, which he got from Shepherd Goldman, a talented student actor and director who helped create the club.

"My son saw that Jerry had the ability to act, recognized within Jerry that he had the potential to be a performer," asserted Shep Goldman's mother, Ruth, Betty Seinfeld's friend and confidant. "Shepherd stimulated Jerry and got him on stage."

After a number of auditions, Goldman, who was a senior, cast Jerry as Sampson Capulet in the club's memorable production of *Romeo and Juliet*, which was the school's one and only sell-out, a box-office smash at the time.

The playbill said:

"*Jerry Seinfeld emerges on stage for his debut. He is enrolled in the drama course in school—both his classmates and his friends know Jerry for his wild sense of humor. He absolutely loves cars and believes in living for today . . . We would like to thank Kal Signfeld Signs and Village Pizza for their sponsorship.*"

But Kal's involvement with the club outlasted Jerry's, who drifted away after only one play while his father continued to make signs and posters promoting subsequent drama club productions, and, of course, Signfeld Signs.

"In the crowd scene in the beginning of *Romeo and Juliet*, the fight between the two families, Jerry got hit over the head with a sword and apparently decided drama wasn't really for him," recalled Ann Pastorini McPartlin, who taught English at Massapequa High and was the drama club's advisor. "Jerry never tried out again after that."

There are some, however, who theorize that Jerry fled the club not because he was killed off in the first act, but rather because of a more personal, complex reason. Already viewed by classmates as a "nerd," he now feared he might be labeled a "queer" like some other club members. In the early 1970s, conservative, provincial Massapequa High had not yet come to terms with gay liberation, which came of age in June 1969—only two years earlier—when the patrons of the Stonewall Inn, a gay bar in Greenwich Village, a short train ride from Massapequa, fought with police who had rousted them. Their reaction was unprecedented, and the ensuing melee sparked the gay movement.

"The drama club had a lot of gay people in it," acknowledged Spence Halperin, its vice president, who played Cardinal Woolsey in *Romeo and Juliet*. "The group identified itself as being kind of gay, kind of out there, welcomed anybody, was on the fringe. And Jerry was definitely not out there in any way, was very down the middle, so that's why he might have left the club."

McPartlin, whose husband, Joe, had given Jerry the Brattinella Award, agreed that because of the club membership's sexual orientation "a lot of guys didn't join—and back then homosexuality wasn't even openly spoken about. We did have guys who left the club because of the gay thing. All Jerry said was, 'It's not for me.'"

Club member Grant King, who became a professional singer, said that at the time none of the drama club students who were gay had "come out, or had even acknowledged to themselves that they were gay," like himself and, he said, Shep Goldman, among others, who revealed their sexual preference after they got into college. "I don't think there was even a question of publicly announcing you were gay at Massapequa High in the early 1970s. I didn't see *anybody* consider it," he noted. "But people talked about people being gay. Everyone thought I was queer. I was once attacked, run off the ice at a lake by a bunch of guys who chased us through the woods and terrified us and smacked us around with their hockey sticks and called us faggots."

Looking back, King, who was Shep Goldman's best friend in high school, saw the drama club as "a magnet for people who felt 'other.' The biggest club at Massapequa High was sports. If you weren't on the football team, or the baseball team, or basketball team, you were considered 'other.' And the drama club drew a lot of people who were creative in some way, and that can be a magnet for gay people as well. The club was also highly competitive—the competition was for attention, and there was all the typical adolescent stuff, lots of gossiping, lots of intrigue.

"Why Jerry left the club was never explained but I never got the impression he was homophobic, or nasty, or intolerant of anybody—but, of course, that's filtered through many, many years. I certainly had my own run-ins with people at that school who made it clear they despised me, but Jerry was never one of them."

Sadly, Ruth Goldman's talented son, who had a promising career in theater and music as a writer and composer, died in the early 90s. Whatever her son's sexual identity, Goldman asserted that Jerry had blatantly snubbed him, and had kept his distance from other club members. "Jerry and Shepherd became friends through the drama department," she explained. "Jerry came to my house for the cast parties. But after that show they were not that friendly. My son was interested in the arts, music, and Jerry's interests were different. Jerry was not friendly with any of these drama club kids. First of all, truthfully, Jerry was not *liked* by those kids because he was *very* standoffish. And if he felt that way [about the sexual orientation of the other drama club members] he shouldn't have been in the drama department."

Shep Goldman felt rejected by Jerry in the wake of *Romeo and Juliet*, Ruth Goldman recalled. "Let's put it this way—they met on the Long Island Railroad one day and Jerry wouldn't even say hello to Shepherd. He just ignored him."

In the mid-90s another member of Massapequa's drama club, Tracey Revenson, who became a professor of psychology at New York University, held a memorial service in her Greenwich Village apartment for Goldman, whom she described as "a remarkable person—mesmerizing, brilliant, creative, funny, sort of the kind of central person everyone gloms onto." Many alumni of the once close-knit drama club attended, but Jerry was not among them. "He wasn't even considered," she said.

Like Ruth Goldman, Revenson said she never had the impression that Jerry desired to be a part of the group after *Romeo and Juliet* closed—most likely because of some of the club members' sexual orientation.

"There was this unspoken, semispoken thing that it was okay to be gay," she said. "It wasn't on everybody's tongues, but somehow the drama club created a sort of setting where anybody could be who they were, like the guys who weren't interested in girls. It was a group of people who would get together and sing Melanie songs—and I never got the sense that Jerry wanted to be a part of that."

Jerry wasn't the only professional stand-up comic who came out of Massapequa High School and the drama club. The other was Marga Gomez, who was also in Jerry's class of '72. Like Grant King and Shep Goldman and others in the club, Gomez said she wasn't sure of her sexual orientation in those days, and "wasn't particularly happy" at Massapequa High, she said, noting, "Most high schools are not good places to be gay." While she acknowledged her close friendship with members of the drama club, she avoided commenting on her view of Jerry's role in the group and his reason for leaving. "I don't really want to go there," she said. An exotic-looking Latina, Gomez studied drama and creative writing at Oswego College, where Jerry first matriculated, and later was reviewed by comedy critics as one of the premier lesbian stand-ups.

Years later, when Jerry became a big name himself, he became the subject of gossip among friends, and within the entertainment industry, that he was either gay or bisexual. Privately, Jerry was concerned about the talk, and efforts were made by his people to make him appear more macho. Even Ruth Goldman, back in Massapequa, had heard the talk. "My niece is an actress in Hollywood and she used to tell me that Jerry was gay, and I said, 'That's such Hollywood crap!'"

—10—

To Israel, with Love

In the summer of 1971, between his junior and senior years, Jerry joined several dozen other New York–area Jewish teens for an eight-week $800 tour of Israel that included living and working on a kibbutz as part of an exchange program. But Jerry was not a happy camper about the prospects of spending his last vacation, before going off to college, in the Promised Land.

"He whined and grumbled about it for two months before he left," recalled a family friend. "He didn't want to go. He wanted to stay home, watch TV, mess around. He felt Kal was forcing Israel on him. It was just like Jerry's feelings about the shul—he could live without it because it was his father's thing. It was Jerry's form of rebellion. Some kids got into drugs. Some kids shoplifted. Some kids joined the Marines. Jerry acted like going to synagogue and visiting Israel was on a par with being circumcised on a moving stagecoach."

But in the end, being the good Jewish son that he was, Jerry methodically and obsessively packed his bag with neatly pressed jeans and T-shirts, his high-top sneakers, extra dental floss, spare toothpaste, and an abundance of soap. It wasn't as if he thought Israel was uncivilized, it's just that his compulsive cleanliness was already firmly entrenched. He also took along a green flowered tablecloth that was a gift for his host family, bade farewell to Massapequa, and reluctantly flew off to Tel Aviv with the others in the group, including his pal David Elkin.

They spent a week or so together touring the country, during which time Jerry, who had switched from eyeglasses to contact lenses, lost one of them. "We all were on the grass in front of his hotel on our hands and knees with a flashlight looking for the goddamn lens, which we never found," recalled Vic Elkin who, along with David's mother, had flown over to visit the boys. "As I remember," he added facetiously, "that's about the most exciting thing that happened on their trip."

Was he ever wrong.

When, in 1998, the news media treated the last episode of *Seinfeld* as if it marked the end of all civilization as we know it, a journalist in Israel, Amir Kaminer, of the daily *Yediot Aharonoth*, jumped on the bandwagon and decided to nose around for something juicy about Jerry at Sa-Ar, the banana-plantation kibbutz in the north near Nahariah, where he had stayed more than a quarter century earlier. *Seinfeld*, which was broadcast a season behind in Israel, had developed a huge following there because of its cynical, ironic, edgy Jewish*y*–New York*y* humor.

A *Seinfeld* fan, Kaminer had once read comments Jerry had made about the trip, which weren't very positive. "I didn't like the kibbutz," he had stated. "Nice Jewish boys from Long Island don't like to get up at six in the morning to pick bananas. All summer long I was looking for ways to avoid work."

When Kaminer got to Sa-Ar, he had hoped to discover that Jerry had lost his virginity that summer—Jerry had once claimed in a *Playboy* interview that he lost it at "nineteen or twenty . . . I'm not positive, to tell you the truth"—but he came up with basically *bupkis*.

The intrepid reporter, however, did find a member of the kibbutz who recalled that Jerry and an American girl named Janice—no one could remember her last name—had become "boyfriend-girlfriend." But that was shot down when Kaminer interviewed Itzhak Ben Shalom, who declared, "Oh, no. They had nothing going on. I think *she* was in love with him. Jerry was nice and quiet, and holding back." Another remembered Jerry as "a nudnick," adding, "Seinfeld was like a cactus—not impressive and no one paid attention to him." Jerry was constantly making jokes that no one understood, according to a woman. "But he laughed all the time at his own jokes." But Jerry's host family, Shoshanna and Emanuel Perreg, had fond memories. "It was really strange for me, for a teenager to give me a present," she said, referring to the tablecloth. "We had many volunteers at the house and none of them *ever* gave me a present." She remembered Jerry as "a really serious guy. We gave him lectures about the kibbutz, values, education."

In the end, after extensive legwork, Kaminer had gotten as much detail as he could for his story, including a fading kibbutz photograph believed to be that of the future famed comedian—a skinny, bespectacled teenager "who seemed to be full of thoughts" sitting at a table during a birthday party.

But there was much more intrigue to Jerry's stay in Israel that summer of '71. In fact, as Kaminer had suspected but could not prove, Jerry had fallen for a girl, the self-described "fucked up" Jewish daughter of a Queens accountant, who became his soul mate and perhaps more.

Jerry and Cathy Ladman discovered in Israel that they had much in common: both were fans of Bill Cosby, Vaughn Meader—like Jerry, she was knocked out by his *First Family* album—Jonathan Winters, Robert Klein, Elaine May, and Mike Nichols, and, like Jerry, had vague hopes and dreams of one day becoming a comedian. And, like Jerry, she felt alienated. As with Lonnie Seiden, who was still part of Jerry's small circle, Ladman fit the profile perfectly: she saw herself as an outcast, having spent "years of being fucked up by my parents. I was very restricted when I was growing up. Which is why I graduated at sixteen—to get the hell out. My father was very strict, overbearing . . . I was never good enough."

In Israel, she acknowledged that she and Jerry formed a strong bond because of their similar interests and feelings.

"He was my first boyfriend, really, and we used to share our dreams of becoming comedians." Ladman was working on another kibbutz near Sa-Ar, but saw Jerry virtually every day. "We both confided in each other, we both were very sensitive," she said. Riding on a tour bus once, Jerry suddenly stood up, acting like their guide, and cracked her up. "On your right," he said, directing his humor at her, "you have my left hand, and on your left, you have my right hand." With that he had won her over, heart and soul. "He was *so* funny," she said, laughing years later about the incident.

Back at home, Jerry told Elkin that he'd met a girl he really liked. "I remember him sort of coming home with somebody from the trip—Cathy Ladman—but he never talked much about her." Jerry was even more secretive with Lonnie Seiden regarding his relationship with Ladman. "Jerry brought me back my 'swucky' from Israel and I still have it—it's an olive-colored wood duck or swan and we called it a 'swuck' because we couldn't figure out whether it was a swan or a duck. But he never mentioned anything about her to me."

That was Jerry's modus operandi for handling relationships throughout his life—juggling them, compartmentalizing them, being secretive about them.

Later, Jerry implied to Caryn Trager, with whom he became involved about a year later, that he and Ladman had been intimate after meeting in Israel. "My memory holds that Cathy Ladman was the first," Trager said. "He met her on the kibbutz and I think that's where it happened. He was very fond of her." If Jerry was candid with Trager, his comments to *Playboy* about losing his virginity at nineteen or twenty were false.

In September 1971, Jerry returned to Massapequa High, seemingly under Ladman's influence. Following her lead of leaving school early, Jerry had decided to take an accelerated schedule as a senior so he could graduate in January instead of June '72. Jerry and Ladman were privately linked on and off for some years to come, and she later credited him for giving her the confidence to become a stand-up comedian and comedy writer.

Jerry also convinced his friend, Chris Misiano, to take a bigger class load and graduate early with him. "We married our schedules together, so we could get out," he said. "We felt like it was exciting to move on, and we had had enough of high school. It was fun and it was okay, but we wanted to grow up."

Jerry's English teacher, Norma Rike, noticed an extreme change in Jerry when he returned to school from Israel.

"After that summer he made it clear to me that he really wanted to get out of Massapequa. He only had one or two male friends that I ever saw him with. Maybe calling him a loner is putting it too harshly but he wasn't with the 'in' crowd; he wasn't a jock and he wasn't with them. By the time he came back to school in the twelfth grade he seemed to be awfully cynical. He found everything wrong with everything. He told me, without being specific, 'People are always out to get you,' things like that. But he was not the most forthcoming student I ever had. He was a very private person. He didn't talk a lot about himself, or about his father or mother, whom I never met, never had a conference with them. Most of all he just thought Massapequa was the dullest town in the world. He told me, 'I can't wait to get out.' He was close to another English teacher, Roberta Schaeffer, and she used to tell me, 'Oh, Jerry Seinfeld's so cynical about everything.' She said, 'He's got a sit-down attitude.'"

While Ladman appeared responsible for Jerry's decision to graduate a half-semester early, she also is said to have played a role in his sudden dark moods. At some point after he returned to school buoyed by having her as a girlfriend, she is said to have suddenly ended the relationship—at least for a time—and he was devastated.

The other reason given for Jerry's decision to leave school and home early had to do with his father. A number of Seinfeld family friends, among them Ruth Goldman, believed that as Jerry got older he found it increasingly difficult—one called it "smothering"—to live under the same roof with his father, to be in his shadow, despite the fact that Kal was extremely supportive about anything Jerry wanted to do.

"Jerry was living with a father who was very funny every single minute of the day," Goldman observed. "Kal was everywhere, involved in everything—his face, his name. It definitely might have cast a shadow over Jerry psychologically. When someone like Kal is *out* there all the time, children like Jerry can't compete, they become withdrawn. It's true that Jerry did come to hate Massapequa, and wanted out. It's not easy growing up with someone who was performing all day as Kal did. That is hard."

Jerry quickly applied to, and was accepted at, the State University of New York at Oswego, located on the shores of Lake Ontario in the frigid blizzard-belt of northern New York. While Oswego was best known as a teacher's college, it also was one of the few state schools that offered some semblance of a radio and TV program. While he was nowhere near the top of his class, Jerry's grades were good enough for him to secure a Regents scholarship, which covered the first year's tuition of $650.

Jerry said a final good-bye to Massapequa High School, which never held happy memories for him. As with the synagogue of his youth that some members say he abandoned after he became famous, there is a Massapequa High faction who have seethed because they contend he has refused to give the school his financial and moral support—unlike the Baldwin brothers, for example, who have donated money and their name to the cause of their alma mater.

"When the Baldwin brothers tried to get Jerry involved in the fund-raising for the school auditorium he always gave excuses," said Al Bevilacqua. "I nominated Jerry for the school's hall of fame, said we've got to get him in. Right at the time they were building up to the last episode of *Seinfeld*, I contacted his agent, his publicist, I said, 'Can you do me a favor? I'm going to fax you something, can you get it to Jerry?' And with that I laid out a letter nominating him to the hall of fame. After weeks of prodding, I got back the message that he can't make it—he's got the show coming up; he's going to Europe. Fine. So the following year I tried again and can't get him, got the same shit. I don't know what it is. His father was such a community guy. So did fame and fortune go to his head? Some people in the community aren't that happy with him."

Linda Roberts of the Massapequa Historical Society said, "The hall of fame is for people who have gone to the high school and made good of themselves. The school has been trying forever to get him in the hall of fame, trying for years, and he never answered a letter, a phone call. He absolutely ignored them."

Unlike most anxious parents about to become empty-nesters, the Seinfelds did not personally escort their son to college. Jerry, who had skipped his prom—he had no date, no girl—left home for the brave new world of campus life without any pomp, little if any circumstance, and no tears. With just a quick wave good-bye, he zoomed off with his friend Cliff Singer at the wheel of his fast, metallic-gold Pontiac 4-4-2, accompanied by Singer's girl Lonnie Seiden and Michelle Cohen, Chris Misiano's steady, for a farewell road trip.

"It was a great weekend," Cohen recalled. "Chris couldn't go, so I went with Jerry and the others. It was a fun, funny time. I remember Jerry in the car doing an imitation of John Wayne doing an imitation of a chicken."

As Seiden saw it, "It was a last hurrah. We were all saying good-bye."

Their first destination en route to Oswego was the small town of Callicoon in New York's "Italian Catskills," where the Seinfelds, in partnership with Kal's sister, Jennie, and her husband, Manny Bloom, had just bought a cottage with a pond on more than a hundred acres. Betty thought the property would make a great investment because it bordered on the expanding lands of the flashy Villa Roma Country Club, a weekend and summer family resort that was to New York Italians as Kutscher's or Brown's in the Catskills was to New York Jews. (Two decades after the very shrewd Betty Seinfeld first spotted the land, she and another Massapequa couple sold the property to Villa Roma for ten times the original purchase price.)

The Seinfelds and Blooms had bought the place from the Grassos—Cosmo and Gertrude. Years later the zany character Kramer on *Seinfeld* revealed during a hilarious episode that his real name was Cosmo, a moniker that had struck Jerry as funny ever since first hearing it during his parents' real estate transaction. He kept the name in the back of his mind and eventually found a way to work it into his show.

On arrival in Callicoon, Jerry and his friends went to a movie—Michelle Cohen still had the ticket stub as a memento thirty years later. Afterwards, they had a celebratory dinner at the Villa Roma, and then returned to the Seinfeld house to spend the night before leaving in the morning for Oswego. And there was a good reason to spend the night at the Callicoon place: Kal and Betty, apparently a couple of swingers at fifty-four and fifty-eight, respectively, had furnished the master bedroom of their new country digs with an enormous undulating waterbed.

"Being Cliff and I were together, Jerry gave us the master bedroom," Lonnie Seiden clearly remembered. "It was the first time I had ever seen a waterbed. Jerry spent the night with Michelle, and Cliff and I actually did it on Jerry's parents' waterbed. I remember Michelle saying that Jerry was trying to get it on with her—his best friend's girlfriend—and she wanted nothing to do with him. The next morning I said to Michelle,

'So?' And she said, 'Oh, puh-leeze!' I said, 'So nothing happened between you and Jerry?' She said, 'No way!' "

Cohen had a somewhat different take on how the evening went. She said that Jerry's come-on "was very subtle. We slept in the same bedroom because there wasn't another place to go, another room. I think there were just two bedrooms. We slept in the same room, but I wouldn't say we slept together. Certainly nothing *completely* happened, but things started to. It's not like in high school Jerry was fighting women off, okay? Chris and I were involved, but Chris never talked about that night, and I don't know if Jerry ever talked to Chris about it. Jerry and I just kept it to ourselves."

The next day after breakfast the triumvirate of Singer, Seiden, and Cohen drove Jerry to Oswego. "We all kissed and hugged him good-bye and said 'see you soon,' " Seiden said, "and that was it."

—11—

Sex, Drugs, and Suzukis

A loner at home, seventeen-year-old Jerry settled easily into the frenzied life of the Oswego campus. While many of his classmates pined for their families and friends, Jerry was ecstatic about his new independence—overjoyed to be far from Massapequa, away from his omnipresent father, done forever with the synagogue and high school he rejected.

Where he once shied away from groups, Jerry was more gregarious, entertaining those around him with his quirky observations. In the cafeteria and student union, where he hung out between classes, he began to draw a small following, including cute coeds from small towns upstate who were charmed by the skinny Jewish kid from Long Island, the jokester with a shag haircut who sported high-top Converse sneakers and neatly pressed jeans.

Roseanne Mecca, for one, a Catholic girl from near Buffalo, said she had never met anyone like him before. "We came from different worlds. He wasn't like the other guys I'd known, or the ones who were at

Oswego, with the ponytails. He had a whole new twist of looking at things, really appreciated the whole dining-hall experience where he was a great observer of people. That's what I was attracted to."

A junior majoring in elementary education, Mecca was two years older than Jerry when they began dating after meeting in the cafeteria a few weeks after the semester began. Mecca fit Jerry's female archetype perfectly—long, straight, dark brown hair, brown eyes, and she was petite, about five feet tall. Years later, when *Seinfeld* aired, friends of Mecca's who knew she had dated Jerry called to exclaim, "You're Elaine!"—because of her looks and mannerisms.

Jerry was infatuated with Mecca and made it clear from the start that he wanted their friendship to become serious. "But I was hung up on a couple of things," Mecca explained. "As far as having strong romantic feelings toward Jerry, I just didn't have them. We'd be okay for a couple of weeks, and then I'd think, 'this really can't work.' We'd have these long, intimate talks about how maybe we should break up. He would talk about it *endlessly*. He was *so* open about his feelings, and in a way that kind of put me off. He was *very* sensitive, and it was like almost too much to deal with."

During those chats Jerry detailed his relationship with Cathy Ladman the previous summer in Israel. "He said he had been quite hung up on her, and that he had been brokenhearted that it ended, and that I was his first girl since her," Mecca said. "Even though he was pushing for a closer relationship with me—a physical one, which never happened because I was not active sexually—he also was kind of afraid to get too involved with me, because he got hurt by that girl. He said he didn't want to get hurt again."

By early spring Jerry's efforts to advance the relationship had failed and his ardor had apparently cooled, but they remained friends. For her birthday in March, Jerry sent Mecca a get-well card, crossing out "Get Well," and writing "Happy Birthday."

Several weeks later, again working his material in the same student dining-hall, Jerry sat down next to a girl who didn't fit his type at all. But their meeting was the start of what became a serious, tumultuous relationship that continued through the mid to late 70s.

A freshman psychology major from Forest Hills, Queens, Caryn Trager had a pretty face and straight, dirty-blond hair, sported aviator glasses over her hazel eyes, had a sweet disposition, a raucous sense of humor, a hearty laugh, wore tight denim bell-bottoms, walked around most of the time in her bare feet, and smoked pot—"kind of a hippie, but not a *total* hippie."

Moreover, at seventeen, she bore a striking resemblance to Barbra Streisand—"without the nose," she pointed out. But to Jerry, the Streisand look was a turn-on because by the time he got to college he was a rabid fan. After he and Caryn started going together, which was immediately after they met, Jerry would proudly point Caryn out to friends and say, "Doesn't she look just like Barbra?" And then he'd do a campy Streisand impersonation.

Physically, Trager was big-boned, weighed about 175 pounds, and in heels, stood a couple of inches taller than Jerry. Emotionally, she was considered by some to be neurotic and to have an inferiority complex, believing that she was unattractive and obese.

Because of their physical and emotional differences, some campus pals who became an integral part of Jerry and Caryn's circle, such as Trager's Cayuga Dormitory close friend LuAnn Kondziela, thought of them as an odd couple.

"Caryn was much taller and bigger and Jerry was very slight and small and shorter, so I remember being surprised that he was going out with her," said Kondziela, who became a nurse and remained friends with Jerry for years after they left Oswego. "Caryn always put herself down, always talked about how she wasn't pretty and that she weighed too much. She'd do that in front of Jerry and he would always pick her up—say nice things about her and tell her don't do that—don't put yourself down. Caryn had low self-esteem and I think Jerry tended to bring out the better part of her. But I always felt they were pretty evenly matched because they liked the same things, did the same things. Personality-wise, they meshed."

Another member of Jerry and Caryn's crowd, Kondziela's roommate, petite, blond Michele DiCarlo Cerrone, whom Jerry nicknamed "Mush" and who also kept up a friendship with him long after college, felt Jerry and Caryn made an odd mix. "I always thought that they didn't belong together because Caryn, who I liked, was kind of nutty. She would lock herself in her room for hours and smoke packs and packs of Marlboros and drink Pepsi. She seemed like the most unhealthy woman I'd ever met. I saw their relationship as a real hot-and-cold kind of thing.

"They were both very intense people, but I could never understand the relationship because they never seemed to be happy. I just remember her being so high and low, always volatile, you never knew what you were going to get. I was pretty lighthearted and I wasn't the brightest light bulb in the case, so I probably didn't get as intensely involved with Caryn because she spooked me on that level, she scared me. On the everyday lighthearted level everything was fine, but when things got heavy-duty and heavy-hearted with her I found myself not wanting to be around her—like the times she used to lock herself in her room and she'd turn her music up—she loved Barbra Streisand, and she looked like Barbra Streisand—and she would turn Streisand's 'Don't Rain on My Parade' up so loud you could hear it from outside the dorm. Maybe she was a case study for Jerry."

From the moment they met, though, Jerry and Caryn were smitten.

"It was an unusually hot day and I was all sweaty and angry and in a foul mood and Jerry sat down next to me and we started talking," Trager recalled. "I felt extremely unattractive and I thought, 'If he's interested in me the way I am, this could be okay.' I couldn't believe someone so cute, so adorable, so bright and funny, could like me."

With Caryn, Jerry revealed his romantic side. He gave her little gifts, nonsensical but endearing things he found—buttons that had fallen off of classmates' jackets—Caryn soon had a whole collection of them; rocks and pebbles he found; and there was a stream of handwritten love notes. "I'm sitting here in the middle of English class," said one. "I'm always thinking of you."

The same extreme sensitivity in Jerry that was a turnoff to Roseanne Mecca was welcomed and admired by Trager, though at times Jerry could be emotional in the extreme. "Jerry's sensitivity was part of the beauty of our relationship," Trager observed. "He was willing to talk about his feelings. He talked about *everything,* which was unusual back then. You couldn't find that unless you were with a gay man. He bared his feelings—Jerry had no trouble crying."

Early on, during one of the couple's many tiffs, Jerry became so emotional that he ran with tears in his eyes to the school library where he was discovered sitting on a first-floor window ledge by LuAnn Kondziela. "LuAnn saw him sitting in front there crying and all balled up, I remember she said, 'like a rhesus monkey.' I mean guys don't do that," Trager recounted. "He was sitting right in the window frame of the library, right in the center of everything. And it went on for a while because LuAnn had to come back to my dorm and tell me, and I went out there and saw him sitting there, kind of curled up and crying, and I talked to him for a long, long time."

Kondziela clearly remembered the incident years later. "Jerry was really upset that they had had a fight," she said. "He was very sensitive and got upset about things like that. I remember talking to him and he just, you know, was despondent, and wasn't going anywhere, and wasn't moving and so I went back to get Caryn. Afterwards I went over to Jerry's dorm to talk to him because Caryn figured he needed somebody to talk to about what had happened, and why he was upset, and why she was upset. He was really broken up."

Others, like Michele DiCarlo Cerrone, had also noted Jerry's sensitive nature. "I never saw Jerry as a sex symbol, and I was never attracted to him in that way," she observed. "But I did see there was an incredible sensitivity there. He was very friendly with girls, with women, with young women at the time, and I didn't see a lot of young men friends around him. I never saw him with a group of guys, but always with a group of women. I just thought Jerry got along with women better than he did with guys—and not as a guy after sex, but more for friendship. Maybe women just laughed a little louder at his jokes, maybe women just understood his sensitivity. He didn't have that rough-and-tumble kind of way about him like a lot of the other guys did."

Not long after Caryn and Jerry began seeing each other, their relationship became intimate.

"It evolved very quickly," Trager stated. "From the first time we were together, we were *always* together. Jerry was very loving and extremely romantic. His friend Chris [Misiano], who became my close friend, had come up for half a semester and was always knocking on the dorm room door saying, 'Don't you guys ever do anything else?' But the sex wasn't very adventurous. We weren't swinging from chandeliers. Jerry was a straight-laced kid.

"From the time I was in high school, even junior high school, I was dying to have sex *over* with. My body developed quickly and the girls I went to school with were very fast in that way. So I was very aware that I was a virgin, and when I met Jerry I wanted all that finished. Jerry was my first sexual partner, but from what he told me, I was not his first woman."

Trager gathered that Jerry had lost his virginity during the kibbutz trip. "When we began our relationship he was still getting over Cathy Ladman," she said. "It was still painful for him." Even though Trager was aware that Ladman was out of Jerry's life, at least for the present, she had twinges of jealousy and curiosity about her. So when she discovered that a friend had once gone to summer camp with Ladman, and knew that Ladman was now attending the State University of New York campus in Albany, Trager decided to size her up for herself from a distance.

"I didn't want to talk to her," she said. "I just wanted to see her, to see who this woman was who grabbed him the way she did. By then I'd heard stories that she was flaky and I loved hearing that people thought that about her. I don't know if Jerry ever knew I went to just look at her. It wasn't a big issue. It was just curiosity."

Jerry was anxious for Caryn to meet Kal and Betty, and when the first semester break came, he made arrangements to squire her home to Massapequa for a visit, which became a near-fiasco.

"He kept telling me, 'Wait until you meet my parents—they're so wonderful, they're so cool,'" she said, laughing at the memory. "Of course the issue of where we would sleep came up at some point and Jerry said, 'Don't worry. It's not going to be a problem—we'll sleep together in my room.' And I thought, 'That can't be. That's not something my parents would allow.' But he kept saying, 'Yes, yes. It'll be cool.'

"Of course, my first introduction to his parents was watching Betty and Kal and Jerry fighting about why I couldn't sleep with him in their home. I waited in another room while they hashed it out. I was all of seventeen and he was now eighteen, but he seemed shocked that his parents were not able to do that. So I slept in the den on a cot. Of course, we crept around all night long."

Aside from that incident, Caryn bonded with Betty and especially with Kal when he discovered she could jitterbug and was an aficionado of his favorite pop music—1940s jive and swing. Caryn had grown up with that kind of music and loved it—her great-uncle was the songwriter Sammy Cahn. "Kal couldn't believe I knew all the songs, but they were like a steady diet in my house," she said. "I came to love Jerry's parents. I felt both of them really took me in. Really cared for me. "

Along with sex, drugs had entered their relationship. While Jerry refrained from taking a toke of Lonnie Seiden's health food store herb back in junior high, and was never seen taking a hit during high school, he smoked marijuana in college because, as Trager acknowledged, "I was a pothead. Once I found it," she said, "I loved it. My friends in college, we were stoned twenty-four hours a day."

But Jerry, a control freak and a very controlled person, wasn't quite as anxious to get high as his girlfriend. "For most of the time that we were together, he did it also," Trager acknowledged. "He'd be fine with it—the laughing, the sex, the eating. But I just don't think he enjoyed it all the time, and I don't think he really liked it, and I don't ever remember him having anything of his own. He might have done it to go along with me because it was me smoking much more than he."

With a steady girl, Jerry now needed wheels. He had spent hours telling Caryn about his passion for fast cars, and now of his interest in motorcycles. They spent many afternoons sitting on a campus hillside overlooking a nearby highway where he taught her to identify passing bikes. After a while, she became savvy enough to tell a two-stroke engine from a four-stroke engine. Because he couldn't afford to buy a car, Jerry opted for a metallic-gold Suzuki for which he and Caryn shopped on Long Island.

"He loved the freedom and feel of a motorcycle," she said.

Their first big trip—to East Hampton, where Caryn had the use of an uncle's weekend house—ended badly. On the back roads of the then less fashionable beach town, Jerry tested out his new machine, zooming in and out of curves, doing wheelies, cutting loose, with his babe on the back wearing a matching white helmet that Jerry had bought her. "They had to be the same color," she noted. "Details like that were very important to him. It was more anal than aesthetic."

As the noise level from his powerful bike rose, annoyed neighbors called the police who arrived on the scene, stopped Jerry and Caryn, and demanded his license. "Jerry went white," Caryn recalled. "He didn't have a license. The cops said, 'If you don't have a license, you don't ride it. We better not see you on it.' We had to put the bike on a truck and my family had to drive him to the city limits and Jerry had to go home.

"Good kids like us didn't do things like that," she continued. "Jerry's a very law-abiding person by nature, not a troublemaker. He wasn't even a person who would ride up the middle of the street when he could. He'd rather wait in line in traffic. He wasn't Marlon Brando in leathers. We were just two Jews on a motorcyle."

The run-in with the East Hampton police, however, stayed with Jerry for years. "He has a tendency to hold benign grudges," said a friend who mentioned Jerry's refusal to come to the aid of his high school and his father's synagogue. When Jerry bought Billy Joel's oceanfront estate in East Hampton for a king's ransom, he told a confidant the story of the motorcycle incident. "You know," he said, "I really hate beaches. I hate this town. But no one can ever run me out of here again."

—12—

Campus Odd Couple

No one in Massapequa who knew Jerry could have imagined him hooking up with someone like Larry Watson as his college roommate. And that included Betty Seinfeld. When Watson arrived unexpectedly for the first—and last—time at the Seinfeld home on Riviera Drive South, Jerry's mother's "jaw dropped and her eyes bulged," recalled Watson, the moment permanently etched in his memory. "I remember her looking at me. She was shocked—*shocked*—to see this black person show up at her doorstep. She didn't invite me in, and I never got an invitation back."

But neither did Betty Seinfeld's reaction come as a surprise to Watson. His first white roommate's mother, a schoolteacher and wife of an upstate New York dairy farmer, stormed over to the admissions' office and declared, "I will *not* have my son living with a nigger!"

At 6'2" and weighing over two hundred pounds, Watson was the biggest, blackest, baddest, most outspoken, most militant, most theatrical and flamboyant dude in Scales Hall, if not in all of the then racially polarized Oswego campus.

Watson could have been from Mars and Jerry from Venus, for all of their differences. Where Jerry grew up in a conservative, lily-white town and graduated from a high school where only three black students had passed through in almost two decades; a town where Jerry never knew a black person growing up except to see the *schwartzes* who came to clean neighbors' houses, and where his father never hired a black employee in all the years he was a popular, civic-minded businessman in Massapequa, Watson was a product of the crime-riddled, poverty-stricken Eleanor Roosevelt Public Housing project in Brooklyn's notorious Bedford-Stuyvesant, and a graduate of a high school named Thomas Jefferson, which was more like a reform school.

But behind Watson's tough, strident, assertive exterior beat the heart of a pussycat. He was a soft, sensitive, extremely intelligent young man who had been told from elementary school on "what I could *not* do academically and professionally, and I wanted to change things by becoming a teacher and kind of heal myself and other students victimized by public school teachers. Some of the early innovative philosophical work in elementary and secondary education took place at Oswego in the late 1800s, so I chose to go to that state school."

Because he had scored so low in his SATs, his high school guidance counselors told him to forget about attending a four-year college, forget about college altogether, that he would not make it through. But he secretly applied to Oswego and proved them all wrong—he was readily accepted and would later go on to graduate school at Cornell, become an assistant dean at Harvard, an associate professor at a prestigious school of music, and have a professional singing career. Unlike Jerry's parents, Watson's mother—"the only soul sister who worked in a kosher bakery for orthodox Jews in Brooklyn"—and his aunts and uncles proudly delivered him to campus, the first in his family—like Jerry—to go to college.

Despite their many differences, Jerry Seinfeld and Larry Watson clicked, and they remained friends through Jerry's rise to stardom. But their "really close relationship," as Watson termed it, was turbulent because of personal and philosophical disagreements that led over the years of their friendship to heated quarrels and long stretches when anger precluded them from communicating with each other.

"It was the end of my junior year when I met Jerry," said Watson, who was two years older than Jerry. "It was just one of those things that we would kind of pass each other in the hall and just start laughing. We just kind of wanted to get to know each other. It was that kind of connection that we had, which was not an easy thing during that period at Oswego because there was a lot of racial tension on campus. I wrote a controversial column on campus issues for the *Oswegonian*, so most people knew me in the context of being very militant and very outspoken and so they stayed their distance—but Jerry broke through that kind of façade.

"So Jerry and I befriended each other and started talking and when my roommate became ill and was not coming back to school I asked Jerry to be my roommate. I wanted to have the best living situation possible. I wanted somebody who was clean. I wanted somebody who I would have a comfort level with, and who I could have control of the room over, who would be the least difficult to deal with. Jerry's humor and kind of free-spirited nature made him an easy choice."

The summer break before they moved in together, Jerry ventured into deepest, darkest Bedford-Stuyvesant to spend a weekend with his new roommate in the Watson's apartment in the projects.

"This was a big thing for this white Jewish boy to come to Bedford-Stuy, which was like Vietnam, a combat zone. White people didn't get caught in Bedford-Stuy day or night without a police escort," Watson said. "My mother cooked dinner and he loved it and then he said, 'Larry, I'd love to see Harlem. I've never been to Harlem.' So I took this white boy into Harlem and we drove up and down 125th Street with me as the black escort. Jerry was like a kid in a candy store—'Look at this, look at that.' He stayed overnight in our little project apartment, and that was the first time I had a white friend stay over. Jerry and I hung out that summer. We'd meet in Manhattan and we just became great friends."

The visit to Bed-Stuy and Harlem was the start of an invaluable sociological and anthropological journey for Jerry, led by Watson and his "brothers" and "sisters" at Oswego, into the world of black culture. Watson did not think of it in those terms at the time, though he put it into perspective years later when he became infuriated—and let Jerry know it—during an angry confrontation over the dearth of African Americans on *Seinfeld*. But all of that was years away.

In the fall, at the start of his sophomore year, Jerry moved into Watson's room, the biggest on the third floor of Scales Hall, and they became tight as roommates do when they live in such close proximity—two single beds separated by a narrow aisle and nightstands. While Jerry's compulsion for cleanliness and neatness and order was hidden from view in the privacy of the Seinfeld home, it became apparent and known in the more public world of the college dorm.

"One of the things that stands out about my time being Jerry's roommate is pHisoHex, a green soap like pHisoderm, that he always smelled of because he washed himself—his hands and his face—*constantly*. Jerry, and our room, *always* smelled like pHisoHex. He was *immaculately* clean, a very clean-cut pHisoHex kind of boy."

In that early 1970s hippie era many college boys wore their head and facial hair long, raggedy, and dirty, but not Jerry, who sometimes spent two hours caring for his face before shaving, and shampooed his hair compulsively. Caryn Trager recalled how he had the ability to "measure out his shampoo so it would last a whole semester. Jerry was very disciplined in that way."

Another college friend of Jerry's, Jesse Michnick, with whom he became close after he transferred from Oswego to Queens College at the end of his sophomore year, said, "In the length of time it took for Jerry to shave, I grew a beard. It was like an application of cold cream, extended hot-water treatments, filling the sink with water and then, finally, the application of the shaving cream. It took hours. A guy wants to get in and out of the shithouse as quick as he can. Jerry made it a labor. It was like painting the *Mona Lisa*."

Unlike other students who brought TV sets, stereos, and other gadgets and toys with them to school, Jerry lived a spartan life with very few possessions—much like his parents always lived, all of which pleased his roommate.

"Jerry brought absolutely nothing to our dorm room," Larry Watson noted. "I have a photo of me and Jerry, and Jerry's sitting on my side of the room and behind us are my pictures of Diana Ross, my pictures of Isaac Hayes, my pictures of the Temptations. It's all my stuff. I brought the stereo. I brought the TV. It was *my* room and that's one of the reasons why I chose Jerry as my roommate. I don't recall him bringing anything, which was fine." As Caryn Trager pointed out, "Jerry didn't have a lot of things."

Friends observed that Watson was the dominant and Jerry the submissive in their relationship, that Watson controlled the room and their living arrangements, their social schedule, and that Jerry went happily along.

"That was my strategy, to have my own room," Watson stated. "I had very selfish reasons for Jerry becoming my roommate. It was like—it was *my* room and anybody who lived there was a spectator.

"I found Jerry humorous, more interesting than funny," he continued. "The term 'clown' is too strong. He was pleasant, good-natured, easy to talk to. There were a lot of white kids at Oswego who were entrenched in positions they hadn't thought through, but Jerry rose above it all and really didn't deal with any of that. With all of my campus political activities I can't remember him *ever* taking a stand on the war, on racial issues—any of that.

"Other white boys on the floor would come to our room and we would stay up until three or four o'clock in the morning struggling and having these major debates and profound political discussions about the issues of the time, and ending up in tears, hugging one another and vowing the next day to do something different. Jerry was never a part of any of that. He constantly told me I was *too* political. Jerry's always—*always*—been so conveniently neutral, so he was the perfect person for the period when his show became so big—this idea of nothingness, being neutral, not having to worry about political correctness, or the role of women, or civil rights—that you could just go back to that illusory world of the *Lucy* show.

"I cannot remember me and Jerry *ever* having a discussion that didn't center around him, which is very unusual because I'm really a dominant figure and most of the time people are listening to me. Looking back on that time, I listened to Jerry and I was bored. We were talking about nothing. Our discussions were not on the politics of the campus, but were really on rather benign, superficial stuff. Jerry and I talked about Diana Ross."

Indeed, the two roomies had compatible tastes in music. They both adored Streisand, were rabid fans of "The Simply Divine" Bette Midler, and Watson couldn't get enough of Diana Ross and the Supremes—"divas" with enormous gay followings whose tunes played constantly on the stereo in the Seinfeld-Watson dorm room. Years later, after he became a professional singer and musician, Larry Watson publicized his musical performances as having "the stage presence and technique of Motown Diva Diana Ross."

When Midler, who made her name playing the gay Continental Baths in Manhattan for an audience of men clad in bath towels, gave her first performance in upstate New York following the release of her smash first album, *The Divine Miss M*, Jerry and his Oswego gal pals—Caryn Trager, LuAnn Kondziela, and a few others—piled into Michele DiCarlo Cerrone's old green Maverick with its black-and-white checkered seats and drove to Syracuse, leaking gasoline all the way to see her show, arriving with an empty tank but getting good spots near the front of the line.

"Oh, Jerry was a *big* Bette Midler fan," Cerrone said. "When I think of Bette Midler I think of Jerry. We always had Bette Midler's music on. Jerry loved 'Boogie Woogie Bugle Boy.' We played that music in our dorm room and Jerry was there and we'd burst into song and we danced. We'd lock our door and learn to polka to Bette Midler music. I found Jerry to be a very theatrical guy."

Jerry did not bring Larry Watson into that circle of white girls; he had a way of compartmentalizing and separating different groups of friends, which became his modus operandi, friends observed over the years. As Watson noted, "When Jerry went off with his other friends, I was not a part of that. When I think back on my friends at Oswego, I knew everything about them. But as much as we were close, I knew really very little about Jerry. I don't know whether I'd say he was secretive, or whether there just wasn't much to know."

While Watson wasn't part of Jerry's girlfriend crowd, Watson integrated Jerry into his circle of "sisters."

"Larry was the quote-unquote ringleader for most of the minority students on campus, and because Larry and Jerry were roommates, Jerry just was always with us," said Carmel Reese Harris, who came to Oswego as an elementary-education major from Syracuse, and later became a banker and married a schoolteacher and minister.

"We took him in—not that Jerry isolated himself from the white students. But Jerry became a part of our crowd and seemed to fit pretty good. We were doing the new dances—the 'Hustle', the 'Bump,' and I was trying to teach Jerry the new dances that were coming out. And then he'd jump up when people were dancing, and try to dance. We had parties and he was always in the middle, laughing and joking and playing and being Jerry. I think everybody pretty much accepted him.

"We were mostly all wearing Afros at the time—I had a pretty big Afro and I was braiding mine, and, of course, Jerry kind of wanted to share everything that we women were doing," Harris continued. "He had curly, kind of long hair, which we used to tease him about—so I ended up braiding his hair, and it was really more fun than anything else. It was the whole thing of accepting him and who he was, and vice versa. Anything we were doing he would go along with. Of all the white kids on campus, I remember Jerry as the only one in our group. He didn't date any of the black girls, but he was more just like a friend. He didn't seem to be interested in having an interracial relationship.

"Jerry impressed me as being a person who didn't have a lot of limitations—racial limitations, which surprises me because of where he came from. I don't mean he tried to take on the black identity—he was Jerry, but he was very down to earth, and very outgoing, and just real, a very real person. As long as you were friendly and accepted him for who he was, he did the same for you. I really think he was trying to learn about us, the whole curiosity of something different, getting away from some of the stereotypes and realizing that people are people, to get the exposure that he didn't get growing up. And it was kind of that way for Larry and the rest of us to some extent.

"Jerry had a much closer relationship with Larry than the rest of us," Harris emphasized. "Larry and Jerry were like two free birds, able to complement and help each other. They did more things together."

Depending on whom one believes—Jerry or Larry—masturbation was one of those things.

In the late 1990s, a writer for Oswego's alumni magazine, having learned that Jerry had matriculated at the college for two years, brought Jerry and Larry together for an interview on the *Seinfeld* set in Studio City to discuss their roommate days.

At one point in her story, Denise Owen Harrigan wrote, "They're talking. Actually they're attempting to talk, but sentence after sentence disintegrates into giggles or escalates into wild hoots of laughter."

The back story for Jerry and Watson's silliness during the interview, which neither Jerry nor Watson saw fit to explain to Harrigan who was writing for a family publication, was masturbation, a subject, like homosexuality, which Jerry dealt with in hilarious episodes on *Seinfeld.*

"I'm certain the famous masturbation episode on his show"—the episode in which Jerry and his pals Elaine, George, and Kramer have a contest to see who can refrain the longest—"came from events at Oswego," Watson asserted. "Jerry was horrified and shocked that some of us acknowledged, or admitted to, jerking off. Jerry and I had an ongoing debate on this subject. And that's why we became hysterical during that alumni magazine interview—because Jerry swore back then when we were roommates that he caught me masturbating one night in our room, and I said, 'You're a liar! It never happened.' So when we went to do that interview and the woman asked, 'What is one of the most memorable things you remember about being Larry's roommate?' Jerry looked at me and burst out laughing, fell on the floor laughing, at which point I burst out and fell on the floor laughing and said to Jerry, 'You bastard! If you even try, I'll sue you!' "

To some on campus, Jerry and Larry were viewed as an interracial odd couple. Because of the racial polarization, there were few, if any, intense black-white relationships similar to the bond that Jerry and Larry had forged. So the tight-knit twosome caused tongues to wag, especially because their circle included an effeminate black student.

But Watson emphasized that any suggestion that he had anything other than "a brotherly, friendly, wholesome relationship with Jerry is untrue, unfortunate, and has no validity."

According to Caryn Trager, "In that little circle they would goof around, play around—they were hilarious together. But I don't remember any big issue about it. I know I never had an issue with it. I never cared about things like that. All my life in my family I'd been introduced to everything and felt comfortable with everything, and I had gay friends. I think Jerry was intrigued by it because he always felt somewhat like an outsider in the world, not part of any group, not part of the popular group."

Watson acknowledged that the effeminate classmate "was a really good friend who was very close to Jerry and myself." But looking back to that time he saw the circle that included Jerry as "straight and narrow kids" who were sexually naive.

"Although we all had notions about what [the friend] was about in a strange kind of way, we didn't know where he would go off to when he wasn't with us, but we never as young kids at Oswego questioned that," he observed. "We as kids at Oswego were so innocent in our own way. We just didn't talk about our intimate relationships. I was never that kind of guy, and Jerry *definitely* wasn't. It was only when I got to graduate school at Cornell and met all of these middle-class white and black kids that all these issues of sexuality came to fruition—where, then, people were graphically talking about and acknowledging their experiences. But at Oswego we quite frankly didn't do it."

—13—

I Just Want to Be Funny

As the young homosexual in the bathroom scene, Jerry Seinfeld was particularly good . . . the acting was among the best seen in student theater . . ."

Most eighteen-year-old aspiring collegiate performers would have been thrilled and exhilarated with such a positive notice, but Jerry was stunned and appalled when he read campus theater critic Kent McKeaver's review in the October 5, 1972, issue of the *Oswegonian* of his debut in *The Word Game*, an avant-garde, improvisational play.

Though it was the first review Jerry had ever received as an actor, and a rave one at that, he was upset that McKeaver had misinterpreted and misrepresented his character as being a gay; Jerry did not want to be thought of as anything but heterosexual, on or off stage.

Yes, Jerry had realistically improvised a young man standing at a men's room urinal as the older man next to him, played by Garry Carbone, turned, smiled, and said, "Hi, how are you? Do you like me?"

But, no, Jerry never considered that he was portraying a homosexual, or, for that matter, that he would be perceived as one; he believed the audience, which had laughed heartily at his responses to Carbone's bizarre come-on, got it—that his character was an innocent caught in some bizarre social discord, and not a men's-room hustler.

"Jerry and I read the review together and we both kind of looked at each other and said, '*Homosexual*?!' recalled Carbone, then a stocky, long-haired, mustachioed junior, an education-turned-theater major from Floral Park, Queens, who, like Jerry, had a natural comedic style. But unlike Jerry, Carbone wanted to be thought of as a serious actor rather than a comic.

"Basically, we weren't even going for homosexuality," Carbone said. "But I guess when people see two guys in a public bathroom, one talking to the other, that's what they think. I said to Jerry, 'How do you think that makes me feel? It means I'm the *older* homosexual.' I played a very strange, kind of nervous, crazy kind of person. And then I wind up having diarrhea and have to go into a stall. It was very raucous and funny. The audience was hysterical."

Barton Ward, the student who directed *The Word Game*, agreed with Jerry's negative assessment of McKeaver's perception of his character.

"There was never *anything* homosexual intended, except that Garry's character was definitely inappropriate and pretty weird and was completely strange and out there," Ward said. "What it was is a guy, Jerry, who comes into an imaginary bathroom, steps up to an imaginary urinal. In front of the audience, Jerry unzips his fly—everything's in mime—and he's standing there, and Garry walks in, walks up to the urinal next to him, and does the same thing. Garry says to Jerry, 'Hi,' and Jerry kind of looks at him, like any guy would, out of the side of his eyes, like 'who's talking to me while I'm standing at a urinal in a men's room.' I mean, men don't do that, and Jerry's character is kind of taken aback, and eventually Garry grabs Jerry's arm and pulls him around, and then runs to the toilet. The scene was all about the inappropriateness of one guy talking to another guy in a men's room. Literally, the audience members fell off their chairs laughing."

By his second semester at Oswego Jerry had seemingly gone beyond whatever hang-ups he had about acting and the personal preferences of some of those who pursued it—the very reasons that were believed to have driven him from the drama club at Massapequa High after his one and only short-lived appearance in *Romeo and Juliet*.

Around the time he bonded with Larry Watson, Jerry had started hanging out with a theater-arts crowd on campus. "He kind of was in that clique, and was toying around with theater," Watson said. "He hung out with a few actors, young white men who weren't part of my circle with Jerry."

Jerry had just started his second semester when he showed up at the Experimental Theater office for an open audition and reading for *The Word Game*. Joe Martiscello, a local high school teacher and playwright, had shown his twelve-page script to Barton Ward when the two were in a musical theater program at Oswego the previous summer, and Ward was knocked out by it and asked to direct.

"The play was Joe's take on the word game that takes place whenever people meet, the kind of set pattern between people of, 'Hi, how are you? Oh, I'm fine, how are you?'" said Ward. "So, I set it as two children meeting and talking; as a young man meeting a young woman, which also featured Jerry; and two old people talking. But the funniest skit of the four of them was Jerry and Garry meeting in the bathroom.

"When Jerry did his reading Joe was amazed," Ward recalled. "Jerry was funny—but I don't remember him cracking lots of jokes. He just hit it off with Garry and the rest of the cast."

Carbone met Jerry for the first time at the rehearsal, but had gotten advance notices. "Everybody had been talking about this funny young kid, and I had done a lot of comedy while I was up there, mostly plays, and I was considered very funny," he said. "But they said, 'Wait until you see *this* guy. He's going to give you a run for your money.' And of course I was very intrigued. At the first rehearsal Jerry said, 'What the hell, I'm going to make believe I'm peeing,' and I just went along. Jerry was very inventive. We just kind of plugged into each other and worked very well off each other."

Despite the positive review, Jerry was disappointed not only in the critic's perception of his character, but also in himself; later, during most of his stand-up career, he had extreme difficulty dealing with, and was highly sensitive to, negative observations of his act by critics.

As a result of the McKeever column, Jerry once again walked away from theater, at least for a semester. "Theater was a real community at

Oswego," Barton Ward said, "but after *The Word Game* Jerry was never really a part of it. After the show was over we didn't talk again."

But Jerry forged a friendship with Carbone, bringing him into his circle with Trager, and taking him on motorcycle rides. "It was weird because Jerry wasn't exactly a rebellious guy in a leather jacket," Carbone observed. "He was kind of matter-of-fact about it. It wasn't like any big statement." And the two hung out with a theater buddy of Carbone's who lived in a house-trailer off-campus where they made inexpensive dinners and laughed so hard that on one occasion Jerry actually spit up his spaghetti.

For a time Jerry toyed with appearing in another improvisational play called *Story Theater,* in which he and Carbone considered and rehearsed singing Frankie Avalon's hit *Venus*, but changing the key lyric to "penis."

"Jerry laughed about it at first," Carbone said, "but after a while he questioned why we were keeping 'penis' in. He didn't see a need to get blue. He said, 'We can be funny without that language. What's funny? It's just the word 'penis.' " While Jerry eventually passed on *Story Theater*, he set the tone for the future, when he always worked clean.

In the communications department, meanwhile, Jerry was among about a dozen classmates who appeared in an entertaining informational video called "Campus in the Making," which promoted student involvement in campus issues.

It wasn't until near the end of Jerry's third semester, in May 1973, that he agreed to be in another play, his second and last at Oswego. *Bringing It All Back Home* was a hilarious situation comedy–like piece of black humor about the Vietnam War, in which Jerry played a pot-smoking junior high school student. It was one of the early plays of Pulitzer Prize– and Tony Award–winning out-of-the-closet dramatist Terrence McNally, whose later portrait of homophobia called *Lips Together, Teeth Apart*, and then *Corpus Christi*, in which Christ and his disciples were portrayed as homosexuals, sparked enormous controversy and even death threats against him.

Jerry agreed to appear as a favor to Carbone, then a senior, who was under the gun to direct at the last minute for the necessary theater credits needed to graduate.

During the rehearsals, Jerry made it clear to Carbone that comedy was his goal, not acting.

"I was in the middle of some drama," he recalled, "and Jerry asked me, 'Why do you want to do this? Why don't you just want to be funny?' I said I wanted to be an actor and felt I had to give all roles a try; that while I knew I could be funny I wanted to play different parts. My thing was to be like Dustin Hoffman. And Jerry said, 'When you think about it, Garry, who do people really like? Who do people really love?' I said, 'I don't know what you mean.' And he said, 'People really love comics and comedians. They remember only the people who make them laugh.'

"That was really the first time I heard Jerry Seinfeld say, 'I just want to be funny,' that he had no desire to do it within a part. He really wanted to just get up there and be funny, and do his stuff, and embellish everything.

"Through the rehearsal period I told him he was funny, but that he had to be funny within the context of the script, which was the reason I cast him. He was nervous about it, insecure about it. He said, 'I really don't want to be an actor.' I said, 'But you're funny in this. You can get it through the lines, through just playing the part because you're perfect for it, and the lines are very, very funny.' But he didn't want to be confined by a script, he really wanted to do a little more of his own stuff. He was saying, 'I want more of *me* to come out.' He wanted to reveal who *he* was and what he felt the world was.

"Even at that early age, before Jerry was even aware that maybe he really could do something, when he was just raw talent, he was very focused on 'I want to make people laugh.' Of all the great comic actors we had at that point, he talked about his love for Abbott and Costello."

Despite Jerry's anxieties, *Bringing It All Back Home* was a success. Jerry hadn't done anything on stage since *The Word Game*, but the audience watched him and remembered him and enjoyed him. He felt good about it. His lines were funny, and he did them funny, and he got his desired laughs. "There was a big difference between Jerry and a lot of the other actors I worked with in college," observed Carbone. "Jerry really had audience appeal."

By his fourth semester, the end of his sophomore year, Jerry had grown more confident socially, was beginning to focus more on some sort of a career in comedy, and was becoming increasingly restless at Oswego, which he felt was "very safe and very cushioned. Maybe I felt Oswego was just too idyllic for me."

The marker event that sparked Jerry's decision to leave Oswego in the winter of 1974 occurred during a visit to the home of Arthur Gittlen, who taught English literature and African-American literature, and was one of the college's most provocative professors—a bald, muscle-bound, radical Long Island Jew who didn't see himself as a white man but as Malcolm X, and was married to a black Puerto Rican woman and had biracial children.

"He was a guy Larry Watson and I were both fascinated with," Jerry said.

Late in the semester Jerry unexpectedly visited Gittlen's home for the first time at the behest of, and accompanied by, Watson. Gittlen was furious and Jerry was "scared to death, about to pee himself," Watson recalled. At the front door, an outraged Gittlen demanded of Watson, "Why did you bring this white man to my house? All day long my wife and my children are gawked at and asked insensitive questions. The only safe harbor to truly be themselves is our home. I screen who I let in this house, and you, Larry, deserve to give me the right to decide who visits us."

Once inside, Gittlen confronted Jerry, demanding to know why and what he was studying at Oswego, and what he wanted to do with his life.

"Jerry said he wanted to be a comedian," said Watson, "and Gittlen just lashed into him. He said, 'You've just sat back on your laurels and been protected by your family putting you through college. If you want to be a comedian get out of this cocoon and get your ass to New York and do what you need to do.'

"He just railed on Jerry, just gave him the business. Jerry was traumatized. On the drive back to the dorm, he barely spoke. It was the first time I ever saw Jerry sad and depressed. The next morning he woke up and said, 'Larry, I made my decision. I'm not coming back to Oswego.'"

Years later Jerry observed, "I was lost. I sensed that what I needed was not there at Oswego. I began to think that New York City would offer more provocation."

—14—

Queens College Days

To get Oswego out of his system, Jerry took a cross-country road trip after his sophomore year, heading west in a used blue Fiat bartered by Kal for some auto dealer sign work. But his destination was not Hollywood, which, like New York, had become a mecca for comedy hopefuls.

With Chris Misiano riding shotgun, Jerry fixed on a most improbable location, a desolate spot that Frank Lloyd Wright once described as "an endless supernatural world more spiritual than earth but created out of it." For a couple of adventurous and contemplative weeks, Jerry explored the deathly still peaks, gullies, buttes, and prairies of the Dakota badlands.

"He had talked incessantly about going there," recalled Caryn Trager. "He liked the whole cowboy thing which he played as a kid. It was a different world for him." There, among the vultures, snakes, prairie dogs, and coyotes, Jerry confirmed to himself that he had made the correct decision to leave Oswego and finish his college education and eventually begin pursuing his career as a comic in New York City. He was still very much a loner, but the otherworldly isolation of the badlands energized and invigorated him for whatever lay ahead. "Jerry was impatient and trying to find his own path," observed Larry Watson, to whom Jerry had written about his trip. "He didn't know yet if it would be comedy or acting. But by this time he saw himself in entertainment, in the public eye."

With Trager about to embark on an adventure of her own—a seven-month junior-year semester in Denmark, marking the couple's first major separation since they met two years earlier—Jerry raced back to New York, the Fiat's little trunk filled with souvenirs, bon voyage gifts for his girlfriend.

Unhappy and upset that Trager was going so far for so long, Jerry felt as if she was abandoning him. "He drove night and day across country and came back early to see me before I left," she said. "Jerry kept saying, 'How can you go and leave your boyfriend? Why are you doing this to me?' But I just trusted what we had. I knew our relationship would survive."

Jerry desperately missed her, writing at least one long letter a day, all of which Trager kept through the years in cardboard boxes filled with mementoes of their relationship. "It sounds like an exaggeration, but it's not," she noted. "He wrote so many letters, over two hundred, that I became friends with the mailman. The theme of his letters was that he missed me. The letters were very detailed about daily stuff in his life, with a lot of comedy shtick. He wrote pages and pages, and he wrote everywhere, even on the envelopes."

When Trager returned, an excited Jerry greeted her at the airport with a new white ten-speed Peugeot bike as a gift. (She didn't know it at the time, but it probably didn't cost him a penny; Kal most likely had gotten it in trade for sign work, but it was the thought that counted.)

At twenty, Jerry, who had come to despise Massapequa and believed he had seen the last of it when he left for Oswego, now reluctantly moved back into his blue plaid bedroom, taking down the Porsche posters and replacing them with pictures of Laurel and Hardy and other legendary comedians, underscoring the change in his priorities; breaking into comedy was foremost, the Porsches would follow. But Jerry's return to Riviera Drive South was, as a close friend noted, "the downside of his decision to try to make it in the Big Apple, educationally and professionally. With no money of his own he was forced to live at home again in the shadow of his father's strong personality."

At the time of his return, the Seinfeld household had a new member—a petite, blond girlfriend of Carolyn Seinfeld's, then twenty-two, who had gone to secretarial school. Carolyn and her friend, who shared Carolyn's bedroom, worked together in the Manhattan headquarters of Amerada Hess, the global crude-oil-and-natural-gas corporation. Jerry didn't take to Carolyn's chum, or their mundane lifestyle, and teased his sister unmercifully. "He thought they were silly girls who didn't have any values, that all they did was work and prepare for work—curling their hair, washing their stockings," recalled Jerry's friend.

But Jerry had enormous compassion for his homely sister and when she began talking about getting a nose job to improve her looks and chances of snagging a husband Jerry asked Caryn Trager to counsel her. Because the Seinfeld family consensus at the time was that Trager and Carolyn would one day be sisters-in-law, Trager took her mission seriously and with great compassion.

Looking back years later, she said, "Jerry and everybody else in the family was concerned over whether Carolyn should or shouldn't have it done. It wasn't my purpose to say, 'Look, you're ugly, get it done.' The whole gist was—'if it'll make you feel better, then go ahead and do it.' "

It finally took family friend Ruth Goldman to get Carolyn Seinfeld to the plastic surgeon. "She definitely didn't have a lot of confidence about herself," Goldman said. "But I told her, 'this is a cosmetic world today.' I recommended the doctor and he did a fantastic job. Carolyn was happy as a lark. It changed her whole life."

His sister's nose job drama wasn't the only disquieting issue that distressed Jerry when he returned to Massapequa. To help pay his way at home he was forced to work part-time at Kal Signfeld Signs, where he showed little enthusiasm for following in his father's footsteps, and disdain for customers, some of whom remembered him as bored, if not rude. Instead of shop talk, he went on incessantly about comedians and show business.

"You should forget about this comedy," Abe Rosenbloom, Kal's head sign-painter, beseeched Jerry, whose response was to roll his eyes and grit his teeth. "What you should do," the old painter harped, "is buy yourself a paint brush, maybe even a hammer, and get a *real* job. Do something respectable. Do something to make a dollar like your father because you'll never make it in that comedy business, kiddo."

While feeling beaten down in the Seinfeld milieu, Jerry enrolled, with great enthusiasm, for his junior year at then-tuition-free Queens College, a gritty, urban campus in Flushing, Queens, last exit on the Number 7 subway line from Times Square. With a large, working-class Jewish and Italian student population from the neighborhoods, the college was more like an extension of high school than an edifice of higher learning, with graffiti and grime rather than ivy covering its walls. But Queens offered what Jerry wanted—decent communications and performance departments, and easy access to the city's potential.

There, Jerry fell into a small crowd of funny, glib, street-savvy theater and radio-TV majors, a couple of whom were serious about becoming actors—Al Pacino and Andrew Dice Clay types—from the blue-collar, pizzeria-and-deli, faux-stone-and-aluminum-siding, row-home neighborhoods of Queens—an Archie Bunker urban version of Massapequa.

The leader of the pack was brawny, mustachioed, prematurely balding nineteen-year-old Mike Costanza, the Dice Clay of the crowd, dubbed by his neighborhood pals "Mr. Excitement"—a class clown who was "the guy to see on campus" for hard-to-get tickets to sporting events, to place bets on horses and numbers, and, as he later acknowledged, for "buying special discount items that had fallen off the back of a truck."

Costanza's first impression of Jerry was that of "a full-fledged, card-carrying nerd," skinny, eyes gaping wide from behind enormous, silver-rimmed glasses. Jerry, on the other hand, was fascinated and mesmerized by Costanza, a printer's son who resembled, talked, and acted like one of the Seinfelds' neighbor Carlo Gambino's heavies, an Italian-American who hung out with what he later described as "aspiring goodfellas." He was an archetype Jerry had never known personally before. Immediately, Costanza became his idol. "My life was like the movie *Bronx Tale*," he asserted years later. "Jerry found that funny."

Of particular interest to Jerry was Costanza's part-time job, which was a bouncer in a gay bar called Flavors, on Lefferts Boulevard in the Kew Gardens section of Queens, a dive operated by an allegedly mob-connected family friend and neighbor known as Uncle Sal. "Flabbergasted" when he heard where Costanza was working, Jerry, in what was a precursor of the kind of observational stand-up he became known for, asked his pal, "What does a bouncer do in a gay bar? What can this job entail? Do you break up fights between drag queens when they start insulting each other? I mean, Mike, is there really any need for a bouncer in a gay place? I have to tell you, the whole thing sounds *very* strange."

Jerry was introduced to Costanza by Jesse Michnick, whose father was a floor-waxing-and-window-cleaning company employee and collector of Nazi war memoribilia. "Jesse," said Costanza, "had set me up as the Second Coming." A media performance classmate of Jerry's with a Catskills sense of humor, a droll style, and love of comedy, the mustachioed Michnick, a drummer for a rock band that played at bar mitzvahs and weddings, lived down the block from Costanza. The two had been grade-school pals, the Jewish kid protected by the tough-guy Italian.

The other member of Jerry's Queens College posse was second-generation Italian Joe Bacino, a handsome, charming, mustachioed, ladies' man—think Tony Manero in *Saturday Night Fever*. An extremely talented theater major, by the 1990s he had won roles as a character actor in such acclaimed shows as *Law & Order* and *The Sopranos*. Back then, though, he was just another neighborhood guy whose mother kept her Queens home so meticulous—plastic covers on all the furniture—that most of the family's living was confined to the paneled basement, replete with full kitchen and bath. That basement was where Jerry tried out his earliest material, with Costanza and Bacino as his audience.

Shortly after Jerry started classes, he took another shot at acting, auditioning for the drama department's hundred-seat Little Stage production of *One Flew Over the Cuckoo's Nest*, winning the part of Martini, the debut role played by Danny DeVito in the 1975 Academy Award–winning, Milos Forman–directed film. "We convinced Jerry to try out," Costanza said. "I said, 'Look, it's about a bunch of nuts. You'll be perfect.' Jerry was already obsessed with performing. But that was the way he was about everything . . . even down to brushing his teeth and flossing after every meal—this was a guy who could take an hour to decide what shirt to wear before we went out. Everything for Jerry was a performance just waiting for an audience."

While Costanza, who had played the role of Chuckles in a campus production of *A Thousand Clowns*, thought Jerry was hysterical hamming it up on stage in *Cuckoo's Nest*—"mugging and inventing bits for himself," he recalled—other performers had a more critical take.

"Jerry was stiff, not at ease," stated Bacino, who played the lead role of Randle Patrick McMurphy, made famous by Jack Nicholson in the film. "He was only taking acting classes and appearing in plays at Queens to help him with his stage presence. He wanted to be a comedian, not an actor."

Sue Giosa, who played Nurse Mildred Ratched in *Cuckoo's Nest*, felt Jerry had absolutely no concept of acting, and she was extremely frustrated by his self-absorbed attitude on stage. "The other characters are supposed to be terrified of Nurse Ratched because of the power she has, and Jerry would say to me, 'But you *don't* scare me!' And I'd look at him and say, 'It's *in* the play, I have the power to do terrible things to you.' And he'd say, 'Yeah, but you *don't* scare me.' I'd walk in and he'd be nonchalant, like, Nurse Ratched's here, big deal. He really didn't grasp what acting was all about. In the play, Jerry didn't stand out at all."

Giosa, who became a professional actress and married a Queens classmate named Glenn Hirsch, who also became a well-known stand-up comic, remembered Jerry telling her that the only turn-on he got from being on stage was getting laughs. "He said laughs were the addiction. He got a big laugh in *Cuckoo's Nest* and he just *loved* it. Years later he said to me, 'That's what started me off—*that* laugh.' From those college days Jerry was very clear he was going to be a comic. He was driven—driven to be successful. In a way it was 'screw relationships, screw everything else.' He was relentless. He would even say it—'I'm going to make it!' Jerry had no humility about it."

While friends like Giosa and Bacino, who were serious about acting, panned Jerry's stage presence and abilities, he put in a stellar performance as a crippled Vietnam War veteran, persuading unsuspecting, patriotic citizens across the country to buy light bulbs. Jerry had discovered the gig while combing the employment classifieds—*Make Money on the Phone! Make Money in Your Spare Time!*—for something—*anything*—to supplant working at Signfeld Signs. Crashing most nights at the Costanzas to be closer to school and away from Massapequa, Jerry asked his pal to join him at the boiler-room operation in a dingy loft on West Nineteenth Street in Manhattan where some thirty other hustlers worked the phones.

"We used to scam people," Costanza acknowledged. "To us it was just a game, just a joke. Jerry was Dave Wilson, I was Mike Davis, and we'd sell light bulbs over the phone. Jerry and I came up with a bit about not having any hands, losing them in the war. We'd see how much we could make each other laugh, and see who could floor the other guy using the most outrageous stories."

One of the best scams, Costanza recalled, involved the following scenario:

"'Hi, Mr. Cohen, this is Mike Davis from Amvet Lighting, you remember us—we're the handicapped Vietnam veterans with the lighting company.' Then, one of us would drop the phone on the floor, step on it like ten times, pick it up and bang it against the desk, and then pick up the phone again and say, 'Mr. Cohen, are you still there? You know, it's hard to get used to these hooks, but I have these two cases of light bulbs for you.'

"Jerry and I, we would be on the floor, hysterical. But we'd each sell two cases of bulbs, make two hundred bucks for the day for two hours' work, and then go out and fool around. We were doing really well and the guy who ran the place gave us our own office. He loved us. He wanted us to hang out eight hours a day. He was a fast-talking manager guy who seemed to be looking over his shoulder all the time as if the place was going to get shut down at any minute."

Like his father, who once hustled fake holy water without qualms, Jerry apparently found nothing unethical about his own bogus pitch to make a buck. "Jerry thought it was a great way to make money," Costanza asserted.

While he was a standout in the bogus role of a Vietnam vet—the war and the draft had ended by the time Jerry was eligible for conscription—he did take one more shot at legitimate theater, joining the Queens College Main Stage production of Japanese avant-garde playwright Kobo Abe's *Friends*—the absurdist story of an office worker whose apartment is invaded by a family that takes control of his life and eventually kills him. On stage Jerry played the role of a journalist, but behind the scenes he was more of a Lothario—jealously pursuing Helaine Seidman, an actress in the play who he knew was romantically involved with his friend and costar Joe Bacino.

"She was an absolutely gorgeous, blond-haired, blue-eyed, model type—considered one of the most attractive women in the drama department, and the goof was she was called 'Miss Super-Lube' because she posed in a bikini in a car magazine ad for a product called Super-Lube," said Bacino. "Jerry actually competed for her affections. She played my girlfriend in *Friends*, and we had a lot of scenes together, got close and dated for about six to eight months. Jerry was very attracted to her, had no success with her and used to make snide remarks about the fact that she was this beautiful Jewish woman and I was this not-much-taller, dark-haired Italian guy. The three of us—me, Mike, and Jerry—were going to go out to this Chinese restaurant in the city and I called her over in the hallway at school and she gave me a kiss on the lips, and we did that little kind of lovey thing and I'll never forget he said, 'Gee, I wonder what you guys have to talk about?' And I remember saying to him, 'Well, Jerry, it's not the talking that matters.'

"Even if he wasn't famous today I would still remember it because it was that kind of cutting, snide comment that said to me he was extremely jealous and felt that he was entitled to the relationship with her because they were of the same heritage, had the same interests. And I guess he probably felt a certain amount of rejection. She never told me whether she ever saw his jealousy or how far he went to show her he was interested. Years later we'd hang out and we'd talk about 'Hey, remember Helaine?' And he would just kind of nod and smile in his goofball way—'Yeah, *Helaine*.'"

—15—

Girl Trouble

While Jerry's stage presence in *Friends* was judged unremarkable and had gone unnoticed—the play received a mediocre review and Jerry was considered by his fellow actors to be "forced," "stiff," and "very nervous"—his unsuccessful pursuit of "Miss Super-Lube" underscored serious problems that were developing with Caryn Trager as they entered the third year of their intense relationship.

In the beginning, and for the first couple of years, the young lovers had had smooth sailing, though there had been a number of emotional endings—should they take time off to see other people, that sort of thing—but those were quickly repaired. "We didn't fight about much," noted Trager, "because we could talk about things. I used to feel that if we would take a step back and agree to sleep with other people, Jerry would never do it. He would feel guilty."

Now, Trager had received vague, undefined warnings from Jerry about their future. On several occasions, while they were standing together on the terrace of her parents' apartment, he seemed distant, contemplative, even gloomy, and when she asked him what he was thinking he replied, "I'm just preparing for the day that you won't be in my life. You know, you won't always be here." When she probed further she got little more from him. "I just know that one day you won't be in my life," she remembered him repeating.

"It was a declaration, a definitive statement, but I never really quite got it," Trager remarked, looking back. "All I could say to him was, 'I'm not going anywhere.' At the time I used to assume he thought that I was leaving him. But years later I realized that he probably knew that he would leave me. What he said stayed with me and has washed over me all these years, but I still feel he was very committed to our relationship."

While the Seinfelds drew Caryn to the family's bosom, the Tragers—her father, Milton, was a successful insurance broker; her mother, Jeanette, an elementary school teacher's aide—weren't as ecstatic about Jerry. "They thought he was too young for me emotionally, even though he was older than me by a year," she said. "They were worried that I hadn't dated all that much, that I should go out with other people. They didn't know what he was going to do with his life, but they knew he wasn't studying to be a lawyer or a doctor, and everyone in our family was very successful, very financially comfortable, so they didn't think he was good enough for me in some way."

Moreover, the educated, erudite Tragers of Forest Hills, Queens, looked down on the blue-collar, unsophisticated Seinfelds of Massapequa. The Tragers *were* a family of doctors and lawyers; an aunt and uncle, for instance, were chic, liberal Upper West Side artsy types who had a handsome weekend home in the Hamptons long before it became fashionable. Their sophisticated world had a lasting impact on Jerry who was both impressed and envious. Years later, with his fame and money, he bought expensive homes in both places because of their early influence on his taste and style.

"Prestige mattered to my family," Trager pointed out, "and I don't think my parents were impressed with his family."

In the five or so years that Jerry and Caryn were together, the two sets of parents met only once, when Caryn's, on their way back from East Hampton, stopped by Riviera Drive South to meet Kal and Betty. It didn't go well. "On the ride home," Trager observed, "I'm certain my parents were thinking, 'Oh, God, look at that house, look how they live. The son's probably the same way.' I didn't appreciate that. I felt they needed to look deeper."

The first major tremor in what was fast becoming an extremely volatile union erupted when Trager came face to face with Jerry's unfaithfulness, catching him in virtual flagrante delicto.

Like many New Yorkers, she had never owned a car or gotten a driver's license. Frustrated and feeling guilty because Jerry always did the driving, usually in his sister's yellow Toyota or the family's old Rambler, she convinced him, but mostly Chris Misiano, to help her get her license. On a sunny afternoon, to celebrate her new freedom, Trager excitedly got behind the wheel of her father's blue Chevrolet Impala and drove to Massapequa from her parents' apartment to surprise Jerry who she knew was at home. "I was so excited," she remembers.

But the visit became a nightmare that has stayed with her for years.

"I rang the bell and Jerry came to the door all flustered and flushed and pale as a ghost. He was stuttering and stammering—Jerry didn't hide those kinds of things well. I don't remember if he was buttoning and snapping, but the first thing he asked me was why I hadn't called to tell him I was coming."

Once inside, where early on Jerry tried to convince his mother to allow Trager to sleep with him in his bedroom, in the house where Trager had since become a fixture, the probable future daughter-in-law, she instantly discovered why her usually cool and contained Jerry was a nervous wreck. Upstairs, in his bedroom, surrounded by posters of Laurel and Hardy, and Abbott and Costello, hid a pretty, dark-haired, exotic-looking girl who was part of his small Massapequa crowd, a girl with whom Trager and Jerry had double dated a few times, with whom they'd seen the famed drummer Gene Krupa in concert, and with whom they'd hung out at friends' houses.

In the midst of all this Sturm und Drang, a scandalous discovery that could have ended their relationship then and there, Jerry had another shocking surprise.

"With the other woman in the house, he actually gave me a gift," Trager said, still unbelieving years later. "It was a gold watch, a Bulova Accutron. I thought, 'How weird! I've just caught him, and in the middle of this horrible situation he's giving me a watch. For what? Years of loyal service?' I was stunned. But I think by giving me the watch, which, by the way, never worked good, he was trying to prove to me that he still loved me.

"I was so stupid," Trager continued. "At that moment I felt bad because I had surprised him. Like an idiot, I was sorry I had intruded. I immediately left, and by the time I got home my heart was broken. It was very painful. He had a lot of making up to do to get my trust back.

"Early on we had said, 'if I ever sleep with anyone else, do you want to know?' He always said he didn't want to know, and I always said I wanted to know, so I was even more upset that that trust had been violated."

Somehow, their relationship survived. Jerry acknowledged his indiscretion, pledging his love for Trager and asserting that he didn't want what they had together to end. She believed him, convinced that he never again saw the other woman, at least under similar circumstances.

In the end, after the initial shock, Trager reluctantly accepted Jerry's cheating. "Back then," she said, "I had this thing that it would be worse if he got really friendly and emotionally involved with another woman. To me, sleeping with someone once or twice wasn't as bad."

Looking back, friends of Jerry's from those college years, including Trager, came to believe that whatever Jerry did—good or bad, moral or immoral, ethical or unethical—was closely tied to the restlessness he felt about his future as a performer, his still unfocused dream of getting into comedy. His actions—good, bad, or indifferent—were linked to his enormous drive, focus, confidence, and conviction that someday he would make it on to a stage and get the laughs he craved.

Despite Jerry's assurances that he still loved her and wanted to continue on with her, Trager's worst nightmare became a reality. Cathy Ladman—whose bond with Jerry had previously sparked Trager's jealousy—suddenly reappeared in his life.

Living unhappily at home in Queens with her mother and "very strict" and "overbearing" father, Ladman, a self-acknowledged neurotic, was floundering, working various irrelevant jobs—cosmetics, advertising—when she hooked up with Jerry again as he started his senior year at Queens College. She still possessed the dream she shared with Jerry in Israel about becoming a comic. But she didn't have Jerry's brand of confidence to make it a reality. Her parents wanted her to follow the path of other Jewish girls in her circle—become an elementary schoolteacher, which she did for a time, or a speech therapist, or, like her father, an accountant. She'd become so depressed that she even avoided watching comedy on television because it was too upsetting. Moreover, she blamed her parents for instilling so many insecurities in her when she was a child that she felt paralyzed about pursuing her dream. But with Jerry it was different. He gave her confidence. "You're one of the funniest people I know," she said he told her. "You'd make a fine comedian. Just do it." Years later, Ladman, who eventually became a successful stand-up and appeared in a number of sitcoms and films, credited Jerry with being her mentor, calling him one of the greatest admirers of her work.

Trager was in her senior year at Oswego and had come to New York to spend a weekend with Jerry—they spent virtually every weekend and all holidays together after he transferred to Queens College—when she discovered that Ladman was back on the scene. They'd gone to a gathering of friends at a Forest Hills, Queens, apartment that Chris Misiano was then sharing with his older brother, Vincent, and there Ladman was.

"My heart dropped," Trager recalled. "I had heard so much about her from him in the past, weathered what I thought was a really long period about her already. Jerry and I were still very much together and she was there and somehow they had hooked up again and I knew she wanted to go out with him again and I couldn't handle it. Here I was far away at Oswego and they were together in Queens. They were all hanging out together now—Jerry, Cathy, Chris, and our friend, Cliff Singer. Suddenly, I felt very much like an outsider because here I was, coming in from out of town, and here they were hanging out together on a regular basis. I felt very uncomfortable, and they were more intimate with each other than they were with me. And that was *my* place, those were *my* people, that was *my* crowd. I didn't trust Cathy, and whether that was right or wrong, I don't know.

"Jerry's thing was it was no big deal, that they were just friends, that they had the mutual career, comedy thing going on."

For the first time, Trager began to seriously question Jerry's sense of commitment. "I felt like he didn't want to be with me anymore and didn't know how to do it—that he really wanted me out of his life. It really rang clear—'you're pulling away from me.' "

She asked him not to see Ladman again and he agreed. But a short time later, during one of their now ever-more-frequent breakups before the final one, she discovered that Jerry and Ladman were together again. Somehow she got hold of Ladman's telephone number and placed a call—"just to hear her voice and hang up, which you couldn't do today because of *69. I did it out of jealousy."

But Trager got more than she bargained for. An answering machine message came on with Ladman's voice—and Jerry's. "They were doing some kind of shtick, some kind of bit together," Trager said. "I felt horrible pain. I couldn't take it."

All of this—Jerry's pursuit of Joe Bacino's "Miss Super-Lube," his cheating on Trager with the girl in Massapequa, the bizarre gift of the gold watch, and the blatant tossing of his relationship with Ladman in Trager's face—makes Jerry sound like a womanizing cad. But that's not the way his Queens College pals, who saw him on a daily and nightly basis, thought of him.

"I don't remember *anything* between Jerry and women," recalled Sue Giosa. "Whereas the other guys were always with women, I heard him once say he was almost emotionally retarded when it came to women."

According to Mike Costanza, Jerry was the only one in their crowd who had not taken out a Queens girl, an ex-girlfriend of Jesse Michnick's, who was reputed to be an easy make. When his buddies asked why he hadn't dated her, Jerry sidestepped the question. "Your doctors," he quipped, "advised me against seeing her." Turning to Michnick and Bacino, Costanza asked for a consensus. "What do you think—could he keep up with her?" With an evil grin, Bacino declared, "The question is—can Seinfeld keep it up?" Jerry's face turned bright red as he tried to laugh along with the boys.

Costanza noted that if Jerry had a choice between taking a bike ride and chasing girls on the beach, he'd opt for the Schwinn. "Once in a while it was supposed to be guys' night out and if someone showed up with a girl, Jerry would get pissed off. 'Why'd you bring her?' he'd demand. He wanted it to only be guys. Jerry liked to be with the guys, but when it was time to double date, he double dated. He compartmentalized things that way."

Michnick's take was that Jerry was always very guarded about his private life. "Even with me," Michnick noted, "there were certain areas that he never got into, that he sidestepped, which were the details of his sexual prowess. He's able to not let anyone see the inner Jerry. When he was with me he was always looking at chicks—'wow, look at that!' He admired the perfect image, the perfect-looking girl, the girl who had not just big breasts but who had it all—the *Ivory Snow* girl. The girls I dated liked Jerry because he had a different approach. It wasn't that macho swagger that the guys from the boroughs of New York have. He was one of those sensitive types. But I never thought that he was interested in going the other way."

—16—

Catskills Forays

While Jerry appeared to be just another drama queen engaged in daily soap opera intrigues involving his complex intimate relationships, he actually did pursue bona fide classroom studies in his final two years, earning credits toward a degree in communications arts and sciences.

Fifty percent of Jerry's courses involved film history, criticism, and theory, all of which bored him. Still, he earned A's and A-minuses. Stuart Liebman, his instructor in some of those subjects, said, "Jerry was a bright student, but by no means unique. One day he showed up in my class, and was like so many of my other students—not distinctive."

Jerry had kept such low visibility on the crowded city campus that Liebman was unaware that he had even appeared in two plays—he had never talked about his performances—or that he had a comedy career in mind. "Knowing what he's become and looking back to that time it's very clear that he's a very private guy," observed Leibman, who became chairman of the Media Studies department at Queens. "He guards his privacy zealously. He did then. He does now. Did I perceive he was a funny guy?

No. Did he crack jokes? There was no way of knowing because in my class, I told the jokes and he laughed."

After college, the two sometimes ran into each other, and on those occasions Liebman was taken aback by Jerry's changed persona. "He was boasting about his stand-up comedy routines, telling me he was starting to make a lot of money. He would still refer to me as 'professor,' but I think he meant it ironically. I would like to think in the early days he was playing it straight with me. Later, it became more tongue-in-cheek."

The classes that Jerry savored were in the Drama department—acting, stage production, and directing—where he developed a close bond with a highly regarded show-business figure, Edward M. Greenberg, who in 1971 came to teach at Queens from St. Louis, where he was the legendary productions director of MUNY, the nation's largest and oldest outdoor theater and city opera. Greenberg had directed hundreds of productions and scores of celebrities; in *Roberta*, for instance, his star was Bob Hope, an idol of Jerry's at a time when many Americans were highly critical of the comedy icon for his hawkish views on the Vietnam War and his enthusiastic support of Richard Nixon. Jerry spent hours with Greenberg after class and even in the evenings, absorbing his professor's show-business knowledge, gossip, and anecdotes.

Greenberg also was a name on Broadway, engaged by Richard Rodgers, president and producing director of the Music Theater of Lincoln Center, to oversee revivals of *The King and I*, *The Merry Widow*, *Kismet*, and *Carousel*. Those years were a hiatus from Queens where he first taught in the 1950s. Like Kal Seinfeld, Greenberg was a Brooklyn boy, born and bred, and there was something about his sense of humor that reminded Jerry of his father's, only refined and sophisticated.

"Ed was very active, directed hundreds of plays, mostly musicals, and Richard Rodgers adored him as a director," said Greenberg's friend and colleague, Raymond Gasper, who at the time of Jerry's matriculation was head of Queens College's Drama department.

"Jerry took what amounted to a special-problems course with Ed," Gasper noted. "Ed basically helped Jerry work on what was to become his first nightclub routine. He gave him advice about what material to use. In a very important way, he got Jerry started."

At Queens in those days, especially in the liberal performing arts area, it wasn't unusual for a student like Jerry to work on performance material. As it turned out, Jerry had submitted a proposal to the Drama department to work on his comedy, Greenberg sponsored him, and at the end of the semester Jerry wrote a paper, for which Greenberg gave him an A.

"It was fun, not like studying medicine or law," remarked Gasper, who was aware of the close bond between Jerry and Greenberg.

Greenberg's widow, the stage actress Sara Dillon, saw her husband as Jerry's first show-business angel.

"There was no class—Jerry was the class," she emphasized. "They had an *individual* work-study kind of thing. It was all about the material that Jerry had started writing. Ed and Jerry would go over it, and Ed would say, 'Well, this is not funny unless it's said another way.' Jerry would come to Ed and say, 'I think this works, and this doesn't.' And they'd see if they could polish it up together. Ed felt Jerry was a very talented boy."

With Greenberg, Jerry dealt with his comedy on a serious, almost intellectual level; comedy, he strongly believed, was no laughing matter. That relationship was a far cry from his comedic shenanigans with the guys in his crowd. When they got together, Jerry acted like a typical college Bozo, choking with laughter over their dopey "nyuk, nyuk, nyuk" imitations of Moe, Larry, and Curly.

Over the years, Jerry and Ed Greenberg kept in touch. "When Jerry first started having little gigs on TV," Dillon remarked, "we'd have to make a special deal to watch, and then Ed and Jerry would get together and talk about what went over and what didn't and why." Naturally, Greenberg had a wide circle of show-business friends, including the actor-comedian Jerry Stiller whom he'd gotten to know through Dillon when she toured with Stiller in *Annie Get Your Gun* years earlier. When casting started for *Seinfeld*, Jerry once again consulted with Greenberg who, it's believed, suggested Stiller for what turned out to be the hilarious role of George Costanza's wacky father. According to Mike Costanza, however, he was modeled on Mike's own dad.

When Greenberg died of a stroke at the age of seventy-one in November 1995, the *New York Times* obit noted from information released by his family that among his many accomplishments was having

Jerry Seinfeld as a student. At the time, Greenberg's family asked that memorial contributions go to Jerry's training ground—the Drama and Theater department at Queens. Via Federal Express, Dillon also received a touching note from Jerry, on *Seinfeld* show stationery with his picture emblazoned on it, expressing his condolences. "It was very sweet, very nice," she said.

As for Drama department head Ray Gasper, the student named Seinfeld made little, if any, impression on him. "I remember his performance in *Friends*, but I didn't know he was in *Cuckoo's Nest*," he said. "Jerry wasn't very active in the college. After he became famous we looked up his grades."

Jerry is said to have earned a B average in acting.

Despite his close ties to his Queens College clique, the extremely discreet Jerry kept his one-on-one studies and friendship with Ed Greenberg a secret, although the instructor was well known to them. As Mike Costanza observed, "Jerry was funny, but he never really let us know he was going to become a comedian until just before graduation. He never told us about that course. It's weird, but, hey, Jerry's weird."

Even though Jerry's tight-knit circle thought of him as "the mild-mannered, straight-arrow" member of the crew, Costanza noticed that whenever they got together "Jerry loved to taunt, tease, cajole, mock, and heckle me." His critiques involved everything from the way Costanza mistreated his car to the junk food he consumed. "Pizza was a major staple for both of us, and Jerry liked to nail me. 'Costanza, you're the first person I've ever known who can eat a whole pizza while waiting for a slice to go.' That was the way his mind worked—observing, analyzing, and dissecting every scene and personality. His observations about my behavior and habits were often right on the mark. Little did any of us know back then that he was collecting material."

On occasion, Jerry and Costanza, who shared some Drama department classes, collaborated in the writing of comedy sketches—a *Godfather*-like take off on a Baskin-Robbins ice cream commercial; a parody of a *Crazy Eddie* commercial—Crazy Eddie was a steep-discount appliance dealer in New York, but Jerry and Costanza had Crazy Eddie hawking sex, not TVs. "There was nothing that either of us wouldn't try to make some funny bit work," Costanza said.

In a class called media performance Jerry wrote and acted in commercials that were videotaped in black and white in the school's then-humble, closed-circuit TV studio. The spots were critiqued by his classmates and their professor—"a guy in an old, brown pinstripe suit with the pipe, the announcer who lost his contract at a major station, took the buyout, and then spent his final days teaching at Queens College," half-joked Jesse Michnick, who bonded with Jerry in that class. "This very straight-laced guy always told Jerry he wasn't going in the right direction, wasn't doing his assignments in an aboveboard fashion. Jerry didn't respect him and saw him as a loser. But Jerry lived for that class because he saw it as a great chance to get his creative juices going."

While he laughed at the professor behind his back, Jerry found he loved writing, producing, and acting in classroom commercials; years later he did it professionally and for huge dollars thrown at him by American Express, and, looking for new worlds to conquer after *Seinfeld* ended its run, Jerry briefly toyed with the idea of opening his own shop.

In the same media performance class, Jerry, along with Michnick, wrote and performed a goofy skit about two guys from Detroit, with all the key words beginning with the letter "D"—delirious, detrimental, delicious, et cetera. It ended with Jerry cueing the Count Basie Orchestra (on $33^1/_3$) to hit it, as he launched into a lip-synch of a Frank Sinatra number, with a little soft-shoe, holding a tennis racket instead of a cane, for good measure. The overdone, nonsensical, albeit lighthearted bit completely stumped his conservative professor who saw no future for Jerry, but the zany routine left the classmates, who watched it on a TV monitor, in stitches, which, of course, made Jerry feel like a million. "For the first time Jerry saw he could do well in front of a camera," noted Michnick, who dropped out of Queens just short of graduation to take a job in the technical end of radio and television, later becoming one of *CBS Sports*' top videotape editors.

They were still at Queens together when Michnick decided to take the test for his third-class Federal Communications Commission license, known as a radio endorsement, which, among other basics, allowed him to work as an on-air disk jockey, the first step to a first-class license.

Jerry had still not made his pronouncement that he wanted to become a comedian, and had even indicated to Michnick, Caryn Trager, and others in his circle that he might want to work behind the scenes in television—Trager saw him as a comedy writer, not a performer. With nothing else to do, Jerry accompanied Michnick to the FCC testing site in lower Manhattan. Bored while riding shotgun, he picked up the FCC manual that Michnick had studied in preparation for the test, and skimmed through it. By the time they got to the FCC offices, Jerry had assimilated enough about basic communications law to be able to take the test, pass it, and be awarded a license to spin, on air, platters by the Stones and Led Zeppelin, two of his favorite groups at the time. But later, when Michnick took him on a tour of a TV production facility, Jerry feigned absolute panic. "Too many buttons. Get me outta here!"

The Catskills, not cathode ray tubes, were more Jerry's speed, and during his senior year, as he became increasingly focused on his future, he began making forays into New York's Borscht Belt, considered the Holy Land of stand-up comedy, which had as its Vatican a kosher resort hotel named the Brickman, in South Fallsburg. There, the high priests of shtick—the Jackie Masons, the Rodney Dangerfields, the Henny Youngmans, and the Shecky Greens, gathered for after-hours services one night a week for the faithful, a motley assortment of lesser known comics and wannabees, off-duty waiters from other hotels, and hotties—waitresses and showgirl groupies who dug the scene.

The Brickman was the last of a dying breed, where a comedian named Davey Carr wore a waiter's outfit, and when an elderly, unsuspecting couple entered the enormous dining room, he'd say "Table for two, table for two," and take them in tow, schlepping them around for a half hour—through, over, and past other tables, as a put-on. Those in on the joke watched, close to choking to death with laughter on their brisket.

During those special comedy nights, the emcee was Catskills veteran Eddie Schaeffer—he played one of the wacky bellboys in Jerry Lewis's Miami Beach classic *The Bellboy*—who'd hop on stage and declare, "We've got us a couple of class acts from New York tonight. The first gentleman coming out to tell you a couple of jokes is nervous. I know because

I was backstage moments ago while he was peeing on my leg." The line was old, but the audience *plotzed*.

In order for Jerry to get into the club and past hotel security, his Queens crowd concocted a scam—Michnick, a smooth operator, pretended he was Paul Colby, owner of the edgy and trendy Bottom Line in Manhattan, and got away with it.

Once inside, Jerry thought he had died and gone to heaven, rubbing shoulders with the likes of Freddie Roman, nee Kirschenbaum, who once wrote the forgettable but hysterical line: "It's the only deli in history with potato pancakes that can lead to cardiac arrest. They are also used as hockey pucks." Jerry and Roman, who later became the dean of the New York Friars' Club, shared similar backgrounds, though they were almost two decades apart in age: Roman grew up in Queens, son of a shoe store proprietor who, like Kal, the sign-painting shop proprietor, had a great sense of humor. By the time Jerry got to see Roman, he was, in the eyes of his peers, big-time: the opening act for Tom Jones in Vegas. But he still clung to his roots in the Catskills where he held the record of performing ninety-six shows in nine weeks.

That was the life Jerry knew he wanted for himself—to be on the road nonstop, unencumbered, getting those laughs.

"Jerry had great reverence toward the older comedians and the great entertainers like Freddie Roman—the guys who made their living day in and day out in Atlantic City, the Catskills, Las Vegas—he was amazed by them," Michnick observed. "He loved Hope and Sinatra when it wasn't popular. He liked all of the great stories he heard about Dean Martin and the Rat Pack. But above all else Jerry loved being among the old comics because that's what he wanted to do—he wanted to be the guy at the delicatessen one day having lunch with those guys. He dreamed he'd be surrounded by comedians and, instead of being at the Carnegie Deli (a corned-beef-on-rye emporium and comic hangout in Manhattan) with Freddie Roman, he'd have his own entourage. Jerry told me, 'One day I'm giving this a shot.'"

While Jerry had quietly committed himself to a career in comedy, he still was open to experimentation with other forms of performance, and one of those was a starring role in his one and only feature film, a low-budget moralistic melodrama based on the premise that the grass *isn't* always greener on the other side.

As everyone in the project agreed later, Jerry was completely miscast for the part.

In fact, he was brought in not because of his acting abilities—it was clear from his stage performances that he was no budding De Niro and had no desire to be—but rather his access to a key shooting location, the Seinfelds' cottage in Callicoon.

The ten-minute, 16-mm, black-and-white film, which never made it to Cannes, was called *Paree*, a creative title based on songwriters Sam M. Lewis and Joe Young's memorable 1919 ditty "How 'Ya Gonna Keep 'Em Down on the Farm After They've Seen Paree?" The farm, in this case, was the Seinfelds' country place.

"The point of the film at the time was sort of juvenile, which is to be grateful for what you have, that the grass always appears to be greener, but that's not necessarily so. That was essentially the moral," said Vincent Misiano, who wrote the script and gave it its title. "It had some slapstick, courtesy of Jerry, and some overly profound, kind of sophomoric philosophy on my part—the kind of stuff that people do when they're trying to be self-important rather than entertaining."

For someone who one day would micromanage one of the most tightly edited and complexly produced half-hour sitcoms on television, Jerry showed a stunning and surprising disinterest and disregard for the intricacies of the filmmaking process.

"He didn't work on the script and couldn't have cared less about the production or editing end of the project," Misiano observed. "Jerry wasn't interested in how you splice film or how you line up shots. He just did the script. That's all he cared about. At the time Jerry was just nosing up to doing his stand-up routine. That was what he wanted. He didn't fancy himself a filmmaker, nor did he have any interest particularly in being an actor. He just sort of humored us by participating."

—17—

I Want to Do Stand-up

Jerry couldn't understand what made classmate Glenn Hirsch so successful—and the more he thought about him the more green-eyed he became. Here they were, both seniors, and Hirsch was already out there *killing* them, while Jerry was killing time in non-career makers like *Paree*. Meanwhile, Hirsch actually was making audiences *laugh*. On stage. In clubs. Sometimes even for money. It was too much for Jerry to bear.

"Glenn was considered the big performance man at Queens College," said Jesse Michnick, "and Jerry *always* talked about him. He envied that Glenn was getting established in comedy, which was the next big step up. It wasn't that Jerry idolized him. It was more like—here's the first guy who was tangible around campus doing comedy, the breakthrough guy, the guy who showed this could be done."

Feeling confident, self-assured, even cocky about his comedic strengths, Jerry believed he had ten times Hirsch's talent. But Hirsch was out there and Jerry wasn't, hadn't even let anyone, except for Ed Greenberg and sometimes Caryn Trager, read, see, or hear his material, which he had been secretly writing and rewriting over the past year or so.

Meanwhile, every time Jerry heard talk of Hirsch getting gigs here and there he became sick with envy, believing he, not Hirsch, should be in the smoky spotlight, getting the laughs, the applause, the adulation.

But despite his aspiration and drive, Jerry's objectives were still quite limited, according to Joe Bacino, who recalled standing with him on campus one afternoon a few months before graduation when Hirsch walked by with his girlfriend and future wife Sue Giosa. "Jerry pointed at Glenn and said, 'One of my goals in life is to do better than *that* guy.'"

"He felt *very* competitive with Glenn," remarked Mike Costanza. "Jerry felt Glenn was out there ahead of him, making a name for himself, and that both annoyed and motivated Jerry. As a result, Glenn became something of a catalyst for Jerry. He wanted to catch up with him—surpass him."

There were major differences, though, between Jerry Seinfeld and Glenn Hirsch, which accounted for Hirsch's early jump out of the gate. Unlike Jerry, who always wrote his material on yellow legal pads with Bic pens and compulsively rewrote ad infinitum until he felt the words were golden, and then rehearsed down to the second, Hirsch was funny in the moment, could get up on a stage anywhere, any place, and leave them convulsed, which gave him an edge. But even back then, Jerry had the magic elixir that foreshadowed his success: a combination of intense focus and an amazing understanding of the *business* of comedy that few veteran comics ever grasped. In many ways, Jerry from the start had the mindset of a comedy entrepreneur, as opposed to Hirsch who had natural ability to make people laugh but little knowledge of, or interest in, the business end, which may be the reason why he was still happily doing stand-up on cruise ships in fall 2001, while Jerry had become a legend in his own time.

As Sue Giosa observed—not completely objectively, because she later married Hirsch—"Jerry didn't stand out at all in college as a comedian—he was more quiet and private, and Glenn was going to be *the* superstar, not Jerry. Glenn was *so* good, *brilliantly* talented. He never had to work to get the jobs. But you don't make the big cycle unless you're focused to go there. You have to understand the business. You have to be driven—and Jerry had the focus and the understanding and the drive and unbelievable confidence, almost blind faith in himself. He was, 'I don't care what *you* think—*I'm* great.' Jerry always had that. He had the mindset of 'I'm gonna make it. This is what I want.' He had that arrogance and that's what gets you there. At the same time Jerry was a good writer. He wasn't one of the crazy guys, not the loud comedian making jokes on the spur of the moment, whereas Glenn was more *out* there."

On a bitter cold evening in January 1976, about a week before graduation, as members of his Queens College posse were shooting the bull, Jerry, who had been listening disinterestedly, suddenly turned to his pals with words they have never forgotten. Costanza swears it happened in his parents' recreation room; Bacino remembers the event occurring in Costanza's Fiat on the Brooklyn-Queens Expressway, as they were returning from having chow mein at Wo Hop's in Chinatown.

Wherever it happened, Jerry's announcement was momentous, at least to their way of thinking.

"Listen, I gotta tell you something," he intoned out of the blue. "I'm thinking of doing some comedy." Shocked, which is how Bacino reacted, is not the word for it. Costanza thought, "If a career in comedy was in his mind when he came to Queens, he sure wasn't sharing it with anyone." The consensus was, "He's jerking us off." Costanza and Bacino looked at each other thinking, "Okay, Jerry, but you're not *that* funny." But, in a serious voice they'd never heard, Jerry declared, "I've been thinking about it for a while. It's what I really want to do with my life. I want to do stand-up. I just didn't want to say anything until I had some material."

Dumbfounded, Bacino responded an octave higher than usual, "*You* . . . have . . . material? *You*? *Seinfeld*?"

"Yeah," Jerry responded. "In fact, I want to go to Catch a Rising Star. I want to go to the open mike night." Still disbelieving, Costanza stuttered, "*You* have an *actual* routine? *You* have some *bits*? Who the fuck are you kidding? Let's hear it funny boy!" More confident than they'd ever seen him, Jerry once again answered in the affirmative. Daring him to show them his stuff, Costanza and Bacino sat on a couch while Jerry stood nervously in front, doing the five minutes he'd drafted and refined. The one that stuck over the years was what Costanza dubbed "socks"—an observational piece based on the real-life question: why do socks disappear after you put them into the dryer? Why is one sock *always* missing when you take your laundry out? Shouldn't there be a closet just for socks? These were metaphysical questions Jerry had first pondered two years earlier at Oswego, when he started washing and losing his socks. "I remember doing our laundry together in the dorm basement," recalled Oswego classmate Tom Daly, "and he'd joke about the socks dancing in the dryer, then disappearing before you could get them out." Socks became a mainstay of Jerry's early stand-up act.

When he introduced the bit for the first time, the two members of his audience—Bacino and Costanza—were unimpressed, couldn't believe Jerry's audacity, but were diplomatic when asked what they thought. "We laughed and said, 'Whoa, that's pretty good,'" recalled Bacino. "But, of course, his routine *wasn't* funny because he didn't have the timing down."

Jerry also tested "socks" and a bit about the discrimination left-handed people face—having two left feet is a put-down; you're not considered a good guy if you sport the nickname Lefty—on his Oswego theater pal Garry Carbone, who had moved back into his parents' house in Queens. "It's the first time I remember seeing him really nervous," Carbone recalled years later. "He kept looking at me to laugh out loud, which I didn't." A die-hard fan of Richard Pryor's type of edgy humor, Carbone was honest when Jerry asked for his opinion. "I felt a little uncomfortable, a little nervous telling him, but I said, 'I don't know, Jerry, it's almost too gentle for me.' He didn't look happy."

A week after auditioning for his pals Jerry became one of the only Seinfelds to earn a four-year college degree, in media studies, and his grades were good enough to leave the hallowed, crowded halls of Queens with honors. Despite his having made his mark to some extent in academia, there was no big family celebration that anyone can remember. He left Queens as he had entered, with no fanfare—there's not a special mention of him in his yearbook or any college publication dealing with the class of '76.

But eighteen years later, on June 2, 1994, after he had become Queens' most famous and illustrious graduate, he was given an honorary degree of Doctor of Humane Letters. The decision to honor Jerry stirred controversy among faculty and administration members who felt it diminished the degree because the comic was not an academic. "Giving a serious degree to a comedian is not a normal thing," argued philosophy professor John Lange, a member of the Academic Senate, unsuccessfully. The idea to make Jerry a doctor was the brainchild of Ceil Cleveland, a cheeky, publicity-savvy, yenta*ish* Queens College administrator, who wrote a humorous campus magazine piece about alumni who had become comedians.

"So it might kill him to maybe one day wear a Queens College T-shirt on his crazy TV show?" she wrote. "Is that too much to ask? Remember, Jer, we have your permanent records in our vaults."

At the event, Jerry, decked out in academic mortarboard and robe, spoke for a mere 110 seconds, which also infuriated some college bigs. "It was a little more casual than I had hoped," said an irate college spokesman. "We hoped he would have taken the opportunity to say more." In his brief speech, Jerry thanked Ed Greenberg "who allowed me to pursue an independent study here in stand-up comedy, and I really feel that worked out pretty good."

The other highlight he mentioned was that in the first year at Queens he got an exceptional parking spot on Kissena Boulevard—quite an accomplishment, since parking on or near the heavily traveled urban campus was all but impossible. However, Jerry had left out one important part of the story. The real reason he was able to get the space was because his pal, Mike Costanza, had given him a bootleg official parking pass soon after they met, to help cement their bond. Remembering the moment years later, Costanza said, "Jerry looked at the pass warily, and then at me, like I had given him a ticket into some forbidden world. The parking situation was a full-blown obsession for all of us." Costanza could never have predicted at the time that their parking pickle would one day also become an obsession with the fictional Costanza character on *Seinfeld*, and be underscored by Jerry in a speech honoring him at their alma mater. But neither was he surprised when both events happened. "Everything we did from those days wound up in his act or his show, in one way or another."

Betty Seinfeld was flown up from Florida in a private jet by her son for the ceremony. Playing the role of the clichéd Jewish mother for laughs she told reporters that she had always wanted Jerry to become a doctor—which was never the case—and declared, "He really *is* a doctor now!"

Jerry, who had presumably coached her what to say, kept the kicker for himself, not to be upstaged by Mom.

"When my parents were pushing me to be a doctor"—again, a comedic stretch of the truth—"I could have at least said to them, 'All right, all right. Just let me tell jokes to strangers in nightclubs for eighteen years, and I'm sure after that they'll make me a doctor.'"

With his college days behind him, Jerry began his long internship.

–18–

Peanuts on the Bar

Jerry landed on his feet running, but fell flat on his face. Bombed. Big-time. That's the only way to describe his very first public performance as a stand-up comic at an open mike night at Catch a Rising Star on the night of the day that he graduated from college. But his timing was, in the parlance of the mid-70s, right on. Comedy clubs were starting to spring up everywhere, or at least venues that allowed aspiring comedians to get up on stage. Even on TV, comedy was king. Just three months before Jerry's debut, an edgy, satirical late-night show debuted on NBC, and *Saturday Night Live* quickly became a runaway hit, the most influential comedy show since Rowan and Martin's *Laugh-In*. Stand-up was entering its golden age. Jerry was in the right place at the right time.

He thought he was prepared. He had put together about fifteen minutes of loose-knit material, his repertoire consisting of observations on dogs, the beach, driving, parents, his socks bit, and a slicker version of the left-handed people routine. For weeks he had rehearsed and sharpened his act in front of a mirror and the members of his inner circle. Along with attempting to improve the bits, he also concentrated on his stage demeanor, which involved practicing comedic gestures, working on intonation. His bed was littered with sheets of yellow legal notebook pages filled with routines that didn't work, with revision after revision of those that did. The man was obsessed.

That morning, graduation was the last thing on Jerry's mind. He called Mike Costanza and told him it was D-Day—Debut Day. Hearing fear in his friend's voice, Costanza volunteered to drive him to the club on First Avenue, on what was then Manhattan's swinging East Side; it was 1976 and the street represented the sexual revolution and anarchy of the city as exemplified by Jodie Foster's hooker and Robert De Niro's maniacal cabbie in *Taxi Driver*, the big film of the year. First Avenue was a hotbed of singles pickup joints and cookie-cutter postwar high-rises filled with available stewardesses and career girls. But for a funnyman, the street was all about dreams—of making it as a regular at Catch a Rising Star. Oddly, though, despite Jerry's goal of becoming a comedian, he had never once visited that scene or been to Catch, which in those days was one of the two premier stand-up comedy clubs in the city, the other being the Improv. He was coming to it—fearlessly but naively—cold.

On the drive into the city from Costanza's house in Queens, in the new Fiat that Jerry had convinced Costanza to buy—Costanza had a reputation for destroying vehicles, and when they came into the showroom Jerry had quipped, "Mike, the cars back up when they see you coming"—Jerry rehearsed his material repeatedly, with Costanza making suggestions and Jerry making still more revisions. To Costanza, the material sounded so familiar, lines and bits he'd heard Jerry deliver when the two had just been joking around. Now he was going to try them professionally. Costanza thought, *Oy*. As Glenn Hirsch noted, "It's different getting on stage in front of an audience than making your friends from the old neighborhood laugh at your jokes."

Lenny Bruce once told a story of how he was working a small town in Ohio when the phone rang at 11 A.M., a call from the club owner's wife inviting him to dinner after that night's last show. "I hope I didn't wake you up," she said. "No, I'm always up at this hour," he said, his voice laced with sarcasm. "I like to brush my teeth, have a Chiclet." Bruce's line underscored the fact that most comedians sleep during the day. So the bookers at Catch were surprised when this new kid named Seinfeld arrived at the club at two o'clock in the afternoon, with Costanza, acting like a de facto agent-manager at his side, trying to reassure and calm him, to get his name on the list for that night's open mike show, many hours away.

With no one else in line, Jerry grabbed the number one spot, which, Costanza later noted, "showed he had no fear, had incredible determination. The first time out and he wanted to be number one on stage."

Many comedians in those days, about to jump on stage for the first time, drank or toked up in order to relieve the tension. Not Jerry, who led Costanza to a coffee shop where he spent the next several hours in a state of controlled panic, rereading his lines and drinking tea with milk. "His nerves and his fidgeting," Costanza recalled, "made the table wobble between us." Jerry watched the minutes count down to show time. His goal that first night, he told Costanza, was "to get on and get off," and that was the real reason he wanted to be first on the bill.

And then it was time.

First up tonight, a very funny guy who, by the way, just graduated today from Queens College . . . Put your hands together for Mister Jerry Seinfeld . . .

The emcee that night was Elayne Boosler, a sexy, boisterous Brooklyn girl, the college-dropout daughter of a tool-and-die-maker father and a Russian ballerina mother. Two years Jerry's senior, Boosler, with her wild Afro hairdo, hot bod, and on-the-mark material, had gotten into the business several years earlier. Then an aspiring singer, she was waitressing at Catch when she met an up-and-coming comic named Andy Kaufman, a Long Island boy like Jerry, though from a more sophisticated and affluent family; his father, like Kal, grew up in Brooklyn, but went into the lucrative costume jewelry business. Boosler had hot material—"when women are depressed they either eat or go shopping. Men invade another country. A whole different way of thinking"—and Kaufman, who became her boyfriend, convinced her that her future was in comedy, not music. Heeding his advice, she became one of the earliest, feistiest, most thoughtful female observational comics of the Seinfeld generation. As she said later, ". . . my group came along, Robert Klein, David Brenner, and Seinfeld, and were a little sharper in our approach, a little more wise-guy."

But on that big first night of Jerry's career she was horrified and embarrassed as she watched this poor schlep make his debut.

He took the mike and just stood there for what seemed like an eternity to him, but actually, in fact, was "several long, silent moments, hesitating and squinting out at the crowd until he finally found his voice," remembered Costanza, who had taken a seat at a back table. "He'd frozen before the first word ever came out of his mouth. I'd never seen the expression that took control of his face. Like the deer suddenly caught in headlights, this was pure, gawking terror. My hair was standing on end just watching him. He forgot so many lines that his routine sounded like one long non sequitur after another. He jumped from bit to bit without so much as a brief pause.

"Jerry didn't have any segues. He had his jokes written on little slips of paper and he would go from one slip of paper to the next. He kept them in his top shirt pocket. And he would just go through one bit to the next. The fact that he didn't have segues was the only thing about his act that was really funny."

As Jerry recalls the horrific night, "I got up and all I could remember were the subjects that I wanted to talk about . . . so I just stood there and went, 'the beach . . . cars . . .' I did about three minutes and I got off . . . and the sad thing is I'm not embellishing the story to make it funny.

"The first time you get up and see all those people looking at you, it's an overwhelming experience. I bombed bad. I couldn't even speak, I was so paralyzed in total fear. I remember thinking, 'Why should everyone be listening to me? What right do I have to come in and have everyone's attention in this room?' I just felt really out of my place. I had thought I was going to be a hit. It was the usual pie-in-the-sky optimism. It was supposed to be a fifteen-minute set. I did about a minute and a half and then just left."

The audience looked like the guests at the Last Supper. Except for Costanza, no one, not a single person, laughed. When he was finished, Boosler took the mike, raised her eyebrows, shook her head, and pointed at the comedian who was beating a hasty retreat. "*That* was Jerry Seinfeld," she jabbed. "*The King of Segues*."

Jerry said, "After I was done, some friends came over and congratulated me. They said it was very important the audience liked me, even though I didn't do anything. I was inconsolable. I walked into the street and just kept walking for miles and miles."

One of those who patted him on the back and told him to keep his chin up, that he did well for the first time out, was Costanza, but a devastated Jerry, he noted, "was too anguished to even speak," except to say he'd made an ass of himself.

While Jerry publicly acknowledged after he became famous that he tanked the first time out, he also has asserted that a month after he bombed he returned to Catch "with a whole new act. This time I had memorized it. The first time I hadn't."

While that may be true, no one in his circle could remember such a quick return engagement after such a disastrous start.

In fact, Caryn Trager, whom Jerry had been running material by when they weren't skirmishing, said it was more like three months before he got back on stage, and that the club wasn't the prestigious Catch a Rising Star, but rather a "seedy Times Square bar that did comedy" called the Golden Lion Pub, at 143 West Forty-fourth Street, an engagement he had on March 30, 1976—a month before he turned twenty-two. "He was very nervous, as anyone might be," she clearly remembered. "He also had no segues and just went from one bit to the other bit."

Costanza and Jesse Michnick called the Golden Lion the "old and dyin'" because the fifteen-table, shoe box–sized club was rundown and usually empty, but it was the first place that paid Jerry for his performance, albeit a pittance. But there was a quid pro quo: he had to bring in paying bar customers in order to get stage time.

"Sometimes it was me, Caryn, Jesse, and Joey Bacino," said Costanza. "We had an entourage and it was up to us to lend the laughter and buy the booze." Most comedians starting out in those days faced the same kind of extortion. "Comedians were nothing more than peanuts on the bar," Glenn Hirsch wryly observed.

That first night at the Golden Lion Jerry received from one of his idols a compliment that stayed with him during his early, struggling years. In the audience was impish Joe Franklin, the venerable and revered local New York TV talk-show host who interviewed the great, the near-great, and the obscure of show business, and aired old comedy film shorts of Laurel and Hardy, and Abbott and Costello. Jerry had grown up watching Joe Franklin's TV show, as had other comedians. Later, Billy Crystal on *Saturday Night Live* did a classic parody of Franklin's campy *Memory Lane* show, "sponsored by matzoh by Streit's, my friends, for the unleavened experience of a lifetime." Crystal's routine helped resurrect Franklin's flagging career and made him a national celebrity of sorts. Michnick, who was in the audience, said Jerry was ecstatic. "Franklin's pallor back then made him look like he was dead for two weeks. He was a fragile kind of guy but he got up and shook Jerry's hand and encouraged him and said, 'Keep it up, kid' and that was a *big* thing for Jerry to hear."

Years later the schmaltzy Franklin, who emceed nights when Jerry performed, said it was Jerry's manners that impressed him the most. "Jerry was very, very, very polite. It was almost like how Elvis Presley used to call everybody 'sir.' Jerry would say, 'Yes, Mr. Franklin, thank you, Mr. Franklin.' If I would say, 'Jerry, even though you're scheduled to go third, you're not going to go on until seventh,' he'd say, 'No problem, Mr. Franklin.' He was never upset, or ruffled, or impatient, which said to me that he was unpretentious and untheatrical. He always wore a tie and a jacket—always neat, always nice, a beautiful kid, such a gentle man. He told me he respected me. To him I was like an icon, a national treasure," said Franklin, who describes himself as a "legend in my own mind."

Besides impressing Franklin as being a good Jewish boy, Jerry also demonstrated how talented he was. "I saw right away that Jerry had to make it because he was *very* funny. He was very deep-thinking, almost like Jackie Mason in terms of being original and picking things apart."

But after all those years of knowing and respecting Jerry—and having him as a guest on his show a number of times—Franklin admitted to having "a terrible secret." He said, "I never saw *Seinfeld* once in my life. I never saw his show."

Jerry's debut at the Golden Lion was the first of many appearances there, and in other clubs like it, places he described as "not so much nightclubs as restaurants with a table missing. They'd say, 'We took a table out over there—just stand there and talk.' They'd have a sixty-watt bulb over your head, so you felt you were in a french-fry heater or something, and that's how I'd do my act."

The Golden Lion fit that description. With its small stage and overhead "french fry lamps," as Jerry called them, he felt the place was more like an interrogation room. But there were even worse places on his climb up.

"I did shows where people didn't even know I was on. I did a show once at a disco in Queens—real *Saturday Night Fever* type of disco—and they turned up the music and said, 'Now it's comedy time.' The dance floor was jammed with people, music was blasting, and the microphone cord was eighteen inches long . . . it was so short I couldn't even turn my face to the audience. But at that point in your career you're afraid to make waves, so I just yelled my act into the crowd; no one heard—complete bedlam. After I was done I went over to the owner to get paid and he said, 'Oh, I didn't even know you were on.'"

—19—

The Dianetics Kid

Like Bill Cosby, whose clean, observational, monologist style set the standard for Jerry's own brand of comedy, and like Ed Greenberg, who gave him the freedom to develop his material in college, Jerry also found a guru for the focus, concentration, and discipline he felt was required to become successful in the knock-down, drag-out world of stand-up comedy in the late Seventies.

Jerry's motivational guide was L. Ron Hubbard, a science-fiction writer, self-styled pop psychologist, and shrewd marketer, who began what he described as "studies of the mind and spirit" in the early 1920s—when he was only twelve. By the late 1930s, Hubbard, who studied engineering at George Washington University, had written a paper about his so-called research, titled *Excalibur*, but decided against seeking publication because, as he asserted later, it didn't deal with any sort of "therapy" but rather was a discussion of what he termed "the composition of life." His then little-known work was the first to use a word that would carry with it controversy by the time Jerry became a believer in 1977.

Hubbard defined "Scientology," the word he coined, as the study and handling of the spirit in relationship to itself, universes, and other life. Man, he asserted, is an immortal spiritual being whose experience extends well beyond a single lifetime, with unlimited capabilities, even if not presently realized.

In 1947, he wrote a tract dealing with something he called "Dianetics," which, he claimed, provided "therapy" that could easily be used by anyone. His unpublished writing was quietly circulated and reportedly had a huge underground following. Hubbard's first article on Scientology and Dianetics—titled "Terra Incognita: The Mind"—was published in the 1950 Winter-Spring issue of the respected *Explorer Club's Journal*; a second, similar paper by him appeared a short time later—in a newsstand pulp magazine called *Astounding Science Fiction.*

Not long after, the bible of Scientology—*Dianetics: The Modern Science of Mental Health*—found a publisher and quickly became a national best-seller. Almost four decades later it was still on the *New York Times* best-seller list.

Hubbard hit gold because, according to the Church of Scientology International, tens of thousands of paying readers of the book began organizing groups and applying the techniques of *Dianetics*, which didn't come free.

"It soon became apparent that many people audited on these procedures were coming into contact with incidents that seemed to occur in previous lives," the Scientology people said, somewhat cryptically. "Although certain officials in the Dianetics organizations attempted to suppress research into this phenomenon, Hubbard refused to allow this." During further research, Hubbard concluded that Dianetics addressed "man's spirit," and he discovered that "you could do things with it from a very practical standpoint that nobody had ever done before . . ."

In 1954, in the same year Jerry was born, the first Church of Scientology was formed in Los Angeles, and by the time the Seinfelds had moved from Brooklyn and settled in Massapequa and Jerry was being schlepped off to Hebrew school at Temple Beth Shalom, churches of Scientology had been formed across the country and around the world, as Hubbard continued to research what he described as the spiritual nature of man. By the time Jerry got involved with Scientology, the church was raking in enormous revenue from its various courses and publications.

Unlike Judaism—which appeared to turn Jerry off, if his half-hearted interest in Hebrew school, bar mitzvah studies, and his abandonment of his father's synagogue are any indication—Scientology is billed as a religion without dogma. The fact that no one in Scientology is asked to believe anything on faith, or at least that's what the church promotes, was a concept that appealed to Jerry's independent nature. "As far as being a strong, religious Jew when we were close friends, Jerry was not," observed Mike Costanza, whom Jerry tried unsuccessfully to enlist into Scientology. "There was *no* Jewishness. He was devoid of *any* religion. Jerry celebrated Christmas. We exchanged Christmas gifts. For years I would say to him, 'Happy Hanukkah,' and he'd say, 'Oh, yeah, when is that?' Being Jewish never seemed a big deal for him."

Moreover, Jerry was attracted by Scientology's promise to improve his life (i.e., help him to make it in comedy by making him more centered and in control) if he applied its teachings to himself and those around him—with the ultimate goal of true spiritual enlightenment and freedom.

Despite its lofty promises, Scientology has been criticized virtually from its inception. Be that as it may, beginning around the same time he broke into comedy, Jerry became a true believer and proselytizer, which caused the more cynical of his friends to wonder how this Jewish boy from Long Island, a kid with lots of *sechel* who was trying to establish himself in the hip, cool, irreverent, and cynical world of stand-up comedy, could become involved with something as seemingly oddball as Scientology, despite its promises to make his life better. Had Jerry, whom his friends viewed as so controlled and contained emotionally, undergone some quasi-mystical experience, or had he had an epiphany of sorts, which led him down Hubbard's mysterious path?

"One day, out of the blue, he said to me, 'I'm into Scientology. You should take some courses with me. It's great,'" said Costanza. "I remember walking him there a few times, but I never went in. He kept telling me it would be good for me, but he already knew me well enough to know I wouldn't buy into it. I grew up in an Italian culture, was brought up a Catholic. That kind of new religion wasn't my thing."

Yet, alone among Jerry's close friends, Costanza wasn't completely bowled over by his sudden interest in Scientology either. From the time they met, Costanza was aware of Jerry's exploration into ideas and concepts that were considered off-the-wall in the provincial cultures of blue-collar Queens and parochial Massapequa, such as his meditation exercises. "It was part of his personality to explore different things when it came to the mind," he noted. "He didn't do drugs. He didn't drink. He didn't use salt or sugar. He didn't eat meat. Very early on he was into that kind of diet. He was always out in front on all that stuff."

In fact, Jerry had traveled the road to Scientology from a deep involvement in Transcendental Meditation and the teachings of the white-haired, bearded, and robed guru known as the Maharishi Mahesh Yogi, who preached that his brand of meditation opened "the awareness to the infinite reservoir of energy, creativity, and intelligence that lies deep within everyone." While still at Queens College, Jerry had become one of millions of adherents worldwide who were practicing the Maharishi's technique in hopes of gaining deep relaxation, eliminating stress, promoting health, and attaining inner happiness and fulfillment.

"He always wanted to see if there was any way to improve himself," Costanza observed. "The meditation, the Scientology—they were all a part of his self-improvement plan. But in all the years I've known him, I never really knew what was in him that he needed to do those things. As his friend—and at that time probably his best friend—I liked him just the way he was, so I never understood why he needed to do anything more to change himself, because he was controlled and focused anyway."

During college, Jerry spent considerable time at the Costanzas, sleeping over many nights, and that's when Mike first learned of Jerry's esoteric interests, causing some humorous episodes in the household.

On one occasion, for example, Costanza's father—a real character who was said to be the model for George Costanza's father, played by Jerry Stiller on *Seinfeld*—found Jerry sitting on the daybed in the basement recreation room, with his eyes closed, in what appeared to be a zombie-like state. When Costanza told his father what Jerry was doing, the senior Costanza shouted, "What meditating? He's out like a light . . . your friend looks like a corpse. Ya better find out what's wrong with him." Roused out of his trance by the tumult, Jerry made light of the situation, telling Mike his father had taught him a new mantra—"Pasta fazool, pasta fazool . . ."

While Costanza rejected Jerry's overtures to join Scientology, Jerry's loyal and malleable girlfriend, Caryn Trager, succumbed to his proselytizing about its glories. "He got into TM and he stayed with it for a long time," she said. "I tried it, but I never got it. It didn't do anything for me. And then he got into Scientology and he thought that would be good for me, too. He dragged me in. I went because of our relationship. But I didn't like it, either. I thought it was secretive, and anyone who was at what the scientologists called a higher level, including Jerry, wouldn't really tell you what went on until you got there. Jerry kept going to higher levels. He stayed with it. But I only took one or two courses. He would have preferred if I'd stayed. He tried to encourage me to stay with it. I was forced to do these very theatrical exercises with other people—you'd have to sit there and touch knees with them. You had to read their books and look up words that no one understood. I didn't get any of it. I didn't get what I was going towards. I told Jerry I didn't want to continue because it was a mystery. They don't tell you why or what. I was in it for maybe six months and then left. It all seemed so stupid to me, but Jerry took it very, very seriously. Back then he always had a very open mind and would look into anything—and experience it. My view was he got into Scientology to become intellectually and emotionally a better performing person. He was searching for self-improvement."

Jerry told Trager and other members of his circle that within the church he was known as a "preclear"—a person not yet clear, a person learning more about himself and life. As a "preclear," he said, he was involved in a form of personal counseling called "auditing"—one-on-one sessions with a Scientology minister known as an "auditor." Auditing uses "processes"—questions or directions given by the auditor to the preclear to help him find out things about himself and improve his condition, and free himself from what the scientologists claim are unwanted barriers that inhibit, stop, or blunt natural abilities; the object is to increase those abilities so that he becomes brighter and more able. (In its literature, the Church of Scientology has a disclaimer, which maintains that during auditing "there is no use of hypnosis, trance techniques, or drugs.")

"Jerry told me his goal was to get clear," Trager stated. "He was dedicated to it. We all thought it was moronic."

Lucien Hold, who managed the relatively new Comic Strip, an Upper East Side club where Jerry began working on a regular basis in the late

70s, said that "while the scientologists suck you in, try to get you more deeply involved," Jerry spoke "very positively" about the classes he was taking, claiming they were helping him to learn more about the science of the mind, and credited them with helping him to organize his own mind.

Hold, who was starting to see Jerry at the club regularly, noted that the longer Jerry was involved with Scientology the odder some of his behavior became. For instance, at one point he got involved in sleep deprivation to heighten his senses.

"He started out writing material an hour a day, then two hours, then it became four hours," Hold said. "The sleep deprivation allowed Jerry to work longer hours. He did this over a period of months. It's not something he did one or two nights, but as a matter of course. If you're able to make sleep deprivation work, and Jerry was able to, it got his adrenalin going, and that's what kept him awake. I guess he was getting four to five hours of sleep a night at most.

"He was looking to optimize every minute of every day. You'd be talking to him, it'd be ten minutes to one, you'd be in the middle of the most engaged part of the conversation—you might be telling him about the death of your mother, it didn't matter what it was—but Jerry would look at his watch. 'Gotta go. Gotta go. Gotta be in bed. See ya.' That's how the sleep deprivation worked. He had to be in bed at the same time every day, get up at the same time every day. He programmed his body to function that way, so he would be up early in the morning to write. By the time his friends were up in the afternoon, Jerry would be free. He could then hang out. He'd already done his work for the day."

Costanza had also become aware of Jerry's odd, new habits. "It all had to do with the Scientology thing," he said. "It had to do with pain—being able to withstand pain, being able to not sleep. It was all tied in. Jerry started talking about how he didn't feel pain anymore. He believed that if you will yourself not to have any pain, you won't feel any pain. I think I was with Jesse [Michnick] or Joey [Bacino], and I turned to Jerry and said, 'What the fuck are you talking about? You don't feel pain?' I said, 'I'll make you feel pain.' Jerry was talking about his Scientology and said he was attempting to get to some new level of consciousness, and this guy Jerry was with, who was going to Scientology with him, put a lighted match under a penny and got it real hot and put it between his fingers and said he felt no pain. Jerry said, 'It really works for me, too.' And I turned to one of the guys and said, 'He's fucking nuts.' The Scientology thing with Jerry was getting a little bit spooky."

Jerry had kept up his friendship with LuAnn Kondziela from Oswego, and during this time period, after Jerry had moved into his first apartment in New York, she came for a visit and was shocked to find Scientology literature scattered around, including a big chart on his refrigerator dealing with the precepts of the church. "I remember thinking that that was very strange, that he could be into that," she recalled. "To me it was really off the wall. I saw all that Scientology stuff in his apartment and I asked him about it and he explained some of it to me, what Scientology was about and why it was worth doing, and I remember coming away thinking, 'That's the weirdest thing I ever heard.' It just didn't mesh with the Jerry I knew at college."

Jerry's involvement in Scientology was not short-lived, and he still boasted about its influence on him in the mid-80s.

"When he told me he was a scientologist, I was like—*what?*" recalled Susan McNabb, Jerry's girlfriend from the mid-80s until the early 90s. "Not long after we became involved he said 'Oh, by the way, I'm a scientologist.' I said, 'Well, that's crazy, you're not a scientologist, you're Jewish.' He said, 'Well, you can be a scientologist and Jewish at the same time. It doesn't really interfere with your religion. It's not like that.'"

McNabb thought Jerry's involvement with Scientology was "weird" and she also got the impression that Jerry enjoyed the shock value of telling people he was into it. McNabb said Jerry's friend, Chris Misiano, had told her that he had gone to Scientology classes with Jerry in New York, but had not stayed with it. "I remember Chris laughing because he said, 'Once you're on their mailing list, they never leave you alone. My parents still get mail addressed to me from them twenty years later.'"

As Jerry became famous, his involvement with Scientology began to surface in questions from journalists, which infuriated him because he felt it was no one's business but his own, an invasion of his privacy, which he has guarded with intensity and stealth. The very publicity-and-media-savvy comedian realized the press's perception of Scientology was not always positive, and figured he didn't need that as part of his public curriculum vitae. When the subject of his Scientology activity surfaced,

Jerry didn't disavow it, but he downplayed it, and even was able to subtly dissuade at least one writer from mentioning it in an August 1990 profile in *Playboy*.

The writer, Stephen Randall, said the subject of Scientology had come up. But, curiously, there was nothing in his published piece about it because, as Randall acknowledged, "I was sort of relieved" with Jerry's explanation that his Scientology connection was minimal. At the same time, Randall said, "The one thing I didn't want it to be was just yet another Scientology story because it tends to color things," even though Jerry had told Randall that he was actually using what he had learned in Scientology in his act. "He told me it taught him, in a sense, how to handle hecklers better, and it gave him those communications skills."

Leaving out the Scientology angle, which would have been the first time Jerry's connection with the church became public, Randall did make brief mention in his story that "the guarded" Jerry practiced yoga, meditated, had a health food diet, and drank mineral water, and he quoted Jerry as saying, "I like that kind of Zen Buddhist atmosphere, very physically clean, mentally clean."

One of Jerry's characteristics that stood out during the time Randall spent with him was his "mind-boggling discipline," which, of course, was something Jerry had sought from Scientology. "Here was a guy on the road, in strange hotels, in strange cities, and he was able to adhere to a rigorous diet. He was very, very picky about what he ate. I remember having one dinner with him in Tucson, in the booker's office, in a strip-mall comedy club, right before he went on stage, and he was able to find a healthy salad. This was a guy who was forced by circumstances to eat a lot of Jack in the Box, but through discipline he never gave in, and only ate things that were good for him. With that discipline, he wrote every day. No matter where he was, he'd write for hours every day with a twenty-nine-cent Bic pen. He did his yoga every day. There was never a time when I saw him lose control, and I'd been on the road with other comedians and seen them get rattled, just be sort of undone by the road. Jerry did not seem to suffer those types of emotions. It was just remarkable discipline. He was a robo-comic." But Jerry's Scientology was never mentioned.

In the spring of 1992, Jerry sat down for an interview with a different sort of writer—the *Washington Post*'s respected and *very* tough-minded

TV columnist, Tom Shales. Shales had done his homework and learned from Roseanne and Tom Arnold that Jerry was a scientologist. Shales quoted Roseanne, who is Jewish and an oddball herself, as saying cryptically, "You can see it reflected in the kind of comedy he does."

Jerry acknowledged to Shales that he had taken "some" Scientology courses, and emphasized that it was long over. "I was never in the organization. I don't represent them in any way. I took a couple of classes. It was great: how to improve certain conditions in your life, your ability to work, your relationships. It was very pragmatic. That's what I liked about it. It wasn't at all what you'd call highfalutin, for lack of a better term . . . that stuff I learned there really did help me a lot."

Shales pointed out that a 1991 *Time* magazine story had been critical of Scientology, but Jerry dismissed the article in his sit-down with Shales as "really poor journalism."

An infuriated Shales noted in his piece: "Look who's an expert on journalism now!"

Referring to *Time*'s reportage, Jerry told Shales, "I don't let that stuff stop me from getting the information I want," adding, "there are things in yoga that I don't agree with and I don't do. I go to get what I need. And that's the way I approach everything."

Jerry also told Shales he'd always taken a lot of meditation and yoga classes, stressing that "I'm very interested in Eastern thought and I like to explore a lot of different ways of thinking. To me, life is like flipping around the TV set, you know? I flip around to get what I want. It's there for me. But I don't embrace things wholeheartedly. I dissect them and take what I want."

A year later, in a *Playboy* interview, Jerry once again asserted that he had taken only "a couple" of Scientology courses, and described them as "fabulous." Asked if he felt he was an unwitting dupe of the church, he offered an emphatic "no." He said, "I've always had the skill of extricating the essence of any subject I study, be it meditation, yoga, Scientology, Judaism, Zen. Whatever it is, I go in to get what I need. To me, these are supermarkets. I go in to get my supplies, then I leave."

A month before the final episode of *Seinfeld* in May 1998, with the media covering the story as if the show's demise was on a par with the end of the world, Shales wrote his own farewell piece, which was highly critical of Jerry, and which again dealt with his Scientology. Shales pointed out that *Seinfeld*'s "hip irreverence" had become "mean and perverse" in a number of episodes over its nine-year run. He wrote, "The most conspicuous example was the show's treatment of Jews and Judaism . . . Some jokes were about Jewish stereotypes and ridiculed them—but other jokes seemed to be just at the expense of Jews." As examples, he mentioned a young rabbi character on the show who played "a platitudinous simpleton with no sensitivity toward human problems," and Jerry's parents, who lived in a South Florida retirement community where everyone "appeared to be made up of mostly Jews who were selfish and mean-spirited."

Shales noted a number of other examples and wrote "somewhere an anti-Semite is probably getting a big laugh out of this, too. Seinfeld is Jewish . . . but the Seinfeld he played showed no respect for Jewish traditions or heritage . . . . The point of all this?" Shales asked his readers. "Try to imagine *Seinfeld* spoofing the Church of Scientology. Even vaguely. Even remotely. Impossible. Off-limits. Judaism, however, a religion as old as recorded time, was fair game . . ."

Shales said Jerry "took pains . . . to defend the controversial sect" and described the courses he took as very pragmatic and helpful.

In 1986, at a time when Jerry had started raking in thousands of dollars a week in the comedy clubs of America, L. Ron Hubbard, to quote the scientologists, "departed his body." But the church's literature in 2000 stated, "He is still with us in spirit, and the legacy of his work continues to help people around the world realize their true spiritual nature."

—20—

Doctor Comedy

While Jerry started to generate stage time in the small clubs that had open mike nights, he, like most comedians in the late 70s, received no regular compensation. "Those bastards," he complained at the time, "give us nothing." But making audiences laugh, rather than making money, was what mattered most to him, at least in the beginning.

Virtually every day he commuted into the city from Massapequa and checked in at the clubs, hoping to get picked by the emcee. If he were lucky, he'd perform closer to midnight than three, but many nights he didn't get on at all after hours of waiting and hoping. "I'd get home at four-thirty, depressed," he remembers, "and my dad would wake up and come into the kitchen and talk to me about it. He was extremely encouraging," Jerry notes. "He was a salesman, and that's a similar type of life. You're really not doing any legitimate kind of work. You're just making a living talking people into things. That isn't much different from what a comic does."

While Kal was openly supportive of Jerry, he privately had doubts and fears about his son's future, which he voiced to friends, such as

Massapequa bicycle shop owner Dominic Totino. "Kal used to come into my store and kibbitz, and I'd ask him how Jerry was doing. He'd sigh and say, 'Oy, he's in New York trying to be a comedian.' Kal told me many times, 'I'm going to back him up as much as I can, but I think he's going to have a tough grind. Hopefully someday he'll make it big.' Kal was very concerned, very worried, because he knew show business was a tough thing to get into. He never said what he would have liked Jerry to do rather than comedy, but he made it clear he didn't want him to follow in his footsteps, even though Kal made a decent living in the sign business."

Mike Costanza acted as chauffeur and valet, shuttling Jerry to and from home and between the clubs, because, as he noted, "His nerves made him an unacceptable risk behind the wheel. He was a frantic basket case."

Even though he was living under the Seinfeld roof, Jerry still had to bring home the bacon to help out and have walking-around money. That involved taking a variety of dismal full- and part-time jobs, all of which he saw as temporary until he caught on.

For a period of time, he worked as a production assistant with Chris Misiano at a TV commercial production house called Flickers, on Fifty-seventh Street in Manhattan. While Misiano was working toward a ground-level job as a grip with hopes of eventually becoming a cameraman, Jerry could not have cared less about the film business, and spent most of his time sweeping cuttings off the floors. During breaks, he'd sit on a ledge in front of the nearby Chase Manhattan Bank branch, watching the office drones pass by, vowing he'd never spend his life in a straight job, even if his budding comedy career hit a dead end.

"The thing about Jerry back then," observed Vincent Misiano, Chris's brother, who also worked at Flickers for a time, "was that he knew what he wanted to do very early on, and had the talent and good fortune to pick exactly the right time. He didn't hold very many other jobs. He sort of went from a couple of short-run, odd jobs into comedy and made a living at it."

After a stint as a shoe salesman, Jerry waited tables for the lunchtime crowd at the busy Brew and Burger on Third Avenue and Fifty-third Street, which was not his cut of sirloin, especially when he discovered the gig could be damaging to his credibility. "He really complained about that job," recalled Jesse Michnick, who had moved upstate to work as an engineer at a radio station. "Jerry would say, 'You get a guy as a customer, who's seen you one evening in a club on a date with his girlfriend, and the guy's sitting up front laughing his brains out. The next day I'm serving him a burger and the guy looks at me and says, 'Haven't I seen you somewhere before? You look familiar. You're the guy from last night!' Jerry said they were always pissed off because they had paid five bucks to get in and thought they were watching professional comedians, not waiters."

More befitting him was the storefront jewelry operation he ran, an idea fueled by Costanza who was making decent bucks selling cheap sweaters he bought from Brooklyn wholesalers. Costanza had four locations in the city in front of busy Arby's fast-food restaurants, for which he had to pay the managers a hundred dollars a pop so they wouldn't call the cops who'd remove him as public nuisance—if not fine or arrest him as an unlicensed vendor. "Jerry came and saw my operation and laughed his ass off," Costanza recalled. It reminded Jerry of their light bulb–selling scam.

Impressed with the revenue his pal was generating, Jerry went to a wholesale jewelry outlet, bought a selection of cheap earrings, bracelets, and rings, and set up shop. Like his father in the early days, Jerry became a street peddler, hawking his junk from a wheeled cart in front of one of New York's busiest and trendiest department stores, Bloomingdale's. While he believed he'd snag bargain hunters entering and leaving the emporium, more often than not he caught the eye of New York City's finest, on the lookout for rogue peddlers. While he never got busted, Jerry was constantly being told to move on, buster. "Here I was running from the police on the streets of Manhattan," he recalls. "A parents' dream come true." According to Costanza, Jerry took those dead-end jobs as a way to motivate himself. "He kept himself in the throes of extreme poverty," Costanza noted. "Most of the time he was barely able to afford socks."

One of the myths surrounding Jerry's career is that he got his big start and spent most of his early years at Catch a Rising Star. But, other than that first botched performance, which became part of Seinfeld lore, he actually made his bones at a new, less-known club called the Comic Strip, which allowed him to drop his day jobs and devote full time to comedy after the club started paying him, as one of its first emcees, thirty-five dollars a night.

The Comic Strip opened on June 1, 1976, just when Jerry was making his first forays in the business, and he was one of the first in line to audition. It was a shrewd business and career move on his part because he quickly became a big fish in a small pond.

Jerry had auditioned for the daytime crew who told him he was good enough to come back and do a more formal tryout for the owners and a small audience. Sensing this was a big opportunity, he was both elated and a basket case at the same time. Needing moral support, he immediately telephoned an SOS to Larry Watson, then a graduate student at Cornell, with whom he'd kept in close contact. "'Larry, could you please, *please* go with me, I'm very nervous,'" Watson recalled Jerry telling him. "'This is my first break where they're going to let me audition ten minutes before closing.' They told him to come by that night when the Comic Strip closed, and they would let him go on and do his act. So I met Jerry at the club and he got up and did his routine at like 1:45. The only people in the club were the owner, two waitresses, me, and him. The owner laughed.

"Jerry came off and looked scared. There was real fear in his eyes. 'Larry, how was I? How was I?' I said, joking, 'Well, personally, Jerry, I don't find that "did-you-ever-wonder-why" kind of observational shit funny, but white people will love you and one day you will be very rich and famous because this will go over very, very well.'"

As Lucien Hold, the club's first manager and later a partner, noted, "The Comic Strip was a wide-open frontier. Jerry showed up at the right time. He was the first guy we picked out of the crowd of comics to emcee, our first homegrown emcee."

Jerry was like a Boy Scout out to win a merit badge: he appeared to the club's owners and manager to be trustworthy, honest, responsible, reliable, one who didn't make waves, didn't cause problems, made no demands, showed up on time, was neat and clean, didn't hit on the chicks, wasn't a drinker or a druggie. He was Mr. Dependable. And that's why he was chosen to be the emcee, becoming the only paid comic in the place.

"There were some perks in the late 70s," Hold said. "Male comics had access to women, could drink for free, there was a certain amount of drugs around. But Jerry was never into any of it. He also was the first comic I knew who wrote every day."

Every club back then had its own personality and mentality. The Improv, owned by show biz entrepreneurs Bud and Silver Friedman, was known to draw comedians who were in, or needed, or were using comedy as therapy—angry, dark comics who saw the world as being, in Lucien Hold's words, "fucked up." That group included the Richard Pryors, the Richard Lewises, and the Larry Davids, all highly talented and creative, but heavily, deeply angst-ridden. David, who went on to create *Seinfeld* with Jerry, would often walk out on stage in front of a packed audience, slowly turn his head, sizing them up, and then say "Never mind," or worse, and walk off.

"In those days," Hold pointed out, "nobody paid anything, so the club had to give something to the performer, more than just allowing him to go up on stage, some kind of personal perk. The perk at the Improv was that your angst will be tolerated and supported."

At Catch a Rising Star, the prima donna was tolerated and encouraged. Certain comedians, like Richard Belzer, the club's big act and emcee in the late 70s, had their egos stroked at Catch; Belzer could walk in, stride past six other highly talented guys who were waiting to go on, and jump on stage. Belzer thrived on it, and loved the place.

"Catch was like the Studio 54 of comedy," he said. "Celebrities would come. There was a period of time when every major celebrity showed up there and everybody who was anybody found themselves on stage there."

Like Studio 54, playground to paparazzi-bait like Andy Warhol and Truman Capote and Liza Minnelli, Catch was a glitzy joint, with a velvet rope cordoning off the crowds waiting to get in for the next show. The club's owner, Rick Newman, who had studied advertising art in college and then got into the restaurant-and-club business, envisioned his new trendy East Side location as a venue for young talent to perform and be seen by agents and network people. Besides Belzer, his talent included David Brenner, Freddie Prinze, Richard Lewis, Gilbert Gottfried, Andy Kaufman, and Elayne Boosler. If any club was responsible, in part, for the stand-up comedy explosion that was to come in the Eighties, it was Catch a Rising Star, which is probably the reason why Jerry, for publicity and public relations reasons, linked himself in later years to Catch rather than the less-known Comic Strip.

Inside, every night, the eighty-stool bar was packed, the 135 tables filled. Catch had become a hangout for all the *Saturday Night Live* people—Belushi, Radner, Aykroyd, Chase. Early on Jerry, wearing sneakers, jeans, and a checkered sports shirt, showed up with Costanza and gawked at the beautiful and famous people showing off the latest in Seventies' nightclub chic, which didn't include Levis and Nikes. But Jerry understood what was happening.

"A little glitter," he told Costanza, "goes a long way." In the background, Bruce Springsteen's "Prove It All Night" blasted on the sound system. Years later Jerry told Rick Newman, "Catch just always made us feel like we were *really* in show business."

Jerry intrigued Newman because he was one of the few comics who brought a cassette recorder with him and taped every one of his performances. "He'd listen to it later, hear what worked and what didn't. He was constantly working on improving his act. In those days, lots of the other comics talked about becoming actors." Richard Belzer, for one, dropped comedy in the Nineties to play a detective on the critically acclaimed TV series *Homicide: Life on the Street.*

"Jerry and Belzer were so different on stage," Newman observed. "Belzer would do an hour on current events with a newspaper, or go through a woman's pocketbook and do twenty-five minutes on what was inside—so quick, so funny. But Jerry prepared, prepared in depth, was tenacious about writing his material, and prolific. He took his work home, fine-tuned it, perfected it. The other thing was he always worked clean while a lot of other comedians were gratuitously dirty. From the beginning he had great potential to become successful."

Compared to the gloss of Catch, the Comic Strip was more down-home.

"We always treated everybody the same," Hold asserted. "Our comics didn't use much in the way of drugs, although we did have some good drinkers. But we liked people who had their heads screwed on squarely. My tendency was to gravitate towards people I considered somewhat normal. There was this theory that you needed to be a little bit crazy to be effective—like Andy Kaufman, who was not normal, like Gilbert Gottfried, who is not normal. That theory always bothered me, so I didn't tend to have as much understanding for them, or desire to work with those kinds of people. The people the Comic Strip gravitated to were very funny guys who were perfectly normal—Jerry Seinfeld, Dennis Wolfberg, Ray Romano—very down to earth, normal human beings, who are very good comedians."

The Comic Strip was the brainchild of a couple of bar owners, John McGowan and Richie Tienken, who later became Eddie Murphy's manager. The two owned a rowdy joint in the Bronx that catered to cops from the local precinct house. One of the bartenders was an impressionist who wanted to do stand-up. One Monday night he went to Catch's open mike night, pulled a low number, which meant he'd probably actually get on at a decent hour, and invited all of his friends, including Tienken and McGowan, to come down to see his act. The show started at ten, but by one o'clock he hadn't been called yet. Pissed off, Tienken, something of a tough guy, went to the manager and asked when his boy was going to make an appearance. "We're good, paying customers waiting to see him," Tienken said. The manager, something of a wise guy, looked around the room and, as if saying "who needs you" declared, "We got plenty of good, paying customers. Don't try to threaten me." Looking at his watch, Tienken noted it was now one-thirty in the A.M. and the place was still packed, the drinks flowing expensively. By the time he got back to his table, he'd decided to open his own comedy club, visions of huge crowds and big money dancing in his head. Tienken's idea was to open as close as possible to Catch, located at Seventy-seventh and First, which he did—four blocks up and one block over, at Eighty-first and Second.

The idea of a major straight-comedy club was something new to New York. The other clubs, the Improv, even Catch, still had other kinds of acts mixed in with the comics—dancers, singers. Pat Benatar, for one, started and worked at Catch. Even at the Comic Strip in its very earliest days there was a performing waitress, a singer named Patti Smith, who went on when the comedic talent was light. But the Comic Strip quickly became all comedy, all the time.

As emcee, Jerry, now twenty-three, wasn't making much money—thirty-five dollars a night—but enough, he felt, to get by, and it was better than picking up tips at the Burger and Brew. While the salary was minimal, the job of emcee gave him two enormous perks: stage time—he was on before and after each act—and considerable power. As emcee, he chose who would go on and in what spot; he essentially produced the show for the evening, and he had the juice to befriend people who could help him, drop his name in the right places with the right people. "The power," Hold noted, "was very tangible."

According to Mike Costanza, Jerry desperately wanted to be an emcee. "That was a big jump up for him pretty quickly," he observed. "First of all he had to be good enough to get on the bill at late-night, then it was to break into the regular lineup, then it was to be in the prime-time spot, then it was to be an emcee. It was a natural progression. But it wasn't the money—it was absolutely the power he wanted. I mean he's suddenly the emcee at the Comic Strip. That was big."

By the end of his first year, toward the end of 1977, a number of Jerry's peers were calling him "Dr. Comedy." He had earned the sobriquet because he could fix any joke that was DOA and give it new life, like Ed Greenberg, his Queens College mentor. He had a preternatural understanding of comedy, the nature of comedy. Those lonely childhood years he spent locked in his room watching comedy on TV and listening to comedy albums did not pass in vain. Comics at the club would make appointments with him to perform microsurgery on their material. A comic seeking medical assistance for a dying joke would say, "So, the girl walks out of the store, blah, blah, blah . . ." And Jerry would take out his scalpel—his ubiquitous Bic pen—and go to work on the material. "Why don't you say 'blah, blah, blah—and then the girl walks out of the store.' That'll work better." Or he'd say, "That word is just not a funny word. Why don't you try this word. This word is a funny word. That word is not a funny word."

Most of the time, he was right.

As Hold, who by 2001 had known Jerry for a quarter-century, observed, "He knows funny better than probably anybody I've ever known in this business. But just hanging out, he's not a naturally funny guy in his own right. He doesn't crack jokes. But those people who are naturally funny don't tend to work as hard as someone like Jerry; they don't tend to write as hard, edit as hard. Jerry was always a *very* serious guy. He can laugh, and tell a story, and make it appear off the cuff, but it's not. He's not like someone who, whenever he opens his mouth, is funny. A naturally funny guy delivers slightly differently every night, like a good actor will do. But from the very earliest days at the Comic Strip I noted that he delivered everything exactly the same—exactly the same nanosecond of a pause, then the punch line, then another slight pause, which might last for 1.2 seconds. Exactly the same all the time. When Jerry goes on talk shows, he always makes sure he's asked certain questions. It's not that he's ill at ease, he's extraordinarily confident—the first hyperconfident comedian I ever knew. But he has to construct what he wants to say—and then he's very funny. He knows how to make people laugh, but it's programmed."

This was underscored one night at the Comic Strip when Jerry and another comedian, Larry Miller, who became bosom buddies after meeting at the club, exchanged material. A naturally funny comedian who can describe in detail the contents of a cantaloupe and leave the audience in hysterics, one April Fool's Day night Miller decided to switch his act with Jerry's. Jerry took Miller's material, nowhere nearly as finely honed as his, made it a little cleaner, but embarrassingly struggled with it, couldn't make it interesting, let alone funny, and died on stage. When Miller got up, though, he took Jerry's material, which was written within an inch of its life, and delivered it with his own panache and natural sense of humor, and killed them.

Hold, who witnessed the performance that night, said, "If I hadn't known before that Jerry was the better writer, Larry the better performer, I certainly knew it after. But Jerry has always known his limitations; he always knew what he was good at, where his strengths lay, and had a very strong ego. So when he did *Seinfeld,* he had no problem putting people on who were funnier than he was because he was so confident in his ability to know what was funny."

—21—

Major Changes

During the best of times they'd wake up together and Jerry would compulsively recite a mantra in his head—the words "Lights. Camera. Action."

"Lights" when he opened his eyes, "camera" when he sat up in a bed they shared whenever they could get together for a night, and "action" when he stood up. The words, coming out of his Scientology and Zen practices, apparently helped motivate him for the day ahead.

But now, with his career seemingly on track, those romantic nights with Caryn Trager were becoming few and far between. Entirely focused on making it in comedy, Jerry determined there was no place in his life for his loving and loyal girlfriend of five years. In 1977, he decided to end it with her, a decision he had been plotting for some time.

"We had many conversations, Jerry and I, about the fact that Jerry knew he was going to have to break up with Caryn, and I told him I just thought they were made for each other," Mike Costanza said. "They read each other's minds. They knew what they were thinking before they said it. They were *that* close. And their many years together had formed a bond that I thought was unbreakable.

"But Jerry's very focused, probably the most focused person I've ever known. When it's time for business, it's time for business, and Jerry knew that he'd be on the road. He knew he'd be traveling, and he just knew that he couldn't have any ties. Jerry's the type of guy who wants everything in his life to have a definite beginning and end—like his stand-up act. At that stage of his life, he looked at marriage as an endless date."

Marriage to Jerry, in fact, was something Caryn had figured was inevitable, even though they rarely talked about it, and were part of a generation that didn't think it necessary or mandatory. Still, their intense, albeit turbulent, years together, and her heartfelt belief in Jerry's strong commitment to her, despite a couple of indiscretions on his part along the way, caused her to envision a bright and wonderful day when the two of them would bond for life under a *huppa*.

"I always thought if we had more time together we would have been married," Trager acknowledged years later, having never married. "When he broke up with me, all everyone said was, 'We thought you and Jerry would be together for the rest of your lives.' No one could see me without him. They couldn't fathom we weren't together."

While Trager was officially a graduate of Oswego's class of June 1975, she required six more credits to get her diploma. Moving back in with her parents in Forest Hills, she finished her schooling in a semester at Queens College, in January 1976. To celebrate her graduation, she and Jerry went on a romantic ski trip to Mont Tremblant, in Quebec, a gift from her parents. To her, everything between them seemed copacetic. Or so she thought. This was at the same time that Jerry's life was undergoing a major change, when he was starting to get stage time at the Golden Lion, Comic Strip, and Catch, where, into the wee hours of the morning, Trager clapped, laughed, and whistled for her man until he finished his set and she could go home, exhausted and bleary-eyed, even though she had to be at work in the morning.

"Waiting on line outside one of those clubs in the freezing rain wasn't my idea of a good time," she said ruefully, looking back. "But my boyfriend wanted me to do it, and I was standing by him. It got to be hard because he got on at 2 A.M. and then we'd go down to Chinatown to eat, and I had to get up and go to a straight job. But I didn't stop doing it because it was hard for me. I eventually stopped because he really didn't want me there anymore. I was probably making a lot of assumptions. There's a concept called 'mind-reading psychology,' and I think I did a lot of that then."

As Costanza, who had sat with Caryn through many of those late nights, observed, "She and I would hang out at the clubs and hold hands watching Jerry on stage, hoping he did well. That's how we were together. She was totally in love with him, totally loyal to him, and would do anything for him."

But Jerry saw that their lives were now at opposite poles. While his world was late-night, hanging out, schmoozing—an electric show business scene—hers was daytime, water-cooler gossip, and mundane, clock-watching, office work. Though she wanted to be a psychologist, her first job out of college was analyzing whether she'd put too much sugar in her boss's coffee, working as an administrative assistant to the director of player personnel for the National Football League, a prim nine-to-five secretarial gig. After a year, she left to work for the blue-collar merchandiser JC Penney as a wholesale-inventory control specialist, a job she secured with the help of Chris Misiano's mother, and one Trager kept after Jerry had started gaining national recognition and making big bucks on the club circuit in the stand-up comedy boom years of the 80s.

The end of their relationship did not occur in one horrific, explosive moment that would live in infamy for both of them. Rather than actually confronting her to say "we're through" and face a possible teary, anguished, name-calling blowup, Jerry took the easy way out: by avoiding her, by seeing her less often, so their breakup was spread out until she got the message, rather than concentrated in one nuclear moment.

"In the beginning he used to run his material by me and wanted my feedback," she said. "At the end he wanted to keep the comedy stuff separate. He really didn't want to be in the relationship anymore, but he never verbalized it. One of either my strengths or flaws is that I do a lot of mind-reading, and what I read was that he didn't want to be there anymore. He was closing me out. That's the way I felt. It was killing me."

But Costanza remembered the breakup as much more heated and concentrated.

"There was," he asserted, "a *big* crying scene and I remember wishing I could help her, and feeling so bad. It's a very bad position to be in for a woman to talk to a guy's best friend and say, 'Why did he break up with me? Don't you think we can get back together?' What could I do? Caryn did not deal with the breakup very well. She was an emotional wreck.

"I was with Jerry right before it happened, and right afterwards. It was a big thing for him, too, *very* big. But he just knew he had to do it. They had gone out on a date when he told her they were through, and Caryn became hysterical. They had to leave the place.

"Caryn was figuring we're together five years now, we'll probably get engaged next year and get married, because that's what it was looking like," he continued. "Even Jerry was looking like he was in favor of that. He wasn't voicing his displeasure with going down that road. But, after the comedy was injected into his life, marriage was not going to happen. He knew marriage wouldn't be fair to him or her. In a way what he did was selfish on one hand and honorable on the other. But it was his life and he ultimately had to do the best thing for himself."

Having finessed the denouement of his relationship, Jerry made another major life change in 1977. He celebrated his twenty-third birthday by moving for the last time out of his childhood room in Massapequa, and into his first apartment, a fourth-floor walk-up at 129 West Eighty-first Street, on the Upper West Side of Manhattan. The four hundred-square-foot unfurnished studio, which cost six hundred dollars a month—most of which his father was paying until Jerry got a roommate—would one day become the model for his much-sweeter-looking, one-bedroom bachelor pad on the *Seinfeld* set.

Costanza, who helped him move over a weekend with a U-Haul rental truck, spent much time hanging out there. Jerry's first place was spartan, minimalist, and practical. The only windows looked down on rooftops. There was a daybed for sleeping, a small, Formica-topped round table for writing and eating, and a hideous, macho leather couch with brass grommets around the edges and the base, next to which sat a wooden magazine rack. The only charming attribute in the otherwise dreary apartment was a fireplace with a mantel upon which Jerry put a jar where he stashed the earnings from his small gigs, something he had learned from his parents who hid their emergency money in the freezer of their refrigerator, money Kal jokingly dubbed "cold cash."

Next to the door of the telephone booth–sized bathroom, Jerry placed a bureau for his neatly rolled socks and folded underwear. On the wall above, he tacked up photos that he had brought with him from his room at home, old publicity shots of George Burns and of the ventriloquist-comedian Edgar Bergen and his smart-ass dummy, Charlie McCarthy.

"I want to be in this business as long as these guys," he told Costanza, pointing at the comedy icons. "They remind me of what I need to do to have the kind of act that it takes to be that good and have that kind of longevity. I want to grow old in Las Vegas."

Costanza thought, "My God, at twenty-three, he's already thinking about achieving immortality as a stand-up."

A big shocker occurred within the first year after Jerry moved in. Living alone for the first time in the big city, and unsophisticated in its ways, he entered a period of deep insecurity and loneliness. Needing love and comforting, he called Caryn Trager out of the blue and asked for a reconciliation. "I was hesitant at first," she said. "It had to really mean something. My feeling was he just needed something familiar to carry him through. But we got back together and it was good again for a while."

Trager moved out of her parents' apartment and into a ground-floor flat—"I could see from my window homeless people peeing"—of a brownstone her grandmother owned on Eighty-eighth Street, between Broadway and West End Avenue, a short walk from Jerry's place. They fell into the same routine. He spent most of the day working on his act at the Formica-topped table, while she was at work at Penney's Manhattan headquarters. At night, she'd hang out in the clubs while he performed, but she sensed that once again he was trying to end their relationship as he grew more confident on his own.

"Slowly, he stopped calling me," she said. "That, for me, was a real betrayal, because I think he wanted me to come back just because he moved into the city and wasn't secure living on his own and he needed a familiar, comfortable thing. For me that was a real breach.

"We broke up for the last time on Broadway, when we ran into each other on one of the islands where the benches are, in the spring of 1978. I said to him, 'It's obvious that you can't do this and you don't want to be with me.' He was saying he was sorry. He gave me the 'you deserve better' speech. I remember saying, 'I can't make you feel any way that you don't feel. There's nothing to be sorry about.' And that was it. It was very devastating."

A couple of years after the split, Jerry and Trager and a few of their mutual friends had a reunion at her apartment and then went to Greenwich Village for dinner, but it didn't go well. "He didn't feel like him anymore to me," she said. "The first thing he said to me was, 'You can't imagine how much money I'm making.' Which was really bizarre because it had never been about the money with him. It was never important to him. He was nervous and uncomfortable."

Jerry's family, who loved Trager, couldn't believe he had dropped her, and she remained friends with them for years afterwards. "The Seinfelds," she said, "felt Jerry was an idiot for breaking up with me." When Jerry's sister, Carolyn, got married in the early 1980s, Trager was one of the bridesmaids. Trager had started dating another man after Jerry, and they double dated with Carolyn and her fiancé, Elliott Leibling. At his sister's wedding reception, Jerry sat at a table near Trager, openly flirting and holding hands with another woman.

For years afterwards there was no contact between them. Eventually moving to Los Angeles, where Jerry had settled, she followed his career during his immensely successful stand-up years, reading the various newspaper and magazine write-ups, but she had difficulty watching *Seinfeld* because many of the bits on the show reminded her of funny incidents and events that had occurred during their years together, and the memories were too painful. The Elaine Benes character, Trager was convinced, was modeled on her. "I was the original Elaine," she stated, "because I was the only girl in his crowd."

Looking back in 2001 on his amazingly successful career, Trager stated, "I've been happy for him. Jerry's a funny kid from Long Island who hit the lottery."

—22—

Strange Bedfellows

With Caryn Trager out of his life, Jerry wouldn't have another serious relationship with a woman until the 1980s. But there were a few females whom he dated for a while, among them Carol Leifer, after he discovered her at the Comic Strip; but their bond was strictly comedy, not romance.

"Jerry was really fun and considerate, not a jerky boyfriend," Leifer noted years later, after she had worked as a writer for several seasons on *Seinfeld*, following a failed marriage to another comedian and a successful stand-up career touring small-town chuckle huts, which she had grown tired of. "It was fun for me to date someone who was already doing for a year what I was, by that time, trying to do for a living also. I liked the perks of going to watch him do his shows. He was such a pro compared to my being such a novice."

After Leifer, Jerry dated two petite, dark-haired, olive-complexioned girls from the boroughs. He told LuAnn Kondziela he was "in love" with one, whom he identified as Donna, asserting that they had been dating for eight months, although none of Jerry's male friends remembered any Donna in his circle. Jerry later told Kondziela he had broken off with

Donna and was now seeing a girl named Camille. "Something about those Italians," he remarked.

Lucy Webb, though, was an entirely different female archetype in Jerry's sphere. She had come to comedy and the Comic Strip, where they met, from her home in Tennessee; at the age of five she was running lines with her mother for a production of *Damn Yankees,* and by twelve was winning acting awards. Dark-haired, with an open, sensuous face and chubby proportions, she closely resembled the White House harlot Monica Lewinsky and had even served as a Washington intern; her boss was not Bill Clinton but the politician who would be Clinton's vice president, Al Gore, then in his first term as a congressman from the "Volunteer State."

By the time Webb started seeing Jerry, she and another young woman were doing stand-up as part of a duo called "Rents Due, Too," playing the Comic Strip, Catch, the Improv, and Dangerfield's.

"Jerry and I had a relationship, and it's something I've never really talked about to anyone," said Webb, who married stand-up-turned-actor Kevin Pollak. "Jerry and I were seeing each other, and I helped him move to California, and I was at his first Carson show."

As with Leifer, Jerry's bond with Webb was strictly show business. Susan McNabb recalled that shortly after she became intimately involved with Jerry in the mid-1980s, they were once headed to Jay Leno's house for Thanksgiving dinner when Jerry, in his drama-queen persona, stopped her at the front door. "I just want to warn you that a girl named Lucy Webb's going to be here and I used to go out with her and I'm just warning you in case it's uncomfortable." It wasn't, she said.

Still another young woman who differed from Jerry's archetype—the dark-haired, big-breasted, New York Jewish or Italian princess—was Texas-born and -bred Karen Jane Greene, four years Jerry's junior. From a well-to-do family, the pretty redhead had graduated in 1975 from the Hockaday School, a tony, private girls' school in Dallas, and had earned a bachelor of arts degree in Spanish from the University of Texas in 1979, before becoming a flight attendant for American Airlines. The two began seeing each other after meeting at one of the comedy clubs in New York, where Greene was based at the time.

"He was working three comedy clubs a night," Greene said, "and earning just enough for cab fare. We dated and at some point became friends and have stayed good friends all these years. We both saw a lot of potential in each other, and we've enjoyed over the years seeing each other succeed."

Many years later, after Jerry became rich, powerful, and famous, he would play an important and curious role in Greene's own continued drive for power and success, no longer aloft in the friendly skies serving Coca-Cola and peanuts, but in the commanding bastion of conservative Republican politics. All of that, however, was still to come.

Meanwhile, Jerry decided to hook up with an intriguing roommate to share the cost of his Manhattan digs.

A talented stand-up comedian, older than Jerry by at least two years, Henry Wallace, the roommate, resembled Larry Watson, Jerry's Oswego roomie—big, theatrical, and African American.

"Wallace broke his neck to meet me back in those days," Watson recalled. "He said, 'I just want to meet you because Jerry has spoken so much about you, and you're Jerry's first black friend.' He was intrigued because I had shared so much with Jerry and that we were close, intimate friends. Jerry carried all of that over in terms of his meeting Wallace. Jerry was sensitive and aware [regarding black culture] and to Wallace it was like, 'How did you get that way?'"

Jerry met Wallace, who used the stage name George Wallace—a riff on the one-time infamous racist Alabama governor—at the Comic Strip when both of them were starting out. The two had come from entirely different worlds—Wallace was a son of the South, born and raised in Atlanta's ghetto. But both shared the same dream, to make it big in comedy.

For Wallace, comedy was his first love but his second career. Before coming to his first audition at the Comic Strip, billing himself as the "Rev. George Wallace," a charismatic pseudo-minister who sported a flowing cape and quoted from what he called "the Big Book of Bell," which was the thick Manhattan phone book, he had been in the world of advertising—not button-down-Madison-Avenue-Brooks-Brothers-style advertising, but rather the gritty, Times Square side of advertising—hustling lucrative ad space on the sides of noxious New York buses. "In advertising I had to sell space, whereas in comedy I have to sell myself," he says, echoing Jerry's feelings when recalling his days peddling jewelry in front of Bloomingdale's.

"George is the guy who talked the Comic Strip into advertising on buses," said Glenn Hirsch. "The Comic Strip was the first to advertise that way, and that's how George became involved in the clubs. When you were a big black man in those days in comedy, you stuck out like a sore thumb, and yet he was a very funny and charming guy who moved up quickly."

Before the advertising game, Wallace, who went to the University of Akron, had peddled rags. "George," observed a member of Jerry's circle, "was sort of a Kal Seinfeld in black face."

Jerry and Wallace had an instant bond, though those around them couldn't quite figure out exactly what it was except for their common obsession with comedy, and a drive to succeed in it. For instance, both had the same philosophy about stand-up, though their styles were entirely different. As Wallace put it, and as Jerry concurs, "There is no sex or drug that makes you feel as good as being funny on stage."

Not long after Jerry rented his place, Wallace moved in. "Jerry and George's relationship grew," said Hirsch, who worked in the clubs where he'd hang out with them. "Comedy makes for strange bedfellows. Maybe it was a security thing for Jerry. But I feel their friendship grew from sitting at a bar in a comedy club together for five hours a night, seven days a week, talking about everything in their lives. To me they were just good friends."

A highly placed entertainment industry player who was closely involved with Jerry's career said, "Jerry and George spent a lot of time together. They shared a small apartment, though they were on the road a lot. George was the most loving, kind, sweet, supportive man who had an older quality about him. It always seemed like he was not Jerry's peer, but more like he took care of Jerry. When I would ask Jerry about his relationship with George, his response was always, 'We're best friends.' It was *always* 'best friends.'"

Nancy Parker, a talented, versatile comedian, a young-Bette-Davis look-alike who did offbeat impressions on command—Dr. Ruth Westheimer, Barbara Walters, Shelley Long, among others—got to know Jerry and Wallace at the Comic Strip in its earliest days. Along with Elayne Boosler and a few others, she was one of a handful of working female comics in the 70s. Unlike most of the other male comics with whom she hung out, Jerry, she noted, didn't show much of an interest in pursuing women, and she perceived little sexuality. "I know Jerry and George were very good friends, and George was a very sweet man who I held in the highest regard," she said.

Most of the comedians lived in the city, but David Sayh was one of the few who owned a car and commuted to his home in Riverdale, in the Bronx, a route that took him through the Upper West Side. More often than not, if Jerry and/or George were around, he'd offer a ride. Like others in the business, Sayh was aware of the gossip.

He had his own questions about Jerry's friendship with Wallace, because of what appeared to him to be Jerry's disinterest in women, not that there's anything wrong with that.

"Show business and groupies go together," he said. "We were all in our twenties and there were a lot of women around who were looking to meet the guys, and the guys who wanted pussy were getting it. Everybody who I knew who was straight was chasing women. It was the 70s. It was there. It was pre-AIDS. The comedy world was burgeoning and there were available women and a lot of guys took advantage of that. Jerry, though, was not known as a girl chaser, and he could have gotten some beautiful women. The talk was, 'What's with him?'"

Specifically, Sayh pointed to what he felt was Jerry's refusal to date a young hottie, a cute Italian girl from one of the boroughs who hung out at Catch and had fallen hard for him. "She was after Jerry, she was dying over Jerry when Jerry was a nobody, before anything happened in his career," Sayh said. "She was coming on to him because she was so in love with him, had a big crush on him, but he wasn't responding. There were some girls who were just looking to get laid—starfuckers who'd fuck anybody who was on stage, a notch on the belt for them. But she wasn't like that toward Jerry. She really liked him, had a genuine, sincere interest in him, but he never would go out with her. I remember her asking me, 'Doesn't he like girls? Is he gay?'"

After Jerry rejected the young woman, Sayh said he dated her for a time, doubling with his pal Richard Belzer, who went out with her girlfriend.

In 1980, when Jerry decided to seek his fortune in Los Angeles, he gave close friends his new temporary address, which was 1000 Westmont

Avenue, apartment 302, in West Hollywood, an area known for its gay and lesbian population.

Jerry's instructions to his friends were to send any mail to the Westmont address, in care of "George Wallace" until, he said, he got a place of his own. Wallace apparently used the apartment when he was working on the West Coast. Jerry also kept the rent-controlled Upper West Side apartment in New York for years, and Wallace continued to share it with him.

When Jerry became intimately involved with Susan McNabb in the mid-1980s, Wallace was omnipresent in their lives, but the very accepting McNabb adored him. However, there were times when Jerry's friendship with him infuriated her, especially when he spent time with Wallace instead of her, including taking a Hawaiian vacation.

"Suddenly I wasn't invited," McNabb said years later, still wondering why. "I was hurt. Instead of me, Jerry went off to Hawaii with George, and gave me no reasonable explanation, and I was absolutely Jerry's girlfriend. Throughout our relationship, Jerry and George were very close. They always shared the apartment in New York. They talked on the phone all the time. They went on that trip to Hawaii together. They spent time together. George was part of Jerry's life. And I always found their relationship a little puzzling, frankly.

"George was the sweetest guy," she continued. "He was always so kind to me. I loved him, but I never knew what they had in common. They just had a connection. They just really clicked. They were very close."

Robert Williams, who was Jerry's agent for nearly a decade and Wallace's for much of the same period, beginning in the early 80s after Jerry moved to Los Angeles, felt the speculation about them, together and separately, was kindled by the fact that they rarely, if ever, were seen with women.

"There was conjecture and after Jerry became much more visible, the rumors started," recalled Williams, who was president of Spotlite Enterprises, a powerful booking agency. "The assumption was, 'We don't see them with a lot of chicks, therefore they must be gay.' Well, welcome to Hollywood. Jerry and George were managed by the same people. They were dear friends and a lot of that speaks to their style of humor. Also,

George in those days happened to be on bills with Jerry where George was the middle and Jerry was the headliner.

"Like any man, Jerry had his liaisons. He had his dinner dates. I saw Jerry with dates, but he didn't put them in the front row, or have them hang off him. When he stopped dating a woman, he'd say, 'She was a lovely girl, but obviously my lifestyle [being on the road all the time] is not right for her.' That's the way he handled it. He was not someone who confided. That was not his style. Jerry's a loner, a private man.

"Jerry was always on the road. He was a sailor who didn't dock, and that led to talk that he wasn't one of the boys. So, what I was hearing was—'How come you don't see him with a chick every night?'—because the comedians on the road were infamous for having a new woman show up every night.' "

The talk about Jerry and Wallace extended to Jerry's old gang from Queens who knew of their close relationship and wondered what it all meant. "I heard that rumor and it took me by surprise," said Jesse Michnick, who discounted it because he believed Jerry was "intrigued and fascinated" with Wallace for other reasons.

"George was from the world of advertising, and he was a person who had an older soul than the younger comedians Jerry was hanging with," Michnick observed. "When he came into the room, the dynamics of the room changed because he had such a commanding presence and had no problem going into some shtick and commanding everybody's attention. He had a booming voice, an aura, and being around that gave Jerry a good feeling. Unlike a lot of people who were struggling in comedy, George knew that he was going to be a comedian, and so did Jerry. George was a professional. He was the guy who Jerry would have at the deli with him."

They remained the tightest of friends for a lifetime. Jerry helped Wallace get the best manager in the comedy business, George Shapiro, and helped to hook him up with Spotlite. Wallace's philosophy has been, "Everybody wants to laugh, and we got something for everybody—old, young, black, Oriental, gay, or lesbian." Like Jerry, he has always worked clean. For one episode of *Seinfeld*, in the fall of 1996, Wallace played a doctor. "We try to keep business and friendship apart," he says of his relationship with Jerry. "You're always gonna feel funny working for your best friend." A couple of years later he talked about the possibility of even touring with Jerry. "Seinfeld will open for *me*," he boasted tongue-in-cheek to a *Washington Post* reporter. "Big letters: George Wallace. Little letters: Jerry Seinfeld."

And Wallace was Jerry's best man when he finally got married, ending his long bachelorhood. But Wallace never tied the knot, though he claimed in a "somber and detached" 2000 interview with a reporter for the *News Tribune*, in Tacoma, Washington, that he'd dated the same woman for twenty years, but did not have time to take the relationship further because he was always on the road—the same excuse Jerry had used for years.

—23—

Faster Than a Speeding Bullet

It was a heady time for a bright, shrewd, driven-to-succeed, on-the-make young comic like Jerry in the New York of the late 1970s. The market for the new breed of stand-up was growing, and real talent, like Jerry's, was limited. The city itself, meanwhile, was in a state of moral, fiscal, and urban decay. Times Square, gaudier and raunchier than ever, had become pimp-and-mugger heaven. The South Bronx was the poster child for the deteriorating cityscape. A twenty-four-year-old deranged postal worker named David Berkowitz, whom the tabloids dubbed "Son of Sam," was running amok, terrorizing the town. Where garbage didn't cover the sidewalks and stoops, street gang graffiti, homeless people, and used hypodermic needles did. On the verge of bankruptcy, the city had to get on its knees like a ragged street beggar and plead with Washington for a handout.

Depressed and stressed, New Yorkers needed to laugh, and Jerry was among a small group of slick, urban comics who were there to fill that need.

By the fall of 1977, he was emceeing the three most important nights of the week at the Comic Strip—Friday, Saturday, and Monday, still at thirty-five dollars a pop. As emcee, he showed a remarkable eye and ear for new blood. In an audition on the night of Labor Day 1977, Jerry first gave Paul Reiser and Rich Hall the green light. "It was very unusual to pass two enormously talented comics in the same night," Lucien Hold noted. "All of them went on to fame. It was quite an alignment of the stars that night."

At the same time, Jerry was getting gigs at other clubs—Catch, the Improv, the Bottom Line—with some regularity. He was walking on air. As he told his Oswego pal, LuAnn Kondziela, "Everything has been just sensational. All things are going well." As a health nut, the one thing he couldn't stand about the comedy club scene was the cigarette smoke, and he jokingly told her that despite his increasingly busy schedule, "I still find time to fight emphysema."

Kondziela had recently started working as a licensed practical nurse with patients in a locked ward and had filled Jerry in on some of the horrors of her job. Using a word he wouldn't be caught saying on stage, and feigning jealousy of her career versus his, he confided, "Sometimes I sit in the bar at the Comic Strip just after I've gotten off, the sound of thunderous applause and screaming oversexed women still echoing in my ears, and I wonder, is it fate? Why must I be a lowly comedian? Why can't I have people shitting on my pants leg and smashing me in the head with arm boards? I suppose we must accept our station in life. I could shit on myself but it just wouldn't be the same."

The stage time, rather than the few dollars he was earning, was paying off. One glorious night Rodney Dangerfield showed up with a bevy of bimbos and a few cronies, caught Jerry's act, and was impressed enough to give him face time on an early HBO comedy special, *Rodney Dangerfield: It's Not Easy Bein' Me*, a showcase for young comics.

While they did entirely different comedy, Jerry admired the veteran Dangerfield. Having watched him work a boisterous club crowd, Jerry was convinced Dangerfield got "his energy just from the buzz," which Jerry felt was a wonderful talent. He felt lucky to have caught Dangerfield's eye because pay cable—HBO and Showtime—was a pioneer in broadcasting stand-up comedy as a staple of its early programming. Shows like Dangerfield's were inexpensive to produce and garnered enormous profits because they were aired repeatedly. For Jerry, it was a way to quickly appear before a national audience, not just couples from the boroughs out on dates. It was an important gig, another notch in his belt on the way up.

For weeks before the taping, he honed his material, now amounting to about ten minutes. Like always, he agonized over every word, nuance, and his timing, and over which routines would play best on TV. His act at that point consisted of old standbys and some new material—from socks in the dryer to the weirdness of computer dating.

But there was no need for him to be so concerned about being on the tube for the first time. The camera would always love Jerry. His squeaky-clean material, observational humor of subjects that everyone from Canarsie to Podunk could relate to, and his clean-cut demeanor were tailor-made for family television. To the trained comedic eye, Jerry had it down. As Glenn Hirsch observed with some envy years later, "Early on he developed the perfect act for television where you don't move from the camera lens and present very funny premises and pictures verbally. I didn't have that ability. I was all over the stage."

Around the time of the Dangerfield special, Jerry got a shot on another TV show called *Celebrity Cabaret.* Like Kal Seinfeld, who as a young veteran had considered fixing televisions as a lucrative business, Jerry had already determined that TV was a place he wanted to be somewhere down the road. Thinking big, he made it his immediate goal to one day get a shot on Johnny Carson, something most young comics dreamed of but few attained.

By the spring of 1978, his plan was getting closer to reality. Networking in the clubs, Jerry had bonded with the comic David Brenner, a clean-cut Jewish boy from Philadelphia, who had quickly earned a reputation as a talented and on-the-mark monologist and observational comic, much like Jerry, though more on the neurotic side. Carson liked him and had had him on the *Tonight Show* a number of times, even using him as a guest host, which gave him considerable influence with the king of late-night television. "Brenner is going to get me on the *Tonight Show*," a confident Jerry confided to Kondziela. "But I still won't do it for at least a year and a half. I'm in no hurry. I have to get much better. It's nice to know it's there for me, though, whenever I want it."

Jerry also had discussed with Glenn Hirsch his dream of appearing on Carson. "Jerry made it clear his goal was to go to L.A. and he had a focus about doing the *Tonight Show*," he said. "He knew making it on Carson would launch him."

Still, at that point, Jerry was working for the laughs and peanuts in the clubs, and though his material was getting stronger and his confidence level higher, there were many bad nights, such as the one at the Raleigh Hotel in the Catskills, the first time he worked in front of an older, Jewish audience. Jerry preferred audiences made up of young, urban professionals and college kids, who understood where he was coming from. Still, he had looked forward to the appearance in the "Jewish Alps" because of the payoff. "There's a lot of money I can make in the mountains," he told Kondziela.

The hotel had sent a van to the Comic Strip to pick up Jerry, Carol Leifer, Costanza, and a juggler whose name is long forgotten. Jerry would have been better off had the vehicle blown a gasket and never arrived in South Fallsburg. Not one laugh, not a chuckle. "It was like one of the worst performances Jerry ever had," recalled Costanza. The Raleigh audience was not Jerry's audience, and he should have known that from the many nights he spent watching the comics at the Brickman. The Catskills audiences liked Jewish humor, which Jerry detested. He thought it was too easy and boring, the same thing over and over. "I have to do what I do," he told Costanza. "It has to come from who I am." On the ride home that night, having bombed badly, Jerry was sullen and depressed. But Leifer and Costanza kidded and joked so much about his horrible debut among the Chosen People, Jerry's people, that he soon was hysterical with laughter. There would be other Catskills dates where he'd be more successful. After one, the Jewish comedic god known as Jackie Mason approached him. "Ta tell ya the truth," the ex-rabbi said in his trademark accent, "I look at you and I can see that you're a funny guy. And, oy, I can't stand it."

Another disaster occurred at Princeton University, a Las Vegas Night fraternity party booking that he got through the Comic Strip. For three hundred dollars, Jerry was asked to perform for forty-five minutes.

Costanza, the chaperone, watched sadly from a corner of the room—filled with rich frat boys and their debutante dates, shooting craps and spinning roulette wheels, getting shitfaced on beer, booze, and pot—as Jerry stood in the middle of this menagerie seriously performing, ignored by absolutely everyone but Costanza. There were many such bookings, but Jerry quickly learned to grin and bear it. It was all, he realized, part of his internship. Stage time was stage time. He took the money and ran.

But at his home club, the Comic Strip, he was fast becoming a celebrity, based, in small part, on the fact that it was there, one exhilarating late night, that a petite but big-haired, big-breasted, sensually lipped, Jordache-jeaned girl from Queens who looked like she could suck the chrome off an exhaust pipe was the first ever in Jerry's young career to ask him for his autograph. "How do you remember all that stuff?" she asked, handing him a pen and her telephone number. Testing his developing show biz charm, a big, toothy smile that would make any Queens *maidel* swoon and think "husband," Jerry signed a Comic Strip placemat menu for her.

But as soon as she turned to leave, disappointed in not receiving an invitation to join him at, or under, the table, an opportunity 99.9 percent of the other comics in the club that night would have jumped at, Jerry tossed her number. Seated at the table during the historic autograph-signing moment, Costanza couldn't believe he rejected such a come-on. Later, he observed diplomatically, "He was just never much interested in the rough-and-tumble of chance encounters."

While he didn't take advantage of the sexual perks, life for Jerry had become a trip. After a night of racing from the Comic Strip to Catch to the Improv, or out to Pip's in Brooklyn, in cabs shared with fellow comics, to do his sets—savoring the laughter, the applause, the adulation, the power, the acceptance—he'd end the night in a booth at a joint called the Green Kitchen, a Greek diner at Seventy-seventh Street and First Avenue, hanging out with the same gang of comedians he crossed paths with all through the night. Exhausted after having done a sixth set in a smoky, boozy room, he'd sit with his comedy pals until almost daybreak, rarely talking, but mostly listening, taking in their war stories, their jokes, their sparring, still high and wound up, not wanting to go home, and laughing, laughing, laughing. For the first time in his life the nerdy kid from Massapequa felt accepted.

Ever since leaving Oswego, Jerry had yearned to have another motorcycle, but with his hectic schedule he no longer had time for road trips. However, he bought a little one-cylinder Yugoslavian bike, and rode it for about a year, but just as quickly sold it. "He realized he could get injured," said Lucien Hold, who lived a block from Jerry and hung out with him inside and outside the Comic Strip. "So Jerry stopped riding. He just rationally looked at it and figured—I could scratch up my face. My career is more important."

Other universes were opening to him, too.

At the Comic Strip, for instance, he charmed a waitress with clout named Hillary Rollins, daughter of Jack Rollins, the coproducer of the films of one-time stand-up Woody Allen. When Jerry mentioned to Rollins that he was a huge fan of the brilliant observational comic Robert Klein, then scoring big-time on college campuses and in concerts around the country, she used her influence to get Jerry, Costanza, and another waitress at the club into a taping of one of Klein's early HBO specials, and then to the aftershow party at Klein's tony Riverside Drive apartment, where Jerry rubbed shoulders with the likes of Paddy Chayefsky and nibbled hors d'oeuvres from silver platters.

Klein's place was not Mr. Donut in Massapequa, or the Greek Kitchen, settings he was used to and comfortable in. As Costanza recalled, "Jerry's eyes were popping out of his head. He felt both thrilled and intimidated being there. We huddled together in a corner because Jerry didn't want to be noticed, didn't want to screw up by making some sort of faux pas. This was what Jerry always wanted—and he was getting a taste of where he wanted to be."

In the clubs with other, more established comedians he didn't want to bring unwanted attention to himself, either, didn't want to screw up this good thing he had going by causing problems. Even when another comic absconded with a piece of his material, he didn't complain, which is on a par with a bank being held up and not sounding the alarm.

Jerry had developed a funny bit about the movie *Jaws*. His premise was that everyone knows that the shark is coming because of the soundtrack. "I don't know why people don't just swim back to shore when they hear the music. They should be yelling, 'Oh, no, cello music. Let's run for it.'" Jerry's audiences laughed. Elayne Boosler thought it was funny, too, and, to Jerry's shock and amazement, she used the bit during an early appearance on the Carson show. But Jerry never complained to her, even as infuriated friends told him to raise a stink. He remained mum because Boosler wielded considerably more clout than he did. His feeling was, why cause trouble. I should be proud she used my material. All's fair in love and comedy.

Glenn Hirsch—who had become hot, doing as many as ten sets a night, making as much as $1,500 in a week and celebrating his success by taking his friends by cab to the White Castle in the old neighborhood—had by now become friends with Jerry, and was intrigued with his persona, the curious persona of a comic who wouldn't bitch when another comic ripped him off. If it had happened to the high-strung Hirsch, he would have ripped the thief a new asshole.

"Jerry just wasn't a guy who made waves," Hirsch observed. "He was very controlled emotionally. He didn't really have highs or lows. Some comedians, if a set didn't go well, they'd get angry, they'd get pissed off, they would attack. In all the years I was with him in the clubs in New York, Jerry never had those situations. I'm sure he got angry, but he never showed it. Of all those guys working the clubs, Seinfeld was most in control of his emotional and physical well-being. He controlled the time he spent at the clubs. He knew he had to do that seven days a week, and if it took him ten years to make it, he would do that.

"He was *so* quiet," Hirsch continued. "In the clubs he would bide his time, wait as long as it took to get on stage. He would do four or five jokes that he knew weren't really funny, but there he was on stage, and he would say to himself—because he told me this—'OK, I can't wait to get to this joke because I know it works.' He was excited about getting to the joke that worked, and he would hone it, and craft it, and cut the fat away. The audiences in that late-70s period liked him because his material was so simple and so easy and so clean.

"But if anyone asked me back then who was going to be a major superstar, Jerry Seinfeld's name wouldn't have come to mind. He just didn't have the demeanor, the personality; he wasn't forceful, he wasn't powerful; wasn't bigger than life. He was very safe and very clean."

By the spring of 1979, Jerry was on his way to becoming a certified road warrior. While New York City and the Comic Strip remained his base, he was often on the road. From regional booking agents—he still didn't have his own exclusive agent—he was getting club dates across the country. In the first quarter of 1979 alone, he had been to California twice, once for three weeks, the second time for a week. He played Washington, D.C., twice, for two and five days; Philadelphia three times. He had bookings in Florida—a theater in Tampa, a college in Melbourne. He played the University of Maryland and a club in Frederick, Maryland. Twice he performed at the Playboy Club in Great Gorge, New Jersey, and did a college concert in Pennsylvania. When he was home, for a couple of days at a time, he'd emcee at the Comic Strip, work the other clubs in town, and do local college bookings—Alfred University, Pace University. He participated in a contest among 150 comedians in New York and placed second; the program aired on Showtime that summer. He also started doing local TV spots, took acting and improvisation classes, and was tested for TV pilots.

It was nonstop and he loved it. Like his comic-book idol, Superman, he was moving faster than a speeding bullet.

But the most important gig of all, the brass ring, he could have grabbed but turned down, at least for the time being. When approached by scouts from the Carson show who saw him on the road, were knocked out by his performances, and offered him a shot, he graciously but emphatically said thanks but no thanks, explaining that he still wasn't ready, that he still needed time to perfect his act.

"Jerry was always very brilliant about accepting and not accepting certain key jobs and moments in his life—like the Carson show," noted Jesse Michnick. "There are people who are great opportunists in life who would have jumped at that offer without thinking. But Jerry's philosophy was—if you're going to be given the golden ticket, you want to make sure you're dressed for the theater; you want to make sure that when you use that ticket, you're ready. Jerry had the greatest ability to say no."

Much of Jerry's decision to put off the golden opportunity had to do with what had happened to his comedian pal, David Sayh, and Jerry wasn't about to let it happen to him. There was talk that Jerry even had Sayh's publicity photo tacked up in a prominent place in his apartment, with a red circle and a line through it. To Jerry, Sayh represented the international icon for comedy failure.

Sayh started life five years before Jerry, as Sidney David Gulkis, in the Bronx, the son of a Jewish electrical-supply-house employee who was active in the electricians' union. In grade school, PS 78, he performed for the first time at an assembly, playing the Nestle's chocolate jingle on the piano. By the time Gulkis entered Evander Childs High School, on which the film *Blackboard Jungle* was based, he stood only 4'11" and weighed seventy-eight pounds. To avoid becoming dead meat he became a court jester to the tough guys in order to protect himself.

Sharp as a whip, he went on to New York University where he turned serious enough to earn a degree in electrical engineering, a career that bored him. The first time he appeared before a big audience, some seventeen hundred electricians and family members, at a union talent show, he bombed because he couldn't remember his routine, froze, couldn't talk, just like Jerry his first time out. The only laugh he got was when he told the crowd as he walked off embarrassed, "I just did this whole thing in the bathroom before I came on stage. If you want to come in the bathroom, I'll do it for you again."

By 1974, having changed his name to Sayh, he had his observational stand-up act together and was getting stage time at the Improv. Two years later, with impressive talent, extraordinary delivery, and dynamite original material, he was named emcee at Catch a Rising Star after Larry David, who was being groomed to replace Sayh's close friend, Richard Belzer, turned down the gig.

On the night of February 26, 1977, a night that has gone down in stand-up history among the cognoscenti, Sayh was discovered by Johnny Carson who visited the club after getting the Hasty Pudding Award at Harvard.

Sayh appeared on Carson on April 27, 1977, and did eighteen minutes. A couple of times the camera cut to Johnny, laughing. When it was over, Sayh, in a daze, walked backstage, had already taken off his jacket, was in the Green Room, when talent coordinator Paul Block grabbed him to do an encore. "I was freaked," Sayh said. "I put on my jacket and they kind of pushed me out again and I couldn't believe it. Johnny got up from his desk, came over to the stage, and shook my hand. It was just the most incredible moment in my life."

Virtually every stand-up in New York who wasn't on stage at that moment, including Jerry, watched Sayh become a made man. He was now as golden as you could get in the comedy business. Wealth, women, whatever, would be at his feet.

Naturally, Johnny invited his discovery back for a second appearance on June 9, 1977. Again, Sayh, using all new material, was a hit. But that changed when the new fall season started and Sayh was on the couch with an all-comedy lineup—Richard Pryor and Rich Little. Sayh was the last act and he bombed, but so had the others. "The audience was so terrible," said Sayh, "that Little actually got up and redid Johnny's monologue as Johnny, in hopes of waking the crowd. At one point Pryor said on the air, 'It's a million dollars' worth of talent here and the producer's sitting backstage saying, "What the fuck's going on?"' And Johnny looked into the camera and said the audience at home is going to be saying the same thing, 'What the fuck's going on?' because he knew it would be bleeped. Before I went on, they tried to get me off the show but they couldn't because nobody wanted to fill my seven-minute slot, the audience was so bad."

Rumors about Sayh's third appearance quickly spread in the tight-knit world of stand-up, especially within his crowd in New York. The word was he'd blown it, had had his big shot, and went down in flames. All kinds of stories circulated: he'd been out the night before getting laid and stoned and was too fucked up to perform well; he used up all his good material and was completely unprepared for his appearance with new bits.

Lucien Hold viewed what happened to Sayh as a lesson in keeping your habits in strict control, as Jerry always did. "He wasn't focused like Jerry," Hold said. "It wasn't that he'd spent the night with a woman. It's that he wasn't as focused on his career, and the promise of his career never recovered from that one set with Carson. One set can make you, and in that particular set it was the beginning of the end of breaking him."

Jerry, the wary, the cautious, the always prepared, appeared to believe every word of what he was hearing.

"He talked about what happened to David Sayh big-time. *Big-time*," said Michnick. "He was obsessed about it. He saw it as Sayh not having enough for his third shot, not being prepared for whatever reason, and it really hit Jerry hard because he was always *very* focused about that first big Carson appearance. David Sayh represented someone he didn't want to emulate. He actually *studied* David Sayh and what happened to him, so Jerry knew when comedians were taking a wrong turn, when they were falling into a certain rut. And he actually put off doing Carson because of the David Sayh thing, saying, 'I don't want this to happen to me.' In Jerry's mind, Sayh was really the definitive example of not being ready. David Sayh was like a warning signal to him."

According to Mike Costanza, "Jerry took very good notice of what happened to David Sayh, and turned down Carson a few times after Sayh bombed. Jerry felt he didn't have enough material to make it past a first appearance on Carson into a second appearance. Jerry wanted Carson bad, but he wanted to make sure he didn't screw it up."

Despite all of the rumors and gossip, Sayh's career didn't instantly die. "I don't think I was unprepared for that show," he said years later, "but the thing about going out and getting laid—that could be very true. There was impact from that show. I knew it intuitively. But one bad *Tonight Show* does not a career destroy."

He had some more appearances on Carson; did the talk shows of Merv Griffin, Dinah Shore, Mike Douglas; toured with Helen Reddy, Roberta Flack, and with Barry Manilow in Vegas where he made a whopping $10,000 a week. In his best year, 1983, he brought in $100,000, but spent it as fast as he made it. By the turn of the century, though, he was working cruise ships and had even done stand-up in a Florida shopping mall. While noting that Jerry himself had some less than winning appearances on the Carson show, Sayh concluded that Jerry's feelings about him "were precipitated by the fact that I had more opportunity and more breaks on this planet than anybody in this business and I failed to take full advantage of them. That's obvious. That's a fact."

Meanwhile, Jerry had decided it was time to take his shot, to make the big move. Unlike David Sayh, Jerry would allow few chances for error in his perfectly formatted, carefully planned, and well-plotted future.

Jerry celebrated the beginning of the comic and tragic, technological and tedious 80s—the Reagan-AIDS-crack-cocaine-personal-computer-"MTV generation"-"Yuppie" decade—by announcing to his friends that he was "going for it"—moving to Los Angeles.

"Yup," he proclaimed to LuAnn Kondziela shortly after New Year's Day 1980, "I'm doing it. The big time, here I come. I hope to do some of the smaller talk-shows this year and Carson in '81," he predicted. "We'll see how it goes. Anyway I've got to stay on the offensive and here in New York I've done everything I wanted to do. I want to be a small fish in a big pond again."

Jerry's California send-off bash was held in March 1980, a month before his twenty-sixth birthday. Following Jerry's instructions, Mike Costanza, one of the party's organizers, made it an all-male affair. "Jerry's idea was to make sure there were no women around," Costanza recalled. "Jerry said that it was to be 'a guys' night out.'"

A stag party was an appropriate choice for the party's venue, which was McSorley's Old Ale House, the oldest establishment of its kind in New York, a charming hangout that opened for business in 1854, whose clientele in those days encompassed a diverse group ranging from Irish immigrant laborers to financiers to powerful politicians like Abraham Lincoln. For Jerry, McSorley's once had an apropos motto for the evening: "Good Ale, Raw Onions, and No Ladies."

Those bidding Jerry good luck, besides Costanza, were his roommate George Wallace, Lucien Hold, Chris Misiano, Jesse Michnick, and his comedian friends—Dennis Wolfberg, Mark Schiff, and Larry Miller. The latter two became part of an inner circle with Jerry and Paul Reiser and a few others along the way who began a tradition of meeting for brunch every New Year's Day, another all-guys celebration.

"For Jerry's going-away party," recalled Hold, "we got a card that was actually a small book, and everybody wrote things in it. Larry Miller had written in kind of a Western, cowboy style, telling a story like a western. When the book was passed to me I read Miller's two- or three-page thing and it didn't have an ending, so I copied the style and went on about my frustration about his half of the story that didn't get finished. It was one of the few times," the comedy club operator noted, "where I made an effort to be creatively funny, which I had always tended to resist when you're around competitors like Seinfeld all the time."

It was a night of nonstop laughter, a raucous mini–Friar's Club roasting of the golden boy who was about to make his way to La-La Land. His climb would be so swift and so successful that no one sitting in McSorley's with him that night over endless mugs of dark brew could have ever imagined it happening the way it did.

At the end of the night, Costanza drove Jerry back to his apartment.

"There was a lot going on in his mind, but there was no trepidation about going," he recalled. "There were no fears, no qualms. Jerry knew he had to go and he had so much confidence by that point in time. We told him we were a phone call away if he needed us to come out there—I would personally come out there. If he needed to come back, then he would come back. But he was psyched.

"We hugged each other and I went home. I was happy for him but I knew that I was going to miss the guy. It was one of the few moments in my life when I knew things would never be quite the same again."

—24—

Welcome to L.A.

It was neither the best nor the worst of times in Los Angeles when Jerry arrived, confident and prepared to make it big. Finally, after four years of careful planning and hard work, his material and style honed to a sharp edge, he was there, in the City of Angels and Carson, a far-out town that was the brunt of so much pseudo-intellectual East Coast put-down stand-up humor, laid out before him in all its glory: the blazing sun over the honky-tonk Santa Monica Pier, the smog over the San Fernando Valley, the manicured palm trees along diamond-studded Rodeo Drive, the legendary white HOLLYWOOD sign in the scrubby hills, the kitschy sidewalk stars on the Walk of Fame outside garish Grauman's Chinese Theatre, the naughty show-window of Frederick's of Hollywood on the seedy boulevard, the humongous show business billboards above and the hookers below on Sunset—all of those La-La Land icons were there for Jerry to gawk at when he pulled in like some rube from the Midwest, or, say, Massapequa, Long Island.

But the world he had come for, the world of stand-up, was in a state of high anxiety and disarray—still reeling from a tragicomic and at the same time extremely positive event of Richter-Scale proportions. In its wake there were the walking wounded, physically and emotionally, at least one horrible death, a fire of mysterious and suspicious origin. And there were the friendships severed, the trusts broken, the loyalties deep-sixed, the egos left battered and bruised.

The L.A. comedy world Jerry encountered that early spring of 1980 was nothing to laugh at, though the healing had slowly begun.

While the comedy club scene was about to explode nationwide, there still were only two key stand-up venues in the city, even fewer than in New York. Those showcase clubs were Mitzi Shore's Comedy Store on Sunset Boulevard, and recently transplanted New Yorker Bud Friedman's L.A. Improv on Melrose Avenue—the clubs where stand-up hopefuls got their first shots, and where the scouts for the movies, the sitcoms, the late-night talk shows came looking for fresh blood—all of the reasons why Jerry had moved west.

Shore was the pioneer, the high priestess, the Yoda, the ruler with a velvet glove, "pretty much the den mother of stand-up" in Los Angeles, observed Argus Hamilton, a stand-up comic and close friend.

In 1972, when Jerry was just a freshman at Oswego, Mitzi Shore, née Mitzi Lee Seidel, and her husband, a Catskills-style comic named Sammy Shore, whose big claim to fame was opening for Elvis in Vegas, which wasn't exactly chopped liver, started the Comedy Store on a shoestring, along with another partner, a comedy writer named Rudy DeLuca. A year later, though, their partnership fell apart when the Shores' marriage went into the toilet, much like Sammy Shore's act. Mitzi demanded $1,100 a month in child support for her four children, the youngest of whom was Pauly Shore, who inherited his parents' comic genes and went on to become a stand-up and an actor. Sammy countered—he'd give Mitzi $600 a month *and* the Comedy Store.

Done deal.

An attractive brunette with a weird, whiny voice, the kind her comedians liked to imitate, Mitzi turned what had been a largely losing proposition into a highly profitable comedy empire virtually overnight, working around the clock. As Pauly noted ruefully years later, "I was given to the comedians like a bastard child . . . She didn't have time."

Despising the term "saloon," the kind of places where her ex-husband had worked for years, she called her place an "art colony," and didn't go for "fuck-you-up-my-ass-eat-my-pussy-fart-jokes," as stand-up Tom Dreesen noted. "Most of us were training for the *Tonight Show,* and the *Tonight Show* didn't use those kind of comedians."

Instead, the Comedy Store was a place for young comic *artistes* to work—laugh geniuses such as Robin Williams, David Letterman, Jay Leno, Jimmie Walker, Garry Shandling, Yakov Smirnoff, and the list goes on and on—who were just starting out. "Mitzi was both soft and hard," Dreesen said. "She learned all the hardships of comedy, all the downs, the heartaches, the pains, the neuroses of comedians. She knew the mind of a comedian."

Ironically, the other major club in town, Friedman's Improv, also got its start because of marital problems. He established the place in 1975, after he and his wife, Silver, divorced, and she was given the New York Improv as part of their settlement.

While Mitzi Shore and Bud Friedman had different philosophies for their clubs, were hotly competitive, and swore blood loyalty from the comics who worked for them—"if you worked the Improv, Mitzi didn't want you; if you worked for the Comedy Store, Bud didn't want you, unless you were the hottest comic in town and then they'd both kiss your ass," declared Dreesen—they shared one thing in common: play, but no pay.

Since the talent could use the venues to work out new material, to get stage time, to be seen by the right people, Shore and Friedman's rationale for not paying a red cent was that they were offering the funnymen a valuable service; therefore—unless you were an established name—you worked gratis, take it or leave it, while Bud and Mitzi raked in big money from the bar and the door.

When young comics demanded money they were given a symbolic finger by the owners.

Their fury grew and finally they enlisted Dreesen, a comic originally out of Chicago, who had made the big time—Carson, a CBS development deal, touring with Sammy Davis Jr., earning $300,000 and change a year working the Vegas-Tahoe-Reno circuit, hanging with Clint Eastwood. They had picked him because of his labor-union experience some years earlier, negotiating with the Teamsters. The guy knew *Robert's Rules of Order* by heart.

Dreesen had seen the light regarding Mitzi one night when he came off the road and she asked him to play the Comedy Store's Main Room, which held 450 people. While there were a couple of smaller rooms in the club, the Main Room was where long-established comedians like Rodney Dangerfield and Jackie Mason worked and got the door—meaning they got all of the admission charge, while Shore got the bar and food receipts. The night Shore asked Dreesen to play the Main Room, which was packed, he shared the bill with neophytes Letterman, Leno, Elayne Boosler, and Robin Williams.

As usual, she gave them *bupkis*.

After they finished their twenty-minute sets, the comedians retired to Cantor's, a well-known deli and hangout on Fairfax.

"Jay Leno comes in and goes, 'Hey, what the fuck is going on? Hey, we packed that room! Those guys [Dangerfield, Mason] get paid. It took five of us to do it, but we still filled the room,'" recalled Dreesen years later. "Jay said, 'We're not getting anything.' And it was Jay who declared, 'This is bullshit. We should be getting paid.'"

Not long after, there was a meeting of about a hundred comedians in a rented hall, the initial group and a new group that included Jerry's roommate George Wallace, prop comic Gallagher, black stand-up Marsha Warfield—a room packed with what Dreesen described as "a group of insecure, neurotic, love-starved wrecks for the most part, ranting and raving that they weren't getting paid."

The group, under Dreesen, was quickly organized. Committees, such as one to handle publicity, were formed, and Dreesen had his first meeting with Shore, who, according to Dreesen, was emphatic. "I'm not paying them," she said. "They don't deserve to be paid." He offered a possible solution: instead of charging five bucks at the door, charge six, and let the comedians split the additional dollar. Her reply, "They don't deserve a penny." When he reminded her that the waiters got paid, the busboys got paid, the bartenders got paid, she still refused to budge. They went back and forth for a month, Dreesen reported back to the membership, and now serious strike talk was in the air. He returned to Shore to warn her that the comedians were now well organized, prepared to walk a picket line, and that if they did, they were going to get a lot of publicity.

"I told her, 'People are paying to see them, so they deserve to be paid.' She said, 'People come to the Comedy Store because of the aura.' I said, 'Mitzi, who but the comedians gave the Comedy Store that aura.' I tried to tell her that getting paid gives comedians respect and self-respect. But she just couldn't see it." At one point there seemed to be a breakthrough: she offered to split the door with certain comedians who worked the Main Room, or give them twenty-five dollars a set. Dreesen reported the offer, which was voted down. That night at Cantor's a young comedian approached him. "Tommy," he said, "I killed them tonight. I'm so excited. I never worked in front of such a big crowd in my whole life." Dreesen congratulated him. "He said," Dreesen recalls, "'Man, it's the most exciting thing that ever happened to me. By the way, Tommy, can you loan me five dollars for breakfast?' I told Mitzi that story and her response was, 'He should get a goddamn job!'"

Strike!

Hundreds of chanting, placard-carrying comedians marched in front of the Comedy Store. Signs read, WORKING FOR FREE JUST ISN'T FUNNY, NO BUCKS, NO YUCKS. The media, including the world press, covered the strike. Up front in the line of picketers were the likes of Leno, in military gear, Letterman, others whom Shore had helped. Later, it was said she watched them from inside, these young comics whom she loved and mothered and saw as an extended family, and cried. But she didn't give in.

Then the violence started.

The comedians had a spy on the inside, a Comedy Store waitress who reported Shore's tactics. At one meeting, with almost two dozen of her loyalists—a number of comics crossed the picket line, among them Garry Shandling—a comic on Shore's side stood up and said, "Comedians need a place to work. They'll probably go over to the Improv." At that point the comedians had not started picketing Friedman's club. Another comic stood up and said, "What if there was no Improv?"

A couple of nights later someone threw a Molotov cocktail on the rear roof of Bud Friedman's club and all but a small portion of the club's front was gutted. Friedman called the fire "devastating." There were no arrests.

Friedman asked for a meeting with Dreesen and pleaded with him not to strike the Improv. He offered to immediately set up a little theater under a tent at the front of the club where the comics could work. Dreesen demanded a deal memo from Friedman, which he signed on the spot.

But the violence continued.

"One night one of Mitzi's thugs went out and punched Falstaff, one of our gay comedians, and gave him a black eye," Dreesen said. "They harassed Elayne Boosler one night. We had a crippled comedian, Jimmy Alec, who had polio in one leg and walked with a limp. When I pulled up in front of the Comedy Store the girls were all crying. A big heavyset guy who worked for Mitzi had Jimmy in the parking lot and was about to beat him up. I ran up and got there just in time. He said, 'I'm whipping his ass.' I said, 'I'm in the right fucking mood now to whip *your* ass. What do you say about that?' He backed off."

In hopes of organizing an actual union, Dreesen met with representatives of the American Federation of Television and Radio Artists (AFTRA). The union agreed to run advertisements in *Variety* and the *Hollywood Reporter*, asking everybody in show business to boycott the Comedy Store, putting more pressure on Shore.

The breaking point of the strike involved a frightening confrontation in which Jay Leno, walking the picket line, could have been killed.

Shore had gotten a court injunction to keep the picketers from blocking the driveway next to the club so she and others could get inside. Dreesen was on the line when he heard the sound of a car engine racing on the other side of Sunset and warned the picketers to not stand in front of the driveway, that the injunction had banned them from being there.

"Just then I heard tires peeling. The car raced in and clipped Jay, who spins around and hits the ground and I heard the thud. I look around and Jay's on the ground and the girls are all screaming, 'Oh, my God . . . Oh, my God.' I knelt down to look at Jay and his eyes seemed to go into the back of his head. I made up my mind right there that when this guy who drove his car in is two feet away from me I'm taking his fucking head off, I'm busting him. He's going to be lying right beside Jay because I'm going to sucker punch him so fucking hard, he's not going to know what hit him. I was that angry and crazed at that moment. As he's coming towards me, the girls are yelling, 'You son of a bitch!'" As the ambulance carrying Jay Leno sped away, Mitzi Shore, who was watching in a state of shock from inside her club, summoned Tom Dreesen.

"That's it," she said, "It's over. I want it called off right now. I want to sit down." Tears were in her eyes, still seeing the vision of Leno's limp body on the sidewalk. Lawyers were called. Negotiations went on until four in the morning and the final deal, satisfactory to all, was that the stand-ups would receive twenty-five dollars a set *and* a split on the take when they played the Main Room, and no retaliation against, or black-listing of, any of the strikers. As he left the bargaining table for the last time, Dreesen remembered Mitzi Shore saying to his back, "I hope this means you'll come back and perform here."

Little did Shore know that the scene she was watching involving Leno and the hit-and-run car was not everything it appeared to be. Just before the ambulance arrived, Dreesen revealed, "suddenly Jay looks up at me and winks. I thought, You cocksucker! And laughed to myself. Under his breath he said, 'Be cool. Be cool.' He was lying there like he was dying. He was playing it for all it was worth."

Leno's little act had helped bring Shore to her senses.

For months Dreesen had put his career on hold; he was mentally and physically exhausted; his home looked like combat central. But now it was over, he thought, and he needed to resume his real life. The comedians, too, were thrilled. There was even talk of calling themselves the American Federation of Comedians. Word of the settlement spread instantly across the land. In New York, where Jerry quietly supported the strike but was not an activist—"he wasn't involved, but he was grateful," said Dreesen—the New York club owners also agreed to begin paying their comics, and virtually overnight stand-ups all over the country were finally being compensated for their work.

The first and only comedy strike in history had been a major success, but not without still more tragedy.

A month after the strike ended Dreesen was headed for a meeting with the TV producer George Schlatter about hosting a new show. As he was leaving his home, Dreesen was approached by a young comedian named Steve Labetkin, who, early in the strike, wasn't sure whether to support Shore, to whom he was a loyalist, or his fellow comedians. In the end he joined the picket line and now felt he had been blacklisted by Shore because he hadn't gotten any work at the Comedy Store since the strike ended. "He was sad-looking and I told him, 'Steve, I have to go to this meeting but I promise you that I won't go back on stage at the Comedy Store until she brings you back.'"

Forgetting about the incident, Dreesen went to the meeting and then on to Tahoe where he was opening for Sammy Davis Jr. His life, he felt, had finally gotten back to normal. He was in his dressing room, fifteen minutes before he was to go on stage, when the phone rang. It was Jay Leno. His voice was cracking and it sounded like he was crying. "Tommy," he said slowly, "Steve Labetkin just committed suicide. He jumped off the Continental Hyatt House toward the Comedy Store."

The hotel was next to the club and Labetkin had landed on the club's parking ramp.

The police found a suicide note.

It read: "My name is Steve Labetkin. I used to work at the Comedy Store."

A few nights later someone snuck into Mitzi Shore's office and placed a publicity photo of Labetkin in her chair. She was freaked. On the first anniversary of Labetkin's death, someone put a dummy on the ramp where he had landed, with a sign around its neck that said, "My name is Steve Labetkin. I used to work at the Comedy Store."

That was the doleful mood that greeted Jerry upon his arrival on the L.A. comedy scene that spring.

Courtesy of Caryn Trager

By fifth grade, Jerry *(front row, second from left)* was a nerdy, bashful loner, but he'd already shown a spark of comic genius, imitating President Kennedy for a class skit and poking fun at an obese classmate, which caused a playground chum to toss his milk and cookies. "I would like to do this professionally," Jerry recalls thinking at the time.

Courtesy of Caryn Trager

Jerry inherited much of his comedic talent from his father, Kal Seinfeld, who used his natural sense of humor as a salesman—from peddling bogus holy water from Lourdes to successfully selling business signs. Father and son share a moment together in front of the family's modest Long Island home, circa 1973.

"Oh my God, Jerry's fly's open." That's how Jerry's first gal pal, petite Lonnie Seiden, remembers him when they met in homeroom in their junior high school freshman year. In their first bit of experimentation, they "did a little kissing, a little feeling," says Seiden, who remembers his "clammy hands" and his breath that "smelled like peppers and eggs."

Courtesy of Lonnie Seiden Javurek

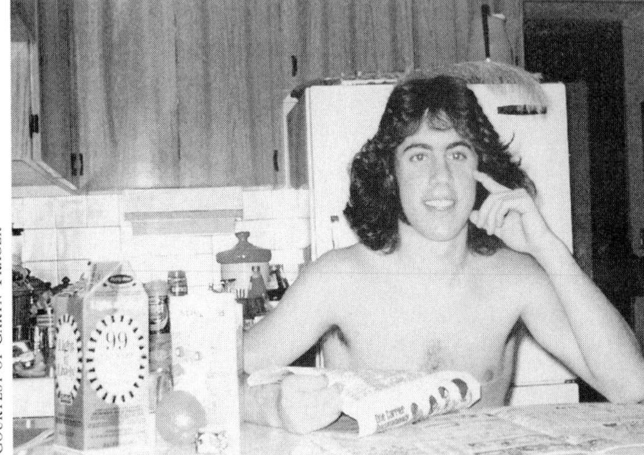

Courtesy of Caryn Trager

Jerry was already a cereal junkie—the milk is just waiting for the fixings—by the time he entered Oswego College in the winter of 1972 *(left)*. Always seeking approval through laughter, he strikes a hilarious pose in a pal's dorm room *(below)*.

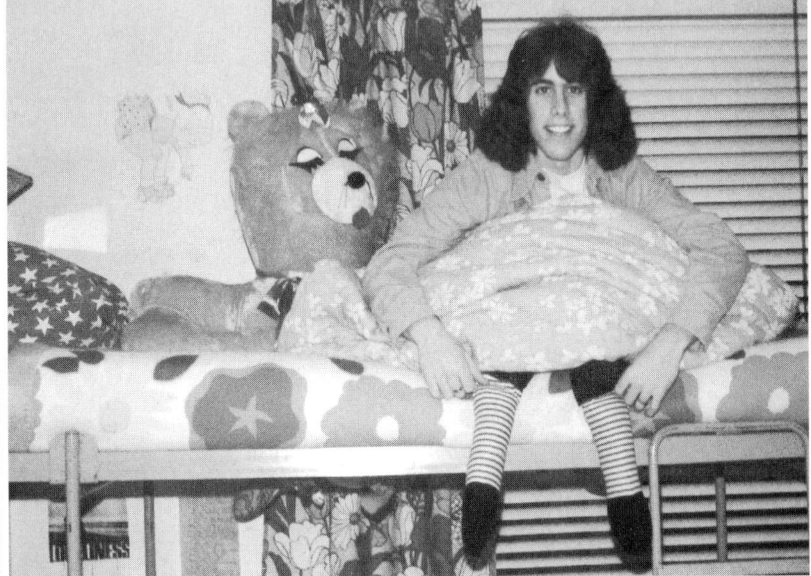

Courtesy of Caryn Trager

COURTESY OF CARYN TRAGER

In freshman year at Oswego, Jerry began a long, tumultuous relationship with psychology major and Barbra Streisand–type Caryn Trager, who smoked pot and was "kind of a hippie, but not a total hippie." Jerry broke up with her when he decided to become a full-time stand-up after graduation. There were joyous times, though, such as a glorious day for the young lovers on the Jones Beach boardwalk.

COURTESY OF CARYN TRAGER

COURTESY OF CARYN TRAGER

Oswego was a racially polarized campus when Jerry struck up a friendship, and shared a dorm room, with outspoken, theatrical Larry Watson *(right)*, who came from a crime-riddled housing project in Brooklyn. Jerry had never known a black person, and Watson introduced him to African-American culture.

Courtesy of Mike Costanza

In his junior year, Jerry declared Oswego "too idyllic" for him and that he needed the "provocation" of New York, so he transferred to Queens College, where he fell in with a crowd of funny, glib guys from the boroughs. Mike Costanza, on whom the *Seinfeld* character George is based, is on the far left. Jerry is sandwiched between the two other members of his Queens posse—budding character-actor Joe Bacino, and Jesse Michnick, who became a top network videotape sports editor.

Courtesy of Mike Costanza

With some pull from Michnick, who was making his bones at a radio station in Watertown, New York, Jerry got one of his early stand-up gigs, and was even interviewed by one of the station's personalities.

© 2002 ABC Photography Archives

Jerry rarely made a wrong move in his climb to the top, but one that he regrets was taking the role of "Frankie" on the popular sitcom *Benson*. Jerry appeared in several episodes playing a governor's joke writer before he was fired in callous Hollywood fashion: "I showed up . . . but there was no script for me." The failure made him even more determined to become the best stand-up ever and to someday do TV again—but only on his terms.

Carson Entertainment Group

Unlike most stand-ups in the late 70s and early 80s, who were in it for the laughs, Jerry thought like a businessman and carefully made his moves, knowing when to say no to an offer. He rejected a number of overtures to appear on *The Tonight Show* until he felt he was absolutely ready, then he trained like an athlete for his first appearance, on May 7, 1981. After the king of late night gave him a thumbs-up, Jerry's career soared.

The Lester Glassner Collection

From the time he was a lonely kid holed up in his blue plaid-wallpapered bedroom, Jerry loved comedy—and his idol from the beginning was Bill Cosby. Jerry was the only kid on the block who had a complete collection of Cosby's albums.

The Lester Glassner Collection

AP/Wide World Photos

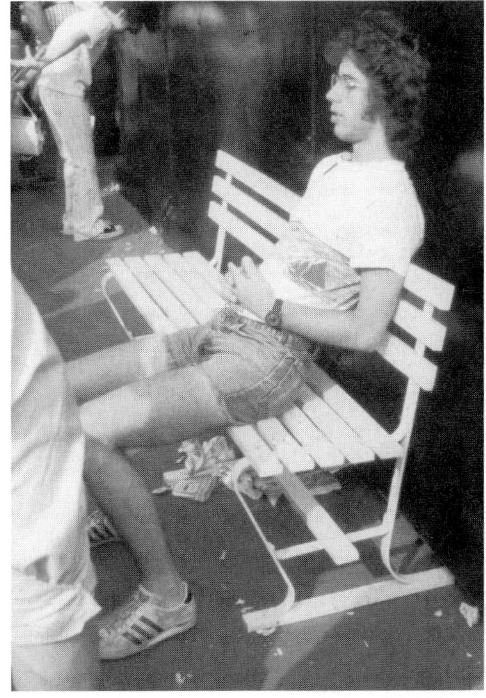

Courtesy of Caryn Trager

Like Cosby, whose clean monologist style set the standard for Jerry's form of comedy, Jerry also had a guru for his concentration, focus, and discipline—and that was L. Ron Hubbard *(above)*, the father of Scientology. Jerry also had an early interest in Transcendental Meditation, and is pictured meditating in summer 1974.

© David Turnley/Corbis

© David Turnley/Corbis

Jerry's close friendship with comedian and actor Mario Joyner was considered odd couple-ish by members of Jerry's circle. One summer, Joyner joined Jerry on a trip to Europe, including Paris, where they're shown camping it up in a café *(above, left)*, and relaxing on the Seine *(above)*.

Getty Images

Spring was in the air and the birds and bees were buzzing in Central Park, where Jerry met and fell hard for a sexy, sophisticated, buxom high school senior named Shoshanna Lonstein, touching off a cradle-robbing cause célèbre. Just from dating Jerry, Shoshanna became a celebrity of sorts, and Jerry benefited, too, even if the press was negative. The upside? The relationship gave him a more macho image.

Central Park also was a place to chill for Jerry and his very close, lifelong friend, the comedian George Wallace. The two met in the late 70s when they were starting out in stand-up and soon became roommates.

BigPicturesUSA.com

Courtesy of Susan McNabb, Photograph by Kenneth Johnson

Jerry began a seven-year roller-coaster relationship with beautiful lingerie and hand model Susan McNabb in the mid-1980s, but his commitment phobia and the fact that she wasn't Jewish made marriage out of the question. McNabb said the Elaine Benes character on *Seinfeld* was originally going to be named Susan in her honor.

After he ended it with McNabb, Jerry began a relationship with sexpot-actress Tawny Kitaen, whose dressing room–trailer was next to his on the studio lot where her show, *WKRP in Cincinnati*, and his show were produced. Her first husband, rocker David Coverdale, of the group Whitesnake, called Tawny, "my whore and my inspiration." Also in awe, Jerry described her to his pal Mike Costanza as if he were "in the presence of greatness."

Courtesy of Tawny Kitaen

Photograph by David Hume Kennerly

As a stand-up, Larry David was the complete opposite of Jerry, who works ultraclean and is considered "normal" by club owners. David, however, was off-the-wall, sometimes coming on stage, looking at the audience, and then walking off, telling the audience, "Fuck you!" Yet, when NBC offered Jerry carte blanche, he asked David to cocreate what became the most critically acclaimed sitcom ever.

Photograph by David Hume Kennerly

The zany coffee shop scenes and chatter were an integral part of *Seinfeld*. Larry David *(standing, right, foreground)* sets up a scene with the beloved gang—Kramer, George, Elaine, and, of course, Jerry.

Photograph by David Hume Kennerly

Jerry's decision in late 1997 to pull the plug on *Seinfeld* while the show was still at the top of its game sparked a media blitz heard 'round the world.
For millions of fans it was the end of life as they knew it. An estimated 70 million viewers tuned in on the evening of May 14, 1998.
Frozen in time: the touching scene after the cameras stopped rolling for the last time.

Courtesy of Mike Costanza

The two Costanzas together—the real-life Mike Costanza, and George Costanza, played by talented Jason Alexander—on the *Seinfeld* set after Mike appeared as a truck driver in a 1992 episode called "The Parking Space." Jerry and Mike's close friendship of a quarter century turned bitter when Costanza decided he wanted to write a book about their relationship.

© David Claborn/Getty Images

Fame has its rewards, and as the money rolled in Jerry bought a fabulous multimillion-dollar house in the Hollywood Hills that was once owned by the actor George Montgomery. Still to come: Porsches galore and palaces of his own overlooking Central Park and in glamorous East Hampton, all from being funny.

AP/Rolling Stone

There was a time when celebrities knew they'd made it if they hit the cover of *Rolling Stone*. Jerry was no exception.

EVERETT COLLECTION

On a par with the cover of *Rolling Stone* is the moment when Barbara Walters comes calling. Jerry shocked friends and family when he told the celebrity interviewer he despised growing up in his hometown of Massapequa.

© DAVID TURNLEY/CORBIS

Jerry poses for Annie Leibovitz, the doyenne of celebrity photographers. When she shoots you, you know you've made it.

SEINFELD STEALS ANOTHER MAN'S WIFE — 3 WEEKS AFTER WEDDING

Womanizing Jerry Seinfeld has broken up a new marriage — stealing the pretty bride from her shattered husband just a few weeks after their wedding day.

GLOOMY GROOM

WHO'S WHO IN THE CAST

OUT ON THE TOWN:

BIGPICTURESUSA.COM

First, it was cradle-robbing with Shoshanna. Then, in 1998 Jerry was accused of wife-snatching, when he scandalously hooked up with a young woman named Jessica Sklar Nederlander, who had just been married to the scion of a powerful theater family. The *National Enquirer*'s story was echoed in newspapers and magazines, on television and the Internet, for months on end.

© Lawrence Schwartzwald/Getty Images

Trying to look her Jackie O best, Sklar in sunglasses savors the attention despite the scandal, and Jerry, wearing his lucky number 9 on his baseball cap, doesn't seem to mind that Sklar's recent groom, Eric Nederlander, publicly declares, "Jerry and Jessica have no respect for decent values. They deserve each other."

© Evan Agostini/Getty Images

Jerry and the stolen bride arrive at the premiere for a film about the life of funnyman Andy Kauffman. Jerry's longtime manager George Shapiro *(right)* is played in the movie by actor Danny DeVito.

© David Turnley/Corbis

Jerry's mother, Betty Seinfeld, who grew up in orphanages and foster homes, doesn't seem bothered by the marital scandal embroiling her successful boy. "Jessica's just right for my son," she declares. "They are made for each other."

© Alex Heining/BigPicturesUSA.com

Flashing her rock from Tiffany's, the proud Mrs. Seinfeld leaves her second marriage ceremony in eighteen months with her beaming groom, long considered a confirmed bachelor.

Photograph by Adam Riesner

Photograph by Adam Riesner

The new bride, who played a role in Jerry's decision to buy and renovate Billy Joel's $32 million home in tony East Hampton, grew up in this modest $18,500 house in Burlington, Vermont *(top)*. Jessica's former brother-in-law, James Meiskin, says he was never invited inside during his years as an in-law. Though Jerry's wife is considered a material girl, her mother, Ellen, and father, Karl *(above)*, are thought of as antimaterialistic.

Not long after Jerry and Jessica tie the knot, she becomes pregnant, and the joyous couple become proud parents of a bouncing baby girl, Sascha, on November 7, 2000.

© Lawrence Schwartzwald/Getty Images

At Jerry's wedding, longtime friend George Wallace *(far right)* toasted Jessica *(wheeling the carriage)* with the words, "If he gives you half the love he's given me, you'll be very fortunate." With the group is a reporter, a friend of Mrs. Seinfeld's.

After the marriage, after the arrival of the baby, Jerry does what he does best and loves the most—he's back on the road, on his own, doing his stand-up, something he says he wants to be doing when he turns 100.

—25—

The *Benson* Fiasco

The dark cloud cast an immediate pall over Jerry's usual optimism. His excitement about finally being in L.A. quickly faded and his hope of getting instantly discovered and catapulted to fame and riches was, for the moment, seemingly dashed. He swiftly discovered that he was, indeed, a very small fish in an immensely huge pond, which didn't feel as tasty as he thought it would. Other than testing his material at the new post-comedy strike minimum of $25 a pop at the clubs—he quickly gave his loyalty to the Improv over the Comedy Store—nothing was happening for him.

Besides accelerating his stand-up career and hopefully getting on Carson, Jerry had another goal: to be chosen to do a TV sitcom, the kind of golden opportunity a handful of other talented—and extremely lucky—young comedians had been given in the recent past.

The first, the idol of all who followed, was Freddie Prinze, real name Frederick Karl Pruetzel, son of a Puerto Rican mother and a Hungarian father. Three months younger than Jerry, Prinze was amazingly just twenty years old when he did his first Carson, which instantly launched him into a starring role in the 1974 laugh-track NBC sitcom *Chico and the Man*. By 1977, during Jerry's first full year as a stand-up, Prinze had already become one of the biggest stars on TV, and a staple of supermarket tabloid gossip. At the height of his popularity—or possibly because of it—Prinze became despondent, dabbled with drugs. Tragically, on the day after his twenty-third birthday, he put a gun to his head and fatally shot himself.

A year after Prinze got his big break as Chico, another stand-up made it to the tube. Like Prinze, Gabe Kaplan, a Jewish comedian from Brooklyn, a decade older than Jerry and Prinze, was a career hero of Jerry's, also a comic who had successfully leaped from the small-stage spotlight to national fame and fortune in ABC's hit *Welcome Back Kotter*, which was considered one of the more realistic sitcoms of the 70s. The cast included a cool Italian named Barbarino, who was played by a young John Travolta.

Another hero of Jerry's who parlayed his odd humor and strange characters into a highly successful series was Andy Kaufman, the seriously strange Long Island boy, who came out of Jerry's New York comedy milieu. From stand-up at the Improv, where he was discovered by comedy agent George Shapiro—who would soon play a pivotal role in Jerry's life—he went on to do his bizarre routines on *Saturday Night Live*, and then to a prominent role as Latka Gravas, the mechanic of indeterminate nationality and fractured English, one of the zany crew working for New York's Sunshine Cab Company on the hit ABC sitcom, *Taxi*, which premiered in September 1978.

That same month, on a different night on ABC, the L.A. Comedy Store's hyperkinetic and manic stand-up Robin Williams, who was so active in the strike, got a sitcom called *Mork & Mindy*, a spin-off from an episode of another successful 1978 show, *Happy Days*. Williams, two years older than Jerry, made *Mork & Mindy* a ratings success, playing an alien named Mork from the planet Ork. His mugging, mimicking, one-liners, and trademark sign-off "Na nu, na nu" (meaning good-bye in Orkan) also launched his successful film career.

Jerry, who knew the history well, and had confidence in his talent, could taste the fame and riches that a network sitcom could bring him. Every day in those early days in L.A. he went from one TV casting call to another in hopes of being chosen. But the competition was immense and intense, and he returned home each night feeling rejected and dejected. Moreover, he missed the grittiness and familiarity of Manhattan.

"He was hurting out there," recalled Mike Costanza who was in regular contact with Jerry during that period. "He was going through New York withdrawal, and he was accepting his career setbacks less easily than any of the bad nights he'd ever had on stage."

Jerry may have been even more down about his new life than his friends suspected. After just a couple of months, he was back east without letting anyone know. And instead of staying at his Manhattan walk-up where he would be seen, he decided to hole up instead at the family house in Massapequa. When Costanza discovered Jerry was at home, he had to lure him into the city. Jerry now let it be known to pals that he considered being in L.A. on a par with being on the road, meaning that he didn't see it as anything permanent. If he failed out there, he'd be back in a flash and pick up where he left off, working the New York clubs and out-of-town gigs. At least he was making a living, enough to support himself without having to deal with all that L.A. hassle.

But not long after Jerry returned to the West Coast in the spring of 1980, he got a break, or so he thought at the time. He showed up to read for a part for the second full season of a sitcom called *Benson*, which premiered on ABC in September 1979, a spin-off of the very hip sitcom *Soap*. That show gave another Long Island–bred, talented young stand-up named Billy Crystal his big career boost as one of TV's first openly gay characters. Benson DuBois, played by veteran African-American actor Robert Guillaume, had worked as the self-assured, intelligent butler on *Soap*, but in the spin-off he was cast as the right-hand man to a governor named Gene Gatling, helping him politically along with looking after his daughter, Katie.

By the second season of *Benson*, the writers and producers, Susan Harris and Thomas Reeder, saw a need to introduce a new character into the mix, a state-house messenger with a sense of humor, named Frankie, who could help the governor lighten up his speeches. Jerry felt the role was custom-made for him, and the casting people thought so, too, and signed him.

Jerry's dark mood vanished instantly. Within his first six months in Hollywood, and without even a single Carson appearance as yet, he had won a role on a network sitcom, and signed a contract with a salary of—

he couldn't believe his own eyes and ears—a whopping four thousand dollars a week, and a guarantee of forty thousand, which he'd get even if he was canned. The show was in production for the fall 1980 season, and Jerry instantly went into taping. He was, in his eyes, on his way to stardom and celebrityhood.

Jerry's pals were surprised and elated that he'd gotten a major TV gig so quickly. One was oddball but very funny stand-up Fred Stoller, who later became a writer on *Seinfeld*, and even played opposite Elaine in at least one episode. But in the early 80s he had worked with Jerry at clubs in the San Francisco Bay Area and at a joint called Freddy's in Bernardsville, New Jersey, a gig they'd gotten through the Improv in New York. Stoller, who was thrilled for Jerry's *Benson* deal, recalled how Eddie Murphy, with whom Stoller was friends, exclaimed, "Wow. That Jerry Seinfeld got a part on *Benson*. He's going to be getting all that money! He's gonna be a star!"

Back home the Seinfelds were having heart palpitations. "My son, the sitcom star," Kal Seinfeld exclaimed to customers, joyfully advertising Jerry's forthcoming appearance on *Benson* on a handwritten placard that he tacked to the side of his truck and drove proudly around town. He also placed a short item in the *Massapequa Post*, which appeared on November 13, 1980. The announcement was wedged between advertisements for a free orientation session for a Trim Line Concepts weight-reduction program at Temple Judea, Irish Cabaret Night at Pattie Daly's Pub, and Italian dinner at Dick Dora's restaurant.

It read:

"Jerry Seinfeld, son of Mr. and Mrs. Kal Signfeld"—always looking to plug his business, Kal used his business-name spelling rather than his family-name spelling—"will make his television debut as a regular on the ABC-TV series *Benson* tomorrow (Friday) night. The 27-year-old comedian graduated from Massapequa High School. Currently living in Los Angeles and New York City, Jerry caught the eye of television producers when he came in second place in the Big Laff-Off, a competition for young stand-up comics, at the Bottom Line last year."

Accompanying the story was a head shot of Jerry looking like John Travolta's Tony Manero—long, thin face, wide-collared disco shirt with the top three buttons opened.

On the air, the mundane dialogue was hysterical—but not in a funny sense, and the laugh-track that followed Frankie's and Benson's lines was canned.

An example of the inane chatter from one of the three episodes in which Jerry appeared that aired on November 14, December 12, and December 19, 1980, took place in Benson's well-furbished office in the governor's mansion. In the scene, Jerry's hair is shorter than the length he wore it for stand-up; he looked preppy in a green sweater, plaid shirt, khakis, and running shoes.

> BENSON: You're a good messenger and a nice kid. I like everything about you—except your jokes. They're not funny.
>
> FRANKIE (pleading): But they are funny.
>
> BENSON: The governor is not going to hire you. He does not need jokes in his speeches. His speeches are jokes.
>
> BENSON (whining): But Benson.
>
> BENSON: Frankie!
>
> FRANKIE: Don't bother to get up. I can throw myself out.

Even back home, with their untrained ear, the Seinfelds of Massapequa exclaimed, "Oy, vey!"

Watching in New York, Jesse Michnick felt sadness and compassion for his friend. "His appearance was disappointing, very disappointing," he observed later. "Luckily he wasn't famous at the time and it was just a bit part. But that experience was pretty disturbing to him. My mom saw him one night on the show and she said, 'If I dropped the fork, I would have missed him.'"

Not only did Benson not want to hire Frankie as the governor's joke-writer because he wasn't funny, but the show's writers decided to revise the story line, sans Frankie. While *Benson* went on to air for seven years, until August 1986, with Guillaume picking up an Emmy in 1985, Jerry was gone as fast as he had appeared, and he was let go in a stereotypically callous Hollywood fashion.

"It happened so fast it was like a car accident," Jerry says. "I flew in from New York the day the show was supposed to start shooting again after the break. I showed up at the studio and went to sit down at the table to read the script, but there was no script and no chair for me. Then the assistant director called me aside to tell me they forgot to tell me I wasn't on the show anymore."

Dejected, Jerry returned to New York. Friends and colleagues said he was furious with the way things went.

"When he came back to New York, we didn't bring up *Benson* to him," recalled Glenn Hirsch. "It was such a terrible experience for him and then to be fired was devastating. We all knew it was not a good spot for him, that it wasn't Jerry. But it was a lot of money and who turns down television when you're a young comedian and somebody says, 'Hey, you wanna be on a network show for four grand a week?' It's common sense that he'd take it. But that's what happens when you're in the wrong place at the wrong time."

Richie Tienken remembered that back in the late 70s he'd been sitting around with Jerry, Paul Reiser, and Larry David, and had asked each of them what they wanted to do in the future. "Paul said, 'I'm not really sure . . . maybe some motion pictures, TV.' Larry said, 'I want to have fun, enjoy myself, do whatever's out there.' And I said, 'Jerry, what about you?' And Jerry said, 'I want to be the best stand-up comedian there ever was.'

"He indicated he had absolutely no interest in movies, or TV, except Carson. So when he got *Benson*, which was terrible, and he was terrible, and he'll admit that, I was surprised. He had put aside his dream because someone waved big bucks in his face. After he was fired and came back to New York, Jerry stopped in to say hello. He was bitching that when he was doing *Benson* he laid off his stand-up work for months, and did nothing but think about acting. Now he was saying, 'Okay, I can go back to doing what I like.'"

To Lucien Hold, at the Comic Strip, Jerry declared, "I'm going to rededicate myself to my comedy because that's something they can never take away from me. I'm in complete command when I'm doing my comedy.

"By that point, a couple of months later, Jerry was getting over it," Hold said. "He saw *Benson* as a learning experience, *and* he'd gotten his money—forty thousand dollars, which was a fair amount of change in 1980. After the embarrassment of the show, he rededicated himself to stand-up. For almost a decade he stayed away from any possibility of doing television."

In the end, bitter about the whole experience, Jerry asserted that he was fired for telling someone else's "crummy jokes."

By the time Jerry returned to L.A. after the *Benson* fiasco, he knew he needed good people to manage and steer his career so that he wouldn't make another mistake.

—26—

Jerry Does Johnny

As a comedians' manager, George Shapiro was one of the best in the business, if not *the* best. It was he, for example, who put one of Jerry's peers, Andy Kaufman, on the map after watching him perform at the New York Improv. Jerry was well aware of Shapiro's reputation—that he was a hustler, but in a good sense; that he was creative, supportive, nurturing—and that he had a partner named Howard West, a tough money-man, a well-oiled closer who made the buyers of their talent pay through the nose.

Not long after Jerry returned to the coast, Shapiro, who had heard of Jerry's talents through the comedy grapevine, sent a junior agent from Shapiro/West and Associates to see him perform, and Jerry was instantly signed.

"Getting George Shapiro was the turning point for Jerry," observed Mike Costanza. "Shapiro became the creator of Jerry's career. He made Jerry, really." As Glenn Hirsch noted, "Having George Shapiro as his agent gave Jerry a lot of clout, a lot of power. He got him into situations that enhanced his career."

With Shapiro behind him, there would be no more *Benson*-style debacles.

Shapiro also agreed to handle George Wallace.

In the 1950s, the wiry, Bronx-born and -bred Shapiro got his start in show business, earning forty dollars a week, in the fabled mail room of the Williams Morris Agency in Manhattan, which was to aspiring young talent agents and managers what Schwabb's drug store on Sunset was to movie-star hopefuls, and what the Improv and Comedy Store in L.A. were to Jerry: a place to be discovered.

As Bernie Brillstein, the founding partner of powerful Brillstein-Grey Entertainment, who worked side by side with Shapiro in the mail room, recalled fondly in his aptly titled 1999 autobiography, *You're No One in Hollywood Unless Someone Wants You Dead, Where Did I Go Right?,* "We competed against each other like a school of baby sharks. Only a few would grow up to be, well . . . big sharks." Shapiro, he emphasized, was one of them, as was West, whom Shapiro had known since third grade at PS 80 in the Bronx.

Over the next decade Shapiro held various jobs at William Morris. For a time he had the good fortune to work as a temporary secretary for Elvis's manager, Colonel Tom Parker, whenever the King's handler came to New York. From Parker, Shapiro learned a lesson that stayed with him for a lifetime, a lesson that Jerry would one day be the supreme beneficiary of, and that lesson was: "Everything is money. Everything depends on money. It's all money. How much money did we make today?"

By the early 60s Shapiro, bringing West along, decided to leave the New York headquarters of William Morris for the agency's West Coast division, heeding Bernie Brillstein's advice, which was, "You're from New York. You're smart. Go to California. They're not real bright out there. You should do very well."

In Hollywood, Shapiro made a rep for himself, handling deals for Steve Allen; introducing new talent like Bill (Jose Jiminez) Dana, overseeing the creation of *That Girl*, the long-running Marlo Thomas sitcom, and discovering a range of talent from Jim Nabors to Andy Kaufman. In *Man on the Moon*, the 1999 film about Kaufman's brief, wild career, the actor Danny DeVito played the role of Shapiro.

By the mid-70s, just as Jerry was about to launch his career, Shapiro and West left William Morris and formed their own operation with an eccentric roster of talent, which included Carl Reiner, father of *All in the Family*'s "Meathead" Rob Reiner; pop-eyed comic-actor Marty Feldman, who played Igor in Mel Brooks's *Young Frankenstein*; Ruth Buzzi, of *Laugh-In* fame; the animator who created TV's *Scooby-Doo;* and an assortment of writers and other creative types.

With a powerful manager in his corner, Jerry quickly hooked up with the aggressive booking agency Spotlite Enterprises that was headed by charming, bright, and savvy Robert Williams. Until the mid-70s, Spotlite handled mainly musical artists, theatrical shows, and disco acts. By the early 80s, when Jerry was welcomed aboard—along with George Wallace—Spotlite had just inaugurated its L.A. operation and was heavily into booking stand-up acts.

"I got Jerry right after *Benson*," Williams said years later, running a new company out of Nashville. "He was not negative at all, but he was far more critical of Jerry Seinfeld the actor, than Jerry Seinfeld the comedian. He was much more critical of himself, and he was very open about the fact that he hadn't mastered acting yet. But Jerry was not a whiner. He took his lumps and moved on. *Benson* was maybe one of the greatest things that ever happened to him, and there's a reason: it made him more focused about becoming a master craftsman at one thing.

"When he came to me, his goals were very absolute: fifty weeks on the road—'fifty-two if I've got to.' He said, 'I want to work every place. Every city.' That's what he wanted out of his career at that time. Unlike many comedians, Jerry was a businessman. When his first entry into television exploded, instead of the artistic part of him saying, 'I'm going to do it again and again and again and prove them wrong,' Jerry regrouped, made himself a road warrior. He had a game plan. He knew in his mind where he wanted to go. He knew eventually he would do TV under *his* terms. That's a businessman.

"He had short, intermediate, and long-term goals," continued Williams, who liked, understood, and respected Jerry, and collected fifteen percent from every gig he did. "His short-term goal was to be a road warrior, to work, hone his craft, and build a fan base in America. His long-term goals were to get his own show, which is not to be confused with another show. He was eventually looking to do the show *he* wanted to do—not a piece of casting. Even if Jerry had been cast as a lead, he would not have taken it after *Benson*."

Meanwhile, Jerry still had one driving, immediate goal to fulfill—doing Carson. In early April 1981, when he was approached once again by a *Tonight Show* envoy, this time veteran Carson comedy talent coordinator Jim McCawley, he was loaded for bear. With the requisite five minutes of dynamite material in hand, he accepted the invitation from the king of late-night television.

Carson was the big one, the one he'd waited for since he was a kid in his blue plaid bedroom in Massapequa, dreaming of making people laugh, the Cosby comedy albums stacked meticulously next to his hi-fi, the Jerry Mahoney ventriloquist dummy on a bookshelf.

For any stand-up in America, it was either the opportunity of a lifetime or a disaster, as Jerry thought it had been for David Sayh, whose post–*Tonight Show* career spiral-down was the reason Jerry had waited so long to say yes, like a reluctant virgin who had seen her best friend become pregnant the first time in the sack.

Tom Dreesen, whose career was propelled by his first appearance on Carson, never forgot the importance of the show.

"No matter where I went in America somebody would ask me what I did for a living and I'd say I'm a stand-up comic. The next question out of their mouth was, 'Oh, yeah, well, have you been on the Johnny Carson show?' If you hadn't been on Johnny, you might want to be a comedian, you might someday be a comedian, but in the eyes of America, you weren't a comedian. Carson was the stamp of approval."

Jerry knew all the fairy-tale endings and horror stories all too well, and for the next month, until his scheduled debut, he trained like a boxer preparing for a championship bout.

"I rehearsed the same five minutes every night, the five minutes I was going to use on the show," he told Lucien Hold. "I ran five miles a day, every day, like a prize fighter, and I did that same routine for thirty days."

A comedian friend who lived in Jerry's West Hollywood building, Jimmy Brogan, who had come out a year before Jerry from New York, often accompanied Jerry around the track at Fairfax High School, where he obsessed about his big shot—what the material was going to be, the order it was going to be in, the best transitions to use from one bit to the other, his wardrobe, his hairstyle.

Jerry allowed for nothing that could possibly throw him off his routine. A klieg light could blow, a camera could go black, jovial sidekick Ed McMahon could go into cardiac arrest—but Jerry was going to be perfect on that night of nights. He was going to be in command, in control. No matter what, the right words were going to come out of his mouth.

McCawley had wanted Jerry to do a routine he had about luncheon meats, thinking Johnny would find it particularly funny, but Jerry decided to do other material, which he felt fit together better. McCawley argued, but Jerry was adamant. "It was brave of Jerry," asserted Brogan, who later became head writer for Carson's successor, Jay Leno. "Jerry just went ahead and did the material he wanted to do."

On April 29, 1981, with a few close friends, Jerry blew out twenty-seven candles on a birthday cake that had been set before him in a restaurant, but he couldn't eat a bite. He was too nervous. On that birthday night, he'd been a professional comedian for five years, but all of it meant nothing to him compared to his appearance on Carson, now just seven days away.

In Massapequa in the week leading up to the show, a proud Kal Seinfeld had been riding around town with another sign on his van advertising his son's latest accomplishment. In crude black letters over garish Day-Glo orange and green paint, the sign-huckster father had written:

JERRY SEINFELD
OF
MASSAPEQUA
WILL BE ON CARSON
SPECIAL

Kal also took out an advertisement in the *Massapequa Post* in honor of Jerry's appearance. It read, in part: "He dared to dream the impossible dream." Later he told Jerry, "It has nothing to do with you being my son. I just happen to enjoy your work."

On the scheduled night, he went around town with a bullhorn, like some crazed Jewish Paul Revere, rousing the citizenry to turn on their

TV sets at 11:30 P.M. "He went up and down the streets—'Watch Jerry Seinfeld tonight, Johnny Carson, NBC,'" recalled Rabbi Spielman.

Wearing jeans, a sweater, and sneakers, Jerry arrived at the NBC Studios in downtown Burbank for the taping, which was scheduled to begin late in the smoggy afternoon of May 7, 1981. After a briefing from McCawley, he went into makeup and then was ushered into one of the guest dressing rooms so he could change for the evening—a handsome, tailored, gray, two-button sports jacket, crisp white shirt, a subtle red-and-gray tie, dark blue slacks, and relaxed tan shoes, which he had brought in a small suitcase and garment bag. In the Green Room there was a bouquet of balloons waiting for him, sent by Mike Costanza and his girlfriend, with a poignant card reading: "From the Golden Lion to Carson—Wow! Kill 'em, Jerry! Much love, Marie and Mike."

The producers and directors had formatted Jerry to go on about an hour into the show, which was taped starting around six o'clock Pacific Coast time, which meant he'd be seen shortly after 12:30 A.M. East Coast time.

In the Seinfeld home in Massapequa, at the Comic Strip in Manhattan, at all the places where Jerry had friends, all eyes were on the TV when dapper Johnny Carson, starting to bald a bit, wearing gray himself, tapping his trademark pencil on the desk, with the faux southern California potted plants behind him, announced, "Would you welcome him please, Jerry Seinfeld. Jerry . . ."

A nanosecond later a seemingly relaxed, smiling Jerry strode confidently out from behind the tan curtain with a squiggly design on it, clasped his hands together as if he'd been doing this kind of thing forever—this show that was seen nightly by millions, many in bed, waiting for Carson's sign-off to flick off the light and go to sleep—and Jerry launched into five minutes of razor-sharp material. He did a bit about TV weather reports. "This is real helpful," he said. "A photograph of the Earth from ten thousand miles away. Can you tell if you should take a sweater from that shot?" He did a bit about a perplexed driver looking under the hood of his broken-down car. "What are you looking for?" Jerry asked. "Whatever's wrong, *you* can't fix it. You stand there looking for something incredibly, obviously wrong—something so simple even you can handle it—a giant on-off switch." Off-camera, Johnny was laughing, especially when Jerry did his riff about *Guinness Book of World Records* heaviest man, Bob Hughes. What would happen, Jerry wondered, if he lost a few hundred pounds? "What would his friends say? You're a rail, baby, look at you!"

The five minutes flew, smooth as cream cheese, and then Jerry was finished. Five years of hard work for five glorious minutes. He took a bow and looked over at Johnny who winked, gave him the high sign, a big smile, and said, "Jerry, take a bow," and Jerry bowed a second time. There was no invitation to join the panel, like David Sayh had gotten, but Jerry had, in those five minutes, joined the elite. He'd gotten the approval of the don of late night. If Jerry had been a low-level punk working for the mob, he now was a made man, like his one-time Massapequa neighbor, Carlo Gambino. However, it wouldn't be until Jerry's fifth or sixth appearance that Carson would give him "the wave" after he finished his number to come sit on the sofa and chat.

Back in Massapequa, the Seinfelds and five neighborhood friends tuned to NBC all night waiting for Carson. "We stopped our mahjong when he came on," family friend Fran Alter remembered. "We were all so excited. We couldn't wait until Jerry appeared. And then, when he did, we were so thrilled. He even said on the show 'my father's the comedian,' meaning Kal. Kal and Betty were so happy. They couldn't believe their eyes."

At the Comic Strip, all eyes were on the television. "I was happy for Jerry," said co-owner Richie Tienken. "Jerry went on and did great, and everyone in the Strip is watching. The show on stage virtually stopped. And when he finished, everyone's screaming . . . that was great!!! . . . holy shit!!!! . . . . blah-blah-blah. So we're all happy for him. I called him in California and said, 'Hey, that was terrific,' and he said, 'I appreciate that.'

"About two months later, he's on Carson again because Johnny really liked him. Again, everyone's gathered around the TV at the Strip, all these comics hoping to get on the *Tonight Show* like Jerry. So Jerry comes on and does great, but now the reaction is, 'Yeah, that's good. Don't know if it was as good as the first time.' Two months later he gets the show again. They all gather around the set. Jerry goes up and does great, from my point of view. But now you get everybody going, 'Well, I don't think he was as great as the first two.' So now they're starting to knock him and I'm looking at them and I say, 'You guys, you rooted for him for years, and now when it comes down to it you're slamming him.' But that's the way the business is.

"After that first Carson, Jerry could go to Vegas and open for someone and make five thousand a week. And it would also be a good plug for us," Tienken continued. "I'm being totally honest, because the next time Carson might say, 'Here's a guy we spotted at the Comic Strip in New York.' And the next time Jerry appeared at the club we could advertise, 'Jerry Seinfeld, direct from the *Tonight Show*.' So everyone benefited."

A few days after Carson, Jerry returned home to a hero's welcome, but nothing fancy. In his honor, Jesse Michnick's mother hosted a small dinner party in her Queens row house to celebrate her son's friend's newfound success. Michnick invited a group of pals from the Queens College days—Mike Costanza, of course, whose girlfriend, Marie, made green lasagna for the gathering; struggling actor Joe Bacino, who brought his girl. Jerry's pal, nerdy, pockmarked comic Larry Miller, who'd also moved to the West Coast around the same time as Jerry, showed up with a drop-dead gorgeous blonde who was the model for the White Rock ginger ale bottle. And the guest of honor arrived with a brunette whom he had been seeing before leaving for California.

Despite Jerry's new celebrity-star status, the evening was just like the old days, lots of sophomoric giggling and joke-telling, made even more hysterical by the fact that Mrs. Michnick, apparently not much of a kitchen planner, had made only a two-pound turkey for all of the guests. Moreover, throughout the entire dinner, "the White Rock girl was rubbing Larry Miller under the table," Michnick said. "My dad was looking over there once in a while and he'd give me a kick, like what the hell's going on with those two. Under the table we're kicking each other, and Jerry's hysterical."

But after he left that night, Costanza noted, life was never the same for the group. Jerry's schedule precluded such play. He was now in the big leagues.

—27—

Comedy Club Bonanza

With Shapiro and Spotlite behind him, Jerry quickly became—next to his pal Jay Leno—the most driven road warrior on the exploding stand-up circuit, a megaton comedy blast triggered because of the decline and fall of another potent entertainment force—disco.

By the early 80s, disco had had its day, a more than half-decade wild ride inaugurated in 1977 by the film *Saturday Night Fever*, which featured John Travolta as a New York bridge-and-tunnel guy with gold chains around his neck, and dancing feet. The disco club craze made its mark in New York with chic venues like Xenon and Studio 54, which attracted the hedonistic, jet-set, nose-candy, exhibitionistic elite, the "Columbian White" of boldface gossip-column names—the Andy Warhols, the Calvin Kleins, the Roy Cohns, the Liza Minnellis, the Diane von Furstenbergs, the Truman Capotes, the Margaux Hemingways, the Margaret Trudeaux, the Rudolf Nureyevs, and the Grace Joneses, among untold scores of others who made the scene until five in the morning.

Out of nowhere, too, came the fuel for the raging fire—the disco music stars: glamorous, glitzy singers like Donna Summers, Gloria Gaynor,

Sister Sledge, Sugarhill Gang, Irene Cara, Earth, Wind & Fire, ABBA, the Village People, and, of course, the Bee Gees—considered the biggest of all—with the *Saturday Night Fever* soundtrack hits "Stayin' Alive," "How Deep Is Your Love," "Night Fever."

There were the dance styles, too—the Hustle, Rope, Sling, Tango, Latin Hustle, West Coast, and Street.

It was a music, fashion, sex, and drugs fireworks display and party that went on all night, every night, and in what was seemingly a split second, this disco thing had spread from New York, where a "National Disco Week" was declared in 1978, to Middle America, where untold numbers of bars in small cities and towns from ocean to ocean, border to border, virtually overnight, became discos, the owners making huge investments in flashy dance floors, lighting, sound systems, bars, backrooms. It was a gold rush while it lasted.

But by 1982 the disco joyride skidded, screeched, slammed into a brick wall, and was pronounced dead at the scene, declared passé by the cognoscenti.

It was at that point that Jerry's agent Robert Williams, the farsighted head of Spotlite, pretty much created the comedy club circuit from the burning embers of the now-dead discos—the circuit that permitted Jerry to work as many as three hundred nights a year, and rake in, at the top of his game, where he always seemed to be, several million dollars annually; the circuit that helped to make Jerry the best of the best and the brightest.

"Basically, there were all these disco clubs across the country that had closed, all those empty rooms, all these places that had done really well for five or six years in the hinterlands that were now dead," Williams explained, "and I saw that, and Spotlite then pretty much created the comedy club scene from that wreckage.

"By bringing in a stand-up comedian, a Jerry Seinfeld, those club owners didn't have to retrofit their places; didn't have to go in and gut and renovate all over again. They just had to simply figure out how to turn on only one spotlight, put a microphone on, and get people in to buy drinks. The magic of the comedy club business was they didn't have any of the hassles of moving a whole sound system in, of three-hour sound checks. That's what sold the whole comedy club concept—ease and simplicity.

"Like many brilliant club owners, they didn't save any money, so when disco died they were cash-tight. Now, with the comedy scene booming, they could pick up a comedian, a name like Jimmie Walker, for $2,500, $3,000 on one of his nights off. Well, here's a guy with enormous TV exposure who's going to be on Letterman, and Letterman's going to mention the next five dates Jimmie Walker's got coming up. So, if you're a little club in Providence, you just picked up $40,000 worth of free publicity, and you paid $3,000 for the act. It was heaven!

"With a comedian, he walks in, taps a microphone, says hello, someone puts a pin spot on him and you have a show. It became a very, very lucrative business."

Overnight, there were ex-discos turned laugh shops running six and seven nights a week, some that could seat two hundred to four hundred people, with names like the Comedy Shop, Punch Line, Laff Stop, Zanies, Funny Bones, Bananas. The originators of the comedy scene started chains and sold franchises, so there were Catch a Rising Star clubs, and Improv clubs, and Comedy Zone clubs. There was even a lucrative chain established in Canada, and soon Williams helped devise the Montreal Comedy Festival, which he made sure featured Jerry, and now stand-up comedy even had an international flair to it.

By the end of the 80s, at the height of the comedy-club boom, there would be some 550 full-time stand-up venues in North America. The funny business had gotten so serious, in fact, that an industry trade magazine was even launched, called *Comedy U.S.A.*

With so many clubs, Jerry reached a point where he was boasting, "Four days is my maximum to go without working." He hated, he said, taking vacations. He couldn't stand for a moment to be away from the road, a good part of that devotion being his desire to obsessively make up for the *Benson* fiasco and firing, to eradicate the guilt he felt for stupidly putting himself in that spot in the first place. "That's when I decided I was never again going to be in a position where somebody could do that to me," he declared, still bitter a few years later. "As a stand-up comic, nobody can fire me like that."

Glenn Hirsch, also represented by Spotlite, was taking the same ride, as were Williams's clients, Leno, Paul Reiser, and Tim Allen, among other soon-to-be big names in the TV, sitcom, and film businesses, who came out of stand-up. "All of a sudden everyone started to work," Hirsch noted. "Comedy was starting to proliferate all over the country. We weren't all together in a few clubs anymore. We hardly ever saw each other. We were traveling all over the country. From the early 80s on, I was in a different city every week, and Jerry was the same way. We had the exact lifestyle. It was very exciting."

The comedy boom was, in many ways, like the rebirth of vaudeville, with performers constantly on the road, going from town to town, the weeks and months flying by. For the comedy-seeking customers, the clubs harked back to another earlier time—the 40s and the 50s, when a couple out on the town could sit at a small table, get bar service, and see an act, a way of life that had all but disappeared outside of the major cities.

Soon, the stand-ups were joined by other contemporary vaudevillians—comedic jugglers, comedic impressionists, comedic singers, comedic magicians, comedic ventriloquists. These were the kinds of acts a pure monologist like Jerry held in contempt.

"The comedy clubs boomed because all of a sudden in America two members of the family had to work," Williams observed. "The world had become so stressful and comedy was a great relief. A couple in their thirties, who didn't want to sit in an arena, could go out for the night, sit and have a drink or two, and have the intimacy of being close to an artist like Jerry Seinfeld, and laugh for ninety minutes, and that laughter released endorphins that were healthy to the body."

In 1982, a year after his first Carson engagement, Jerry returned from the road to make his first appearance on David Letterman's new *Late Night* show on NBC. Jerry invited Mike Costanza and Jesse Michnick to accompany him to the studio and wait in the Green Room while the show was being taped for airing that night. But the evening became a disappointment when, just before Jerry was scheduled to go on, one of the producers came backstage and told him he had been bumped; another guest was running longer than planned. When the show ended, Letterman strode into the Green Room looking serious, and profusely apologized to Jerry, vowing to reschedule him as soon as possible. Not long after, Jerry appeared on Letterman and was invited to "do panel," meaning sit on the sofa and chat, the first time Jerry had experienced the national TV late-night star treatment.

While the viewers wouldn't have known it, Jerry was a nervous wreck. He only felt comfortable with material and situations that *he* controlled, and Letterman was a loose cannon who didn't always follow the questions and subjects on the index cards that his producers had prepared for him from the pre-interview with Jerry. Afterwards, Jerry asked Costanza, who had again accompanied him to the studio, "How did I do? Honestly, what do you think?" Costanza said he told him what he believed to be the truth. "You killed, Jerry. You killed them out there."

Jerry's only response was, "Really?"

Later, Costanza analyzed the situation, deducing that Jerry, after six years as a professional performer, and despite his apparent self-confidence, still needed "all the emotional reassurance" he could get "as if he didn't quite believe any of this was actually happening."

In the fall of 1982, Jerry had another major career first, an appearance at famed Carnegie Hall, the most famous concert hall in the United States, where George Gershwin, Duke Ellington, and Igor Stravinsky, among dozens of others, had their world or American premieres, beginning in 1891. Now a pop singer by the name of Laura Brannigan was appearing there, having had a big hit in the summer of '82 titled "Gloria," and Jerry was her opening act.

Jerry had gotten backstage passes for Costanza and Joe Bacino. In Jerry's dressing room, waiting for him to go on, Costanza asked Jerry how much he was being paid for the one night. "You don't want to know," Jerry responded teasingly, knowing that Costanza would keep pushing for an answer. "I'm your friend," Costanza said, patting Jerry on the shoulder. "I want to be sure they're taking care of you." Costanza may have been sorry he asked, so stunned was he by Jerry's response: "They're paying me twenty-five thousand dollars." Usually quick with a comeback, Costanza, who was then trying to make ends meet running a small deli for which Signfeld Signs had painted the "Bologna, 75 Cents a Pound" placard as a gift, was at a loss for words.

The audience loved Jerry's show, and backstage after his performance he was standing talking to a circle of people that included Costanza, Bacino, and Bacino's wife. As they were chatting before going off to have dinner in Chinatown, an old haunt from their Queens College days, a teenage boy approached Jerry. "Can I ask you a question?" he said.

"What is your theory on comedy?" Watching this tableau, Bacino was stunned. "It was so weird for me that someone was seriously asking Jerry Seinfeld what his theory of comedy was," he said later. "Jerry responded very seriously. I don't remember what he said, but he began to really pontificate, to talk about what his theory was, and where it came from." Watching Jerry, Bacino thought to himself, "This is *not* the Jerry I knew. He seemed so full of himself."

Larry Watson, who had always considered his relationship with Jerry to be more like that of close and loving brothers, also experienced a difference in Jerry that put him off. Watson had been planning a trip to L.A. and called Jerry at his West Hollywood apartment—the two were in regular contact by phone, exchanging updates about their lives—and when Watson mentioned his visit and the possibility that he might need to crash with him for a few days, Jerry responded enthusiastically.

"So I arrive in L.A. and I call him from the airport and I say, 'I'm here. Let's hook up,' and he said, 'We'll have lunch tomorrow.' The next day I wake up and call him and we make plans to meet in West Hollywood, and while I'm getting ready he calls and says, 'Larry, I'm so sorry. Something just came up and I've got to go to an audition, but we'll meet for dinner.' Okay, no problem. I called him a half-hour before our date and he says, 'Larry, oh, my God, I don't believe this, I just got called for a late meeting. I have to go. We'll meet tomorrow for lunch.' The next day he calls and says, 'Larry, I just got called for another audition. This is how it works out here. We'll meet later.' Finally, we meet for dinner and the first thing he says is, 'I've just been invited to this great party, and you know I need to be on the scene, so I gotta run.' I said, 'Jerry, you can stay on the scene.' I said, 'You're full of shit.' There was no response because Jerry cannot be confrontational—except if it's for something he wants—if it violates him in some way. I went back to Boston and we didn't speak for like two years."

But Bacino and Watson, in their own ways, had taken a narrow view. Jerry had now become a polished performer, a talent who earned big money, who was being sought after for deals, opinions, and soon for interviews. He was still the same Jerry from Queens College, but also different, headed down the path of fame, wealth, and celebrity, and that naturally inspired envy and misunderstanding in old friends.

Later in 1982, the Queens College pals got together again, this time for Jesse Michnick's wedding, which took place at a synagogue in Brooklyn, with Jerry serving as one of the ushers and Costanza as the best man. Jerry shattered the solemnity of the ceremony when he whispered to Mike, who cracked up, "If you're the *best* man, why is she marrying Jesse?"

—28—

Love and Death

For an ultimate road warrior like Jerry, marriage, let alone a simple relationship, was an alien concept. He had no personal life, which he savored. He had no commitment to a woman, no one to interfere with his schedule, no birthday or anniversary presents to buy, no social engagements to fulfill, no conjugal obligations to meet. His existence was much like his father's had been years earlier when he was a door-to-door peddler, only with more money.

"Jerry's life," Robert Williams acknowledged, "was the job. He had absolutely no private life. He *was* vaudeville."

But, in the Orwellian year of 1984, the year Jerry turned thirty, which for him was a major emotional barrier to cross, the *concept* of marriage became a serious consideration; it finally seemed like a good idea.

"Turning thirty just blindsided me, and I thought I had better hurry up and do the things I needed to do," he says, recalling that period of his life, "so I got engaged."

What he didn't reveal was that the stars were in their proper place on his astrological chart, which told him it was the time to get hitched.

It all started some months earlier at the Hollywood Hills home of Jay and Mavis Leno, where Jerry, in his rare free time, often hung out, playing video games, trading road warrior war stories, gawking at Leno's growing motorcycle collection and, in general, just chilling. They'd become best of friends, shared similar backgrounds, and their birthdays were just a day apart—Jerry's on April 29, Jay's on April 28, but Jay was four years older.

Like Kal Seinfeld, Jay's father, Angelo, was a salesman, of insurance, and, like Kal, Angelo had a great sense of humor and was a master storyteller, talents his son greatly admired; at one point, Jay even considered becoming a funny insurance peddler. Like Betty Seinfeld, Jay's mother was an immigrant, from Scotland. Amazingly, both Jerry and Jay once got the same type of comments from teachers in elementary school. Jay's wrote, "If James used the effort toward his studies that he uses to be humorous, he'd be an A student."

Like Jerry, Jay wasn't involved in extracurricular activities and wasn't good at sports. At Emerson College in Boston, though, the comedy bug bit him, as it did Jerry at Queens College. Jay and a classmate started picking up small gigs, billing themselves as "Gene and Jay, Unique and Original Comedy." Eventually Jay got stage time at the Improv in New York, where, like Jerry at the Comic Strip, he earned a reputation as a comedy craftsman. In 1974, six years before Jerry, he moved to Los Angeles, where the two driven and ambitious stand-ups met soon after Jerry's arrival in the wake of the ugly comedy strike.

Jay had met cute, petite, dark-haired Mavis Nicholson, an aspiring comedy writer, outside the restroom at the Comedy Store on Sunset, and five years later, in 1980, they were married, and from the beginning theirs was an unconventional union: Jay was on the road virtually all the time, and when he wasn't he spent time with his guy friends. "If spending a day off working on cars was the kind of thing that bugged her," Jay once said, "we wouldn't have gotten married."

But Mavis didn't mind; she had her own interests, one of which was serious astrology, and that's where Jerry and the idea of marriage came together, through an astrological chart Mavis did of Jerry Seinfeld, a Taurus.

For Jerry, who had been a scientologist, who had dabbled in Zen and Transcendental Meditation, and who practiced yoga every day, believing in astrological charts didn't involve a big leap of faith. "It was kind of fun for him," recalled Susan McNabb, who became Jerry's girl not long after the marriage plan went awry. "He found astrology interesting, but thought it was funny that Mavis was an actual astrologer because Jay always made fun of things like that, because Jay thought it was a bunch of baloney."

Mavis did Jerry's chart once or twice a year, and even tape-recorded it for him so he could listen to it on the road. In early 1983, she made a startling prediction. Based on her latest reading, Jerry would become engaged within the year. "Jerry was like 'No way,'" Susan McNabb said. "Jerry said, 'That's never going to happen.'" For years he had told friends he didn't see marriage in the cards for himself. "He was always saying that his lifestyle was not conducive to settling down and having a family," asserted LuAnn Kondziela.

All of that was about to change, at least temporarily.

Jerry was on the road, staying at a hotel in San Francisco, and was having some problems with his room. He called downstairs, asked for the concierge, and when he opened the door you could practically hear Cupid's arrow go *Boooiiiing*!

She was a pretty, tailored, coiffed brunette who prided herself on her long, red, perfectly manicured fingernails. She had a bubbly, willing-to-please personality, and a gentle demeanor, which was well suited to a successful executive career in hotel management. She was also Jewish, which didn't hurt when Jerry introduced her to his parents, who were immediately smitten with her.

They began dating immediately, which wasn't easy, since Jerry was on the road most of the time, so it was for the most part a long-distance romance—they never lived together. They went together for about six months. In February 1984, Jerry popped the question and she accepted. So far, Mavis Leno's astrological reading of Jerry's chart was right on the money.

On a trip to New York with his fiancée, Jerry broke the good news to Mike Costanza. "Basically, Jerry just got infatuated with this girl," Costanza said. "She was in another room and he said, 'I have somebody for you to meet.' He tried to like set this whole thing up and surprise me, and I wasn't prepared for this at all. Jerry said, 'Sit down.' I asked why. He said, 'I'm very serious about her.' I said, '*You're* serious about someone.' He said, 'Yeah, I'm so serious we got engaged.' I said, 'Come on, stop fucking around!' He said, 'No, I'm really serious—and *don't* use that language, she's in the other room!' I said loudly enough for her to hear, 'Oh, your fiancée is in the other room, is she,' you know, always kidding around."

In private Jerry told Costanza that one of the things that had most impressed him about her was that she didn't know who he was at first, and accepted him for being Jerry Seinfeld. Period. "He liked the idea that she wasn't impressed with his fame."

At the same time, Costanza noted, Jerry was "petrified that he was going to go through" with marrying her. "He said, 'It's like being on the roller coaster, and you can hear that chain clicking when you're on the way to the top . . .' "

Back in L.A., after the engagement announcement, Jerry had Mavis Leno do his chart again, and she revealed still another shocker.

She predicted the engagement would end, that there would be no marriage, Jerry told McNabb later.

And that's precisely the way it happened.

The story Jerry put out was that when he told his fiancée that he had to work New Year's Eve she freaked out and said she wasn't going to spend that special night sitting in the back of some smoky club alone, nursing a glass of champagne, listening to him tell jokes to a bunch of drunks.

"That's when Jerry realized she just wasn't the girl for him," Betty Seinfeld once told an interviewer. "Jerry needs a woman who will support a family life and his career."

Jerry's sister, Carolyn, thought Jerry's decision to marry had to do with the fact that she'd recently had a baby, and Jerry felt he had to keep up and "get on with life," which included marriage. She speculated that Jerry probably would have gone through with the marriage "to not upset" his fiancée and her parents, but as time went on "he realized it wasn't right. It was very difficult for him to end it. Very difficult."

So difficult, in fact, that Jerry, for the first and what's believed to be the last, time in his life went into therapy. "He felt maybe he should talk to somebody," McNabb said. "He went for one therapy session, and then he was like, 'Okay, I'm fine. I'm cured.' "

A number of Jerry's friends wondered whether he had circulated the New Year's Eve scenario because it sounded believable, but that it wasn't the whole enchilada. Lucien Hold, for instance, heard the tale and privately questioned it at the time, and in the final analysis thought it might be "apocryphal."

"The fact that she wouldn't spend New Year's Eve in a club watching him perform is not why Jerry ended the engagement," asserted McNabb. "The New Year's Eve thing was a tiny element of it, kind of a little problem, but that's not why the relationship ended. It wasn't like, 'Okay, you're not going to be there New Year's Eve, it's all over.'

"He basically freaked out when he turned thirty. He told me that he thought he should settle down and he said, 'I proposed to the first reasonable candidate who came along. And then I realized I don't want to get married. I'm not ready to get married.' It had nothing to do with her. He shouldn't have asked her to marry him. After they broke up, he spoke very highly of her. He kept her photographs in a shoebox. He took all of the blame for the breakup, that it was his doing, his fault. He felt terrible about hurting her, was very remorseful."

There may have been another reason for Jerry's decision to make the bond of holy matrimony for which he wasn't ready, and didn't want, and that was to please his parents who were getting on in age and would have liked to see their son with a wife and kids, especially Kal Seinfeld whose health was failing.

Around the time Jerry's career started taking off in the late 70s, Kal Seinfeld had, after a lifetime of hard work, begun to slow down. He approached Arthur Farb, who ran an engraving business next door, and asked him to manage Signfeld Signs because Kal didn't have anyone else he could trust to run the shop while he was out selling. When Farb explained he already had too much work on his plate, Kal, still the hustler, didn't miss a beat. "If you can't manage my business," he said, "why don't you buy my building? It's a lucky building. I've had good fortune in it. You won't go wrong."

Farb couldn't afford the down payment, let alone the $67,000 asking price, but Kal made him an offer he couldn't refuse.

"Kal was very quick on his feet," Farb noted. "He said, 'I'll lend you the money for the down payment, but you can't tell the bank.'" Farb admits it was a shady deal. "It was definitely under the table," he said years later. "The fact that he didn't want the bank to know about it is not completely aboveboard. The bank would have seen that I didn't have the money and not approved the loan. The down payment amounted to six thousand dollars, but the money never changed hands. However, it was recorded that I gave him the money. Over the next three years I paid him two thousand dollars a year plus interest for the loan. Once we closed on it, it became my building. I'm in business thirty years, so I would say it's been a lucky building."

On the day of the closing, the Farbs were at the lawyer's office on the button, but Kal was late, which was often the case. Mrs. Farb recounted, "When he finally showed up, I said, 'You must not go by English time, you're late.' And Kal looked at his watch and said, 'You're right.' Then I looked at his watch and the numbers were in Hebrew and backwards. It was a joke watch so he could make an excuse to people when he was late. It was typical Kal, always trying for a laugh."

After the sale, Kal, still needing to be active, but with less responsibility, rented another sign shop in town, moving from one end of Broadway in Massapequa to the other end. "His sense of winding down was *not* to stop working, but to start unloading property so that he wouldn't have to pay heavy taxes," Farb observed. Around the same time, the Seinfelds bought their first condo, in Fort Lauderdale. Later, Jerry joked in his act that his parents had retired and moved to Florida because "it's the law." Moreover, Jewish condo life in South Florida became a staple on *Seinfeld.*

By 1984, the year of Jerry's ill-fated engagement, Kal, who now was in the process of selling off his sign company and going into full retirement in Florida, began having trouble with his eyesight. The doctor's prognosis was not good.

"They found a tumor behind one of his eyes," said John Egan, a long-time employee of Signfeld Signs. "It was cancerous. The eyeball was removed and a glass eye was put in."

Arthur Farb heard about the cancer and the surgery from a girl who worked for Kal. "She called me up and said that he had an operation and he lost the sight in one eye and she said, 'Don't be disturbed when you see him. I'm trying to forewarn you.' So that was the beginning of what was to become the end."

On October 20, 1984, Jerry had come off the road to join the family celebration of Kal's sixty-sixth birthday, and the consensus was that he was beating the cancer that had taken half of his eyesight. But three months later, on January 31, 1985, Kal Seinfeld was rushed to Bethesda Memorial Hospital in Boynton Beach, where the Seinfelds had bought a new condo. The operator, the hustler, the finagler, the lovable sign king of Massapequa, the born clown who had passed on his genes of fun and laughter and comedic brilliance to the son who adored him, died suddenly in the hospital emergency room at 6:30 P.M, his death apparently related to the illness that took his eye. Jerry was heartbroken. At Jerry's side, George Wallace telephoned Larry Watson to tell him, "Jerry's devastated. It's a real blow."

Arrangements were immediately made with a funeral home in nearby Sunrise to fly the body to New York, where funeral services were held on Sunday, February 3, followed by burial at Montefiore Cemetery in Queens.

"Jerry spoke at the funeral," said Lucien Hold, who had been invited. "He said he got his sense of humor from his father. I'd never seen Jerry so emotional. When the coffin was being lowered into the ground, he broke down and cried and sobbed. It was really intense. It was interesting to see Jerry lose control like that, because he was always *so* controlled. His crying shouldn't have been surprising, but it was; it was testimony to how powerfully I always thought he was in control of his emotions, so those tears were a surprise. I'd never seen him, or known him to be vulnerable. It was very touching."

—29—

Sex, Money, and Accolades

Think "Robert Palmer Girl." Think the bevy of slender, leggy, leather-clad, brunette, mannequin-like models in Palmer's slick, erotic mid-1980s MTV videos "Addicted to Love" and "Simply Irresistible." That's the best way to describe Susan McNabb when Jerry met her at a Beverly Hills party on May 2, 1985, three days after his thirty-first birthday.

And what an exotic birthday package she was: slicked-back raven hair; sensuous brown eyes; fair skin; red, red lips; black leather miniskirt that revealed slender, shapely legs sheathed in black lace stockings that seemed to go all the way to heaven—she stood 5'8" tall in flats, but teetered at six feet in spiky heels. And underneath that trendy L.A. uniform was a body to die for, a svelte one hundred pounds sculpted as if by a modern-day health-club da Vinci into an eye-popping 34C-23-33 form.

Despite the Hollywood "look," the lingerie model and aspiring actress—five years Jerry's junior—was an old-fashioned girl, with old-fashioned values, from a good Christian family, a contemporary, conservative Aphrodite from Asheville, North Carolina, a homey town where the Blue Ridge and the Great Smokies meet, the daughter of an advertising agency executive father and a real estate broker mother, both of whom owned long-established businesses in town. A 1981 graduate of the University of Tennessee where she majored in English literature, McNabb considered going to law school, but then thought, "Hey, modeling's more fun." Deciding to cash in on her looks after graduation, she moved to Charlotte, North Carolina, where she got an agent and started doing fashion shows, print work, and commercials, before moving on to bigger, more lucrative modeling assignments in the larger market of Atlanta.

In July 1984, at the urging of a hometown girlfriend, a hairdresser who had moved to L.A. six months earlier, McNabb decided "to go for it"—the same words Jerry used in his L.A. bon voyage to LuAnn Kondziela—and moved west with the consent of her parents. Like Jerry, McNabb hoped to break into TV and/or the movies. She and her friend—who became Jerry's haircutter for a number of years—her friend's sister, and the sister's five-year-old daughter shared a nice two-bedroom apartment on Hayworth Avenue, in West Hollywood, not far from Jerry's junior one-bedroom on Westbourne, which was basically a bedroom–living room combination, the two sparsely furnished spaces separated not by a wall, but a step up.

McNabb loved comedy, being around comedians, and often went to see them perform at the Improv. So, when she looked across the room at that Beverly Hills party and spotted Jerry standing alone, she thought she recognized him, but couldn't quite place him. "It's like I know this guy," she remembered thinking, "but in L.A. you think it's probably a celebrity because there were other actors at the party, like David Keith." Looking around, though, she spotted a group of stand-ups—Jerry's close friend, Larry Miller; another pal, Jeff Cesario—and immediately it came to her. "Oh, these are the comedy guys from the Improv. I had seen Jerry on stage at the Improv and I thought he was absolutely brilliant, but I didn't remember his name."

She made her way through the crowd and introduced herself, telling Jerry she'd seen his act. They chatted, talked about the Improv and mutual friends they had there, because she had gotten to know some of the other comedians, and at the end of the evening Jerry asked her for her phone number.

"I was interested in him from the beginning because I thought he was really funny, very smart," she said years later, one of the few women, like Caryn Trager, who escaped being publicly identified with Jerry after he became an *über*celebrity, beginning when *Seinfeld* hit. "I was impressed with his talent and that first time when we met he was very attentive to me, was obviously interested in me—I'm a girl and I can tell when men like me—and I was flattered. He wanted to know about *me*. He was very polite and very much a gentleman."

Like so many in Jerry's sphere—male and female—McNabb concluded from the start that he was not a player.

"Jerry was not a ladies' man, he was not a guy who knew how to seduce, not a smooth operator, not a womanizer."

A few years earlier a babe like McNabb wouldn't have given Jerry a second glance: too nerdy, too uncool. But now, with growing celebrity and sophistication, that was changing, and he snapped her up.

The next day, Jerry made the call and they had their first date that night, meeting for drinks at Joe Allen's in the trendy Melrose district. Jerry picked her up in his black Saab 900 Turbo coupe—later featured on *Seinfeld*—his first fancy car, which he'd bought recently with a loan from a bank that gave him the first credit he'd ever had, at the age of thirty-one. "We talked. He made me laugh, but it wasn't a constant running shtick. We had a couple of beers and he brought me home and right inside the front door of the apartment he kissed me good-night. The next day he went to San Diego for a job."

For the week he was booked there, he called McNabb every night after he got off stage. "I can't wait to see you again," he told her. "When I get back can we see each other?" He told her he liked her, and she said she liked him. While she was dating different people at the time, she wasn't in a serious relationship. "I remember my roommate being excited for me because I thought I had met somebody I really liked."

Jerry was not in the best of shapes emotionally when he and McNabb began their seven-year roller-coaster ride, during which time they had eight separations, all based on the fact that he didn't want to get married even though he acknowledged, right from the start, that he was in love with her.

"He was," she declared, "commitment-phobic, and remained so. He broke my heart eight times."

Jerry's emotional shakiness had to do with his father's death, which had occurred four months earlier; he was still grieving, still in deep pain, and talked about his father constantly.

At the same time, Jerry had still not recovered fully from his broken engagement, which he discussed at length with McNabb, even showing her photographs of his former fiancée. "Unlike her," McNabb made it clear, "I was happy to sit in the back of the bar *every* night when Jerry worked. When he wasn't on the road, he was at the Improv, sometimes at the Comedy Magic Club in Hermosa Beach, trying out material, and I'd often accompany him. I loved watching comedians. I was so enamored with the whole comedy club scene. If I went out with him I spent part of every evening sitting in the back of a comedy club, and I was happy as a clam."

Their relationship became intimate soon after they started seeing each other. Always obsessive about cleanliness and health, he was now even more careful and vigilant about his body, especially when it came to sex, especially with "the gay plague," as it was being called, raging around him. The beautiful young men of West Hollywood were in a state of high anxiety, their numbers being depleted by the fast-spreading horror. In July, two months after Jerry started seeing McNabb, the actor Rock Hudson, looking ravaged, appeared on a TV show with his dear friend and one-time romantic costar Doris Day. The news broke that he was dying. The screen idol, who made women swoon, was secretly gay and his quick and tragic death forced the AIDS epidemic out of the closet. Hudson's last lover, Marc Christian, one of the pretty boys of West Hollywood, lived not far from Jerry.

For couples in general, gay, straight, and in between, it was a scary time. Jerry and McNabb agreed that he would always use condoms, and their sex was, as McNabb diplomatically put it, "very traditional. We did regular stuff." There were no trips to the sex shops on Santa Monica for adult toys, no visits to the kinky clubs on La Brea, or to the sleek homes in the hills above Hollywood, where couples still gathered to exchange mates. "I'm kind of old-fashioned," she said. "We were both conservative, not kinky at all."

Jerry's naturally intense fear of AIDS and his extreme hygiene came as a relief to McNabb. At the same time, she also thought "psychiatrists would have had a field day" with his compulsive cleanliness.

Being on the road sometimes for six weeks at a stretch, Jerry had desires like any other young man, but as a control freak, he was able—she was confident—to keep them in check. To keep himself occupied when not on stage, he wrote, did yoga, an exercise regime that he could do daily in his hotel room without the use of any equipment.

"I trusted him thoroughly," McNabb said. "We had an agreement that we wouldn't see other people. I always knew where he was. He would get back from the show every night and call me before he went to sleep. He was always faithful."

Despite his intense and loving feelings for McNabb, Jerry still preferred to be on the road, still was happiest when he was performing, still thrilled when he was booked all the time, which was increasingly the case thanks to his aggressive management and booking team, and he never complained about the constant flying, the airport hordes, the getting in and out of taxis. In fact, he thrived on it.

Compulsively packing and unpacking the appropriate wardrobe; getting to and from, and in and out of, airports in the most expedient way possible; getting to and from the clubs he was playing and the hotels where he was staying; properly budgeting his time to allow for travel snafus; developing new material and sharpening the old—these were the priorities of Jerry's very monastic life for more than a decade until he took on *Seinfeld*, essentially his first desk job because it entailed little or no travel, a subject on which Jerry could have written the book.

He had it down to a science.

Meticulous and precise, he always traveled with just two carry-on bags, one for each hand, the weight evenly distributed. He never relied on airline baggage handlers and storage because, if his luggage didn't arrive with his change of dress, he'd be, well, fucked. So he personally packed everything he needed, nothing extra, nothing heavy, and he weighed each bag before he left home. After months of searching, he had found the perfect compact travel clock, the end-all and be-all of bath and shaving kits. He traveled light, like a mountain climber; a reason and a thought behind everything he brought with him.

Even for a gig of several weeks, he brought only one sports coat, one pair of slacks, three shirts and three ties that he could mix and match; three pairs of socks, a pair of shoes or sneakers—white Nikes, always Nikes, with the shoelaces cut by him to a custom length, so if they loosened while he was on an O.J. Simpson–like run through an airport he wouldn't trip over them—and three pairs of boxer shorts, color light blue, which he had discovered while shopping at Neiman-Marcus and bought in bulk, his favorite underwear. His travel methods and wardrobe were as tightly formatted and regimented as his act. On the road, he arranged to have everything cleaned every day so his outfits were always fresh, and when he got home—instead of taking his laundry to the dry cleaners—he amazingly threw everything away and bought new clothes.

When he started seeing McNabb, that changed; she took the clothes he planned to discard, like the expensive sports jackets, which normally would have wound up in the trash. In December 2000, almost a decade after their final break, she still had them in her closet. "I have lots of his old clothes that I still wear—really expensive, well-tailored jackets that he would have just thrown away," she said. "And my dad and brother wear his clothes, too. One time I hit the ceiling because he threw away a brand-new, very expensive wallet. He was mad at himself because he bought brown instead of black, so he went back and bought the identical wallet in black and threw the brown one in a trash can. I said, 'Are you insane? Go give it to a homeless person, but just don't throw it in the trash.' He said, 'Oh, oh, you're right. I didn't think about it.'"

Jerry could now afford to buy and toss thousand-dollar sports jackets and two hundred-dollar wallets with abandon. McNabb recalled a letter he had written to her from Greensboro, North Carolina, where he was appearing. "'Can you believe it? I'm making four grand a week here. I never thought I'd make this much money as a comedian.' And we were going, 'Oh, my God, I can't believe people are paying that much for you to go on stage.' He just thought he had hit the peak."

But Jerry was being overly modest—or secretive—with McNabb regarding his income.

By 1986, as he was becoming a headliner, moving out of clubs and into the larger, more lucrative venues of concerts, the money was rolling in, thousands, even tens of thousands, a week. "This was back when a club owner was guaranteeing him thirty-five thousand dollars a week," Robert Williams noted.

The way the system worked was that the club owners paid a well-known performer like Jerry on the spot, many times in cash, and Jerry would return from the road with a wad of money and McNabb would think, is that a roll of thousand-dollar bills in his pocket, or is he just happy to see me?

Despite his growing income, Jerry was not a financial kind of guy; he didn't read the *Wall Street Journal*, knew little if anything about stocks and bonds, at least then. For a long time, he didn't even have a savings or investment account. "When I met him," McNabb said, "he had forty thousand dollars in a non-interest-bearing checking account. He liked money, he had fun spending it, but money," as she saw it, "wasn't that important to him."

With his newfound riches, Jerry was like a kid in a candy store. For the first time in his life he could buy anything he wanted—the most high-tech music system and television set, the latest, coolest watches—Tag Heuer, Rolex—with all the bells and whistles, gadgets, and buzzers. The Porsches, the first condo, the house in the hills, the apartment on Central Park West, the Hamptons estate—the really big-ticket items were to come.

Like his parents, who hid their spare cash in the freezer of their refrigerator, Jerry had his cache of cash—literally thousands of dollars—stashed away where he thought no one could ever find it, in a unique, secret location in his apartment.

Inside his bedroom closet, buried under the countless pairs of Nikes and Levis 501 button-fly jeans, was an embroidered bag with Hebrew lettering that held his tallis, the prayer shawl he wore at his bar mitzvah. That bag had now become, in effect, the Seinfeld First National Bank, his personal depository, where he hid the wads of moolah he brought home from the road.

When McNabb became aware of the situation, "I was like, 'Jesus Christ! This is crazy.' I'm not a money whiz, but I knew there was a better way than a checking account and hiding the money in the closet in what I called his little Jewish pillow."

She convinced Jerry to open an account at Merrill Lynch, and that was when he first decided he was going to have to invest money and earn some interest. Merrill Lynch wanted a cashier's check for five thousand dollars to open the account, so Jerry and McNabb went into his closet and counted out the right amount of cash and stuffed it into her purse—it was too fat to put in his wallet or pocket. "We got to the office and I'm pulling all these twenty- and fifty- and hundred-dollar bills out of my purse and they're looking at us like we're drug dealers and wouldn't take the money unless we gave them a cashier's check, which we had to get down the street," she recalled.

As part of his act later, Jerry dealt with the issue of finances, asserting that he lost money on every investment he had ever made. "People always tell me, 'You should have your money working for you.' Well, from now on, I've decided I'll do the work, I'm going to let my money relax."

Jerry and George Wallace, who now had his own place nearby on Fountain Avenue but shared the same maid, began to notice that cash and other items were disappearing from their apartments. The thefts stopped after they fired the maid.

Jerry tried as best he could to manage his personal financial affairs, like bill-paying, but his hectic schedule, his long road trips, precluded him from doing a thorough job. For instance, he'd write a check to the gas company for three times the amount he thought he would owe, and hope that when he got home his gas wouldn't be shut off, that there would be a bill for any balance due.

Later, as more and more money rolled in, Jerry decided to hire an accountant.

"I remember at the time warning him about whether this was a good idea, because he knew nothing about this woman," McNabb said. "I'm like, this is scary, giving this person permission to write checks on your account—it's not like he went over everything with a fine-toothed comb. It's like he would charge thirty things a month on his American Express card and if he got the bill and there were thirty-one things, he wouldn't question it. He'd be like, 'Oh, yeah, I probably bought that.' I was concerned about the security regarding the accountant, and I remember expressing that to him and he said, 'I don't have anything to worry about. She's a single mother.' Sure enough she ripped him off."

Jerry was taken to the tune of about fifty thousand dollars, but declined to prosecute, and turned over all of his financial and money management to his sister. The bigger he became, the fewer people he trusted who were not a part of his tight inner circle.

Even though Jerry had developed a national following after his first Carson appearance, it took six years before he started to receive national media exposure even from the supermarket tabloids. The celebrity-news industry as it now exists, 24-7, had not yet exploded. Tabloid television, with screaming celebrity gossip–driven shows like *Extra*, *Hard Copy*, and *Inside Edition*, was just being conceived—*Entertainment Tonight* was the first of the all-celebs-all-the-time genre in the very early 80s. The Internet, which would spawn millions of celebrity Web sites and chat room star gossip, was still a tool for serious academics in the mid-to-late 80s. (By the mid-90s, however, an entire newsgroup called alt.tv.seinfeld, that dealt with every nuance of Jerry's life and show, and scores of Web sites run by *Seinfeld*aholics were in operation on the Web.)

But it wasn't until 1987 that Jerry, despite the reputation he had earned as a top comic, got his first substantive mention in a publication read by millions, albeit in doctors' and dentists' waiting rooms across the land, for the most part.

That's when the then still relatively staid newsweekly *Time* suddenly, somehow, discovered him and the world of stand-up, declaring in its August 24 issue, like a befuddled research scientist announcing the finding of a frightening new life-form, "They're virtually everywhere . . . nurtured in a rapidly growing nationwide network of comedy clubs . . . on TV they get nightly exposure . . . as well as on their own specials for cable networks . . ."

That fall Jerry had his first solo shot on HBO with an hour-long special called *Jerry Seinfeld—Standup Confidential*, which was the brainchild of George Shapiro and Carl Reiner, who helped put the program together. Among other bits, Jerry discussed the power of the Swiss Army knife, which he asserted kept the postage stamp–sized country neutral or at peace. "Back off. I have the toenail-clipper right here."

The show received mixed reviews. A *Chicago Sun-Times* critic, for instance, noted that while the camera "caught Jerry's nuances—especially his extendable chimpanzee-like lower jaw, which juts for emphasis on a joke . . . viewers had to sit through a series of skits that seemed unfinished." On the other hand, Reiner, who started in the comedy game on *Your Show of Shows* with Sid Caesar and Imogene Coca during television's golden age of the early 50s, had a more positive take: "He's not just a comedian; he's a real thinker," Reiner asserted. "When you see him, you're not seeing things you've seen or heard before."

A year later, the same Chicago critic, seeing Jerry perform at the latest club in the Catch a Rising Star chain—a 250-seat room in the Hyatt Regency, in the Windy City suburb of Oak Brook—raved that he was now "a master of irony and observation, fluid and fresh without resorting to abuse or gimmicks."

Noting that Jerry was among the standouts on the comedy club and concert circuit—there were by 1987 at least 260 full-time venues across America, and the number more than doubled by the end of the decade—*Time* described him as the "suburban-preppie cousin" of Jay Leno. "The two are similar in style and subject matter, although Seinfeld has a softer edge." At that point, Leno was on a roll—about to star in a prime-time special for NBC, and be a once-a-week substitute for Carson, which made Jerry, Leno's buddy, a shoo-in for spots on the show.

Meanwhile, Jerry kept plugging away, playing the clubs, telling an inquiring journalist, "I aspire to improving as a stand-up comic. No one has ever been *too* good at it." In addition to his observational material, he'd developed a new comedic concept that he dubbed "human cartooning," which sounded like something from his Scientology days. Human cartooning involved putting thoughts in the heads of animals and inanimate objects, such as a bit he did in which Hopalong Cassidy's horse says to his anxious, apparently lost rider, "Just chill out . . . I know the trail."

The same week as his HBO special aired—it was repeated four times in one month—he was in New York to work the Governor's Comedy Shop in Levittown, a short drive from the Seinfeld home, which the family still owned. The last time he had played the club, doing a routine on Cub Scouts, family friend Marlene Schuss had stood up and made Jerry laugh when she announced that she had been his den mother.

Despite the fact that he was now raking in four and five figures a week, Jerry continued to keep his rent-controlled Upper West Side walk-up, where he stayed when he was in town. Little had changed since he first took the place a decade earlier, sharing it with George Wallace: it was still modestly furnished with a cheap couch, a couple of chairs, a desk, and a giant color poster of Manhattan. The only books were those on comedy, Zen Buddhism, and, as Jerry was a rabid Mets fan, one tome on baseball.

Jerry, steering clear of his complexities and eccentricities, described himself as "a very simple guy" to a *New York Daily News* writer, who had been won over by the nice-guy persona Jerry was perfecting with interviewers. "He wouldn't seem out of place clutching a briefcase and wearing a business suit, racing for the 5:35 out of Penn Station," the scribe wrote in a piece promoting Jerry's cable special.

By mid-1988, Jerry had made some thirty appearances on the Carson and Letterman shows, he'd been one of the stand-ups on HBO's Sixth Annual Young Comedians' Special, was voted America's Best Male Comedy Club performer in a poll of his peers, was honored as the Funniest Male Stand-Up Comic at the American Comedy Club Awards, and was raking in steady five figures a week—$10,000 and up—headlining at high-roller joints, such as the Cascade Showroom, the premier downstairs club of Caesars Tahoe, or the annual Gator Growl at the University of Florida.

At this point in his career, having outgrown the small rooms, he mainly was doing big clubs, concert halls, and amphitheaters, and had signed to do his second HBO special, slated for the air in early 1989.

By his own estimation, he was working stages across America three hundred nights a year, and when he wasn't on the road he was working on new material, which he tried out at open mike nights at either the L.A. Improv, or, when in New York, at Catch A Rising Star and the Comic Strip.

In 1988, *People* magazine, in its first full-scale profile of Jerry, declared, "He has, in short, arrived." The weekly that celebrates celebrities described him as "part visual comedian, part puckish social commentator . . . a tonic to audiences weary of sexual epithets and ethnic jabs."

—30—

L.A. Story

When things were good between Jerry and Susan McNabb they were very good, and when things were bad, they were very, very, very bad. Though they slept together—there were romantic overnights at his place, or at hers, and she'd occasionally meet him on the road in Florida, Chicago, Nashville, and New York when he felt lonely and in need of some TLC and R&R—they never lived together.

"In eight years," she noted, "I never had any of my things at his place."

Apparently taking a moral stance as an excuse for his fear of commitment, Jerry told her, "I don't think living together is right until marriage." Yet, about two-thirds of the way into their relationship, he made a proposal that shocked McNabb.

Would she, he asked over dinner, be interested in being the mother of his baby?

"He was absolutely serious," she stated, still confounded years later. "He wanted to have a child, but didn't want to marry me. He wanted the benefit of a real relationship without the responsibilities. He wanted the love part of it, the fun part of it, but not the commitment part of it. Children are fun and wonderful and it would make sense that he would want one, but he couldn't cross that boundary into commitment land."

McNabb told Jerry no way. He said way. But she argued that she wanted, needed, and deserved more: to wake up every morning with her baby *and* husband. "I told him, 'I don't want to live alone and have Jerry Seinfeld's child and not be married to Jerry Seinfeld, and not live with Jerry Seinfeld.' He just sort of listened, and said, 'Okay.'"

When things were good, he bestowed a number of gifts upon her, nothing big, nothing fancy, which wasn't his style, being the minimalist kind of guy he was, and besides, as McNabb herself acknowledged, "I'm very low-maintenance, and he liked low-maintenance."

While he never gave McNabb diamonds or pearls, keys to a ragtop, or a love nest at the beach, he did dig into his tallis bag and bought her small tokens of his affection for special occasions, like birthdays. "He used to love to take me shopping," she said. "We would go to the mall and he would see a pair of shoes or something that he thought was really cute and he'd buy them for me, like these really cute flat black boots that I just loved and wore for years and years. And he bought me these silver-metallic lace-up shoes—remember we're in the 80s. He never was really into high heels and sexy clothes—he never would think of buying anything frilly like lingerie, but liked me in things that were cute. He wouldn't know Manolo Blahnik from JC Penney."

Over the years, he also bought her a "really pretty" gold bracelet, not engraved; a "really beautiful" silver-and-turquoise Native American necklace; a "not horribly expensive" LaSalle watch on their first Valentine's Day together; and a VCR so she could tape him when he was on TV.

For her thirty-third birthday—just before he blew her off—he guilt-splurged, spending more than five thousand dollars during a shopping spree at the Beverly Center, L.A.'s trendy, ultimate "Valley Girl" mall, on the edge of Beverly Hills.

Knowing that Jerry "never had a free day to do anything but work," McNabb decided to save him time by going to the mall the day before the big event for "my preshopping"—selecting the clothing and shoes that she wanted, and asking the salesgirls at a trendy boutique called Ice and a kicky shoe emporium called Privilege to hold everything for twenty-four hours because "my boyfriend's bringing me here tomorrow night for my birthday and he's going to buy a lot of this stuff." The evening in question was "like a whirlwind" for McNabb and Jerry because, as usual, he was pressed for time, so they wound up running from store to store, where she tried on the things for him.

"I got pretty much whatever I wanted, but my girlfriends were laughing at me because they said I should have gone to Beverly Hills, to extremely expensive designer stores where one outfit costs five thousand dollars, and spent thirty thousand dollars instead of whatever I spent because Jerry had plenty of money," she said. "But that's just not me. I don't know if he appreciated the fact that I wasn't a gold-digger—that I spent fifteen hundred dollars on a leather jacket at the mall rather than ten thousand dollars on a leather coat in Beverly Hills—but I'd like to think he did. I wasn't doing it to save him money. I knew he was making a pile. I just did it that way because that's me.

"When Jerry was my boyfriend, he was the best boyfriend I ever had," she avowed. "That's why I was in love with him. He was a loving, wonderful man when we were together. He took good care of me. He made me feel loved and accepted and cherished."

As with Caryn Trager, who became an accepted part of the Seinfeld family but didn't get the brass ring—marriage—McNabb, too, became a familiar face within the family, which gave her the feeling that one day his phobia would be cured and she would become Mrs. Jerry Seinfeld; after all, she believed, neurotic, commitment-phobic men like Jerry don't usually introduce their women to the family unless they're serious.

When Jerry's mother and a woman friend drove to L.A. from Florida in a new black Mercedes-Benz convertible he had bought for Betty with his growing riches—she was there to attend Jerry's appearance on the *Tonight Show*'s twenty-fourth anniversary special—Susan and Betty Seinfeld met and instantly bonded. It was the first of a couple of get-togethers they had during the course of the relationship. "She was *so* adorable, the cutest lady," McNabb said. "I loved her."

Jerry had even confided in McNabb about his mother's tough childhood, about her life in the orphanages, about how she and Kal were older when they met and "sort of past the age when everybody expected them to marry," that she was a "Syrian Jew and not really as easily accepted by his family because she was a little different, but that she and his father fell in love." And Susan even accompanied Jerry to Long Island for a visit with Jerry's sister and her husband and son.

At the same time, Jerry got to know the McNabbs, a friendly, open, cheerful family, who welcomed him for the first time when he was touring clubs and amphitheaters in Florida. At the time, the McNabbs were leasing a house in Fort Myers, located on Florida's west coast, on the Gulf of Mexico—the city is the location of Mangoes, the winter home of carmaker Henry Ford—as opposed to Jerry's usual stomping grounds, the more Jewish Gold Coast, on the Atlantic, where Betty lived in a plush retirement community.

"We all went to see Jerry's show," Barbara McNabb, Susan's mother, recalled years later. "Jerry got us all in, even my ten-year-old Corey. We enjoyed the show. He used clean language because so many of them seem to think they have to talk dirty to make it funny, and Jerry didn't and I appreciated that. The next day he came to the house and I cooked dinner—he wanted a southern dinner, fried chicken, bread, southern things, 'cause he's not a southern boy and he wanted to try the food."

The McNabb's house was situated on the banks of the serene Caloosahatchee River, where the family could walk out in the yard and fish. "Jerry relaxed a bit and wanted to stand out there and do some fishing," Barbara McNabb said. "So he just stood out there, it seemed like for hours, but caught nothing, and he seemed to enjoy doing nothing, just like on his show. He was casting at one point and he caught the hook on some bushes and broke my little boy's fishing rod, and Jerry felt really bad, and bought Corey a new rod. We absolutely got along fine with him. He was a nice, very polite fellow."

On trips to the West Coast, the McNabbs visited with Jerry, who gave old pairs of his many Levis to Susan's brother, and hung out with Corey when he could.

Susan's mother was certainly aware of her daughter's love for Jerry and the inherent problems in their relationship, but she wasn't one to meddle.

"Susan talked about him all the time," she said. "We kept the long-distance lines going. I liked Jerry just fine. His main problem was commitment, and Susan would once in a while say that they had a falling out, and then, after a while, they were back together. She's a smart girl and if she were in a bad relationship she'd get out. I had confidence in her and trust in her to do the right thing."

McNabb's first competitor for Jerry's affection was not another woman—though that would happen soon—but rather a car.

Next to comedy, Jerry's first love was fast, foreign automobiles, which he fantasized about as a teenager, flipping through the pages of car magazines. His first sports-type car was an inexpensive Fiat. By the late 70s, however, he was advising friends who were car-hunting to consider the very pedestrian but well-engineered Honda Accord—"well made, nice-looking, economical, a good investment," he told his friend LuAnn Kondziela. "I think it would be a great choice, it's worth the money. The only negative is you're supporting the Japanese whale-and-dolphin killing."

Sounding like an op-ed page writer for the *Wall Street Journal*, he added, "The VW Rabbit is a good car too. They're making Volkswagens in Pennsylvania and Volvos in Wisconsin now. National economies are being increasingly blurred into world economies. Volkswagen is as much a part of American life as Chevrolet."

Accords and Rabbits aside, his dream car was the Porsche, the brainchild of a brilliant German automotive engineer named Dr. Ferdinand Porsche, a puppet of Hitler during the Third Reich. Porsche had personally convinced Der Führer of the merits of an innovative aluminum-body car he had designed, which could help the Fatherland gain worldwide supremacy on the Grand Prix circuits. In 1945, Porsche was arrested as a war criminal and imprisoned for a time.

The fearsome racecar he designed with Hitler's stamp of approval was the forebear of Jerry's first Porsche—a 1987 sleek, black 944 Turbo; a 2.5-liter, 217-horsepower rocket that did zero to sixty in about six seconds. With the bucks rolling in, he could afford to pay cash—a cool forty thousand or so big ones. Soon he'd have an airport hangar full of Porsches in Santa Monica, and still later a warehouse full of them in New York.

"He did a lot of obsessing over getting that perfect first Porsche," recalled McNabb, who sat with him as he examined spec sheets. "Every detail was so important to him. The day it arrived at the dealer I was picking up Jerry at LAX, which I always did. After he came off the road, we'd always rush straight home and rip each other's clothing off because he had been out of town for a long time, but this particular time he said, 'Do you mind if we stop at the Porsche dealer on the way home?' and I thought, 'Well, the honeymoon's over. I've suddenly been preempted by a car.'"

The car was beautiful, and the first weekend that it was in Jerry's hands, he and McNabb did what she characterized as the wildest, craziest, nuttiest, weirdest thing in their years together. They got in the Porsche and headed north on Highway 1, the Pacific Coast Highway, with the sparkling ocean on the left, and drove and drove with no destination in mind, no reservations anywhere. They finally stopped in the beautiful central coast town of Monterey, where they got a hotel room and spent the night. "He took two days off to celebrate the arrival of the car, and that was a big deal for him because he performed every night at some club. Doing that was pretty wild for us."

Jerry's love affair with the 944 didn't last very long. Less than a year later he decided to trade up for a faster, sleeker, more sophisticated and far more expensive model—a black Porsche 911.

"The 944 was a perfectly good car," McNabb said. "He loved it so much and had put so much care into ordering it. But at this point his money was increasing so rapidly that suddenly he could buy a *fleet* of Porsches. He bought the second Porsche, and then a third. Soon, every time he picked me up he'd be in a different one. He just loved them. They were his toys. One time we were in a parking lot and a woman tapped the Porsche—one of his Porsches—and he was very upset. I asked him if he got her insurance information and he said no, he'd pay for the damage. He said, 'It doesn't matter. I'm rich.'"

Along with the purchase of the thoroughbred cars came Jerry's first real estate investment. He had spent about a year looking, often with McNabb assisting, and had for a time considered buying a duplex with another comic, Jimmy Brogan, but eventually decided to go it alone. Jerry settled on a quarter-million-dollar condo, unit nine, in a new building at 1222 N. Kings Road, just above Santa Monica Boulevard. With the down payment, the place cost him peanuts—twelve hundred dollars a month—compared to what he was generating in income.

With his move to a large place of his own, McNabb got her hopes up that his fear of commitment would drift away like heavy smog, that she'd move in, and that they'd live happily ever after. "I was," she said, "praying and hoping he would change, that I was going to live there with him."

It had been seven years since Jerry had moved to L.A. and settled in West Hollywood. While there were so many other nicer locations for a bachelor entertainer with big bucks like himself to buy property—near the ocean, in the hills, at the marina—he preferred to remain in what was considered an iffy neighborhood. His new abode was like the last, a short walk south from the frenetic gay scene on Santa Monica Boulevard, and close to George Wallace's place, and the apartment of his new best friend, another black comic from New York named Mario Joyner, who traveled and played with Jerry.

Besides the AIDS epidemic and other sexually transmitted diseases that were ravaging the community, West Hollywood in the late 80s had gotten a reputation as a dangerous, edgy place: street crime was common—violent muggings, burglaries, male prostitution, drug dealing, bar fights, fatal lovers' quarrels, bloody drag-queen cat fights. The area had been considered questionable as a place to live since the mid-70s when gay actor Sal Mineo was brutally murdered, possibly by rough trade, during a robbery outside his home, not far from where Jerry had just bought his place.

Still, Jerry adored the neighborhood, and found his life there idyllic, though if he had had a choice he would have preferred to be living back east, in New York.

His condo was contemporary—sharp angles, clean lines, bright white walls, slate-gray carpeting—with living space on three levels. On the bottom was the master bedroom and a bath; the middle level was the foyer, a guest room, and a bath; and the main level had a large living room, dining area, and the kitchen. Above, there was a small loft and a balcony. The building itself had few if any amenities—the ubiquitous L.A. pool and/or tennis court was missing but there was underground parking. The view? A similar-looking apartment building next door; Jerry's neighbor, the guy on the other side of the balcony, had a monkey, who screeched every so often, but it didn't bother Jerry, who was so rarely there, being on the road most of the time.

In fact, there were still finishing touches that needed to be completed by the builder when Jerry took possession, so he placed McNabb in charge while he went on tour for a month. It was the only time she had actually lived in one of Jerry's places beyond an overnight stay. "While he was away I let the tile man and the other workmen in and I got it all cleaned up and looking beautiful so that when he came home it was all done," she said.

Jerry was particularly upset that one of the shiny black-marble fireplace tiles had a slight chip in it and he demanded that it be fixed. Instead of dealing with the obstreperous workmen, though, McNabb went to the hardware store and bought a tube of Super Glue and made the simple repair herself. Jerry was certain she had performed a miracle because he couldn't see where the break had been. "He thought I was a genius. He couldn't believe I had done it. I never told him it was Super Glue." The other problem McNabb was left to deal with was the entryway tile, which was dark gray to match the carpeting, and Jerry was adamant that he wanted a certain shade of gray grout between the tiles. Speaking to him on the road by phone, McNabb complained that it was hard to communicate the color to the tile man because he didn't speak English. "Just show him the part in your hair," he suggested. It worked because McNabb's hair during her relationship with Jerry had started going prematurely gray, and the color of her part was precisely the shade Jerry wanted for the grout.

Jerry was excited about having his own place, and furnished it sparsely in contemporary bachelor style: a black-leather sectional sofa in the living room, prints on the walls of his Porsche and of the New York Mets; a Scandinavian-style dining table, chairs, and buffet; a state-of-the-art TV and stereo system; and a king-sized bed, for which he spent considerable time researching mattresses to find the perfect one that suited him.

The orderliness, the neatness, the sparseness, Jerry once said about the apartment, kept his mind uncluttered.

Guy friends who visited the place were intrigued and humored by its coldness and sterility. Jay Leno, for one, compared the minimalistic living room to "a hospital with a stereo," and was deliberately sloppy around Jerry, knowing it would drive him up the wall. But Mike Costanza, who probably knew Jerry best, came to visit a few months after Jerry had moved in, and wasn't surprised by the setting, agreeing with Jerry's observation about keeping his mind uncluttered.

"It's all part of his Zen thing," Costanza remarked. "Jerry wanted no clutter, no material things, nothing to get in the way of his comedy and writing. He felt a lot of stuff in the way was a distraction."

For Susan McNabb, Jerry's move didn't change their relationship. He was even more driven and ambitious regarding his career, and even less inclined to commit to her, and that's when things got very, very, very bad.

During one of their conversations he told her, for the first time, "I will never marry someone who's not Jewish." Naturally, she said, "I wasn't very happy to hear that. That was a problem for him. It wasn't for me. I was angry. I said, 'You know, Jerry, that's prejudice and I'm surprised that it would come from you because you're the least prejudiced person I've ever known.' I mean he had no prejudices towards any race, religion, color, whatever. And I said, 'I'm really surprised that you would feel that way because that's judging me based on something I have no control over.' He said, 'You're right, but that's how I feel.' "

She offered to convert to Judaism, but he told her that wasn't the same as *being* Jewish.

A stung and hurt McNabb told her mother about the confrontation.

Years later, Barbara McNabb said that while her family was Christian, "I had no qualms about Susan marrying a Jewish boy. She had dated other Jewish fellows. What Jerry said about his refusal to marry a gentile was more typical of what people of the Jewish faith would say. What Jerry said to Susan was just being upfront with her."

What infuriated Susan McNabb even more about Jerry's anti-shiksa wife philosophy was the fact that he wasn't a religious Jew. Most of his closest and dearest friends—George Wallace, Jay Leno, Mike Costanza, Chris Misiano, among others—*weren't* Jewish; he didn't go to temple, rarely celebrated holidays, though she recalled that he fasted once or twice during Passover, which was probably more of a Zen experience rather than a Judaic tradition. Long before McNabb knew Jerry, Judaism had seemingly meant little to him as evidenced by his feelings about his father's synagogue and his disinterest in his Hebrew and bar mitzvah studies.

However, Jerry, who as a child in elementary school was a target of an anti-Semitic outburst by a playground bully, did have an abiding sense about his faith that manifested itself in curious ways.

In one instance, a very close female friend, who may have even been Jewish, once made a tasteless crack about Jews, the kind of remark that wasn't emblematic of any deep-seated anti-Semitism, but just a crude comment that someone might make without thinking. Rather than deal with it with her, Jerry was so offended that he ended their long and close friendship, an action that those in his circle thought was a bit extreme.

Even more extreme, though, was Jerry's method of keeping an eye on what Jew-haters were up to. He did that by subscribing to a scurrilous, hate-filled newsletter published by the Ku Klux Klan, which looked like it was badly mimeographed in some extreme-right-wing nut's basement.

Susan McNabb, who handled Jerry's mail when he was on the road, discovered it in a stack of his letters, bills, and car magazines. "I remember thinking, What the hell is this all about? This is really weird. Why is he getting this crap?"

When she confronted him, he told her he liked to keep an eye on anti-Semitic groups.

Meanwhile, Jerry's remarks to McNabb regarding his feelings about not marrying a gentile naturally stung deeply, and resulted in another one of their major separations.

It was during the split that Jerry began seeing another woman.

McNabb, who had trusted Jerry implicitly, who had always believed he was faithful to her, was having lunch with a country singer friend and his manager at the Mustache Cafe across the street from the Improv when she eyeballed him and the woman—curly, dark hair; young; trim, fit, built; a Southern California cutie who jogged on the beach and loved to surf at 6 A.M., skied, played volleyball; a tempting and tanned package—leave a restaurant, his arm around her slender waist, and get into his Porsche. "I was devastated," McNabb recalled. "It was extremely painful. Even when we were broken up, I was in love with him. Every time he called, my heart would jump. I was pretty crazy about him."

Chalk the other woman up to Jerry's teeth, those big, Chiclet chompers that are an integral part of his trademark look. Jerry had great teeth and, as McNabb knew well, "went to the dentist regularly." And that's how Jerry discovered his new dish—McNabb's competitor in the battle for Jerry's heart and mind.

As it turned out, Dr. Stephen Buka Effron, Jerry's tooth guy, who had a thriving practice in Santa Monica, was one of those "dentist to the stars" types. Like Jerry, many of the patients who climbed into his chair and read the trades while he made their mouths beautiful were comedians and other show business types. Effron also was the father of a doll named Stacy, who had just graduated from college, with the goal of getting into the publicity game, and her photograph was displayed prominently in a gold frame on her daddy's expensive office desk.

Now a free man since his latest break with McNabb, Jerry took one look and was smitten. Better still, Stacy Effron was Jewish.

Somewhere, somehow, between dental appointments, Jerry and Stacy Effron, about a decade younger, hooked up, and Jerry, who would later gain a national reputation as a cradle robber, was her first big relationship out of college.

The first mention of her appeared in the "Extra" entertainment section of the September 3, 1987, *New York Daily News*, one of the earliest big-city profiles of Jerry. In it, he was quoted as saying that back in L.A. he had an apartment, a black Porsche, and a girlfriend named Stacy. He didn't identify her further. While McNabb was able to maintain her anonymity, Effron's name became a staple of tabloid and *People* magazine stories dealing with Jerry's love life, whether real or imagined.

McNabb received a couple of notices involving her relationship with Jerry, but her name never became public. A year or so after they started seeing each other, Jerry went to San Diego to perform in his first big concert, and a publicity interview was arranged with a reporter for the city's *Union-Tribune* newspaper. While the writer questioned Jerry, his eye was on the stunning woman who was waiting for him nearby. A few days later Jerry awakened McNabb in their hotel room and said, "The whole article's about you!" While he was exaggerating, the writer had indeed mentioned that Jerry, the nerdy kid from Long Island who used to wear braces, well, look at him now, he's squiring a beautiful model. "I loved it," McNabb said later, "because he described me as a splendidly cheekboned brunette."

On another occasion, Howard Stern was interviewing Jerry on his morning radio program and kept pressuring him to talk about his love life. "I heard you're seeing some model," the King of All Media said. Jerry's response was, "I'm seeing an old friend."

Other than those two instances, the media never got wind of Jerry's very hidden relationship with McNabb.

Soon, McNabb came to realize who the other woman was in Jerry's life, and so did Effron, and both women, separately, came to the conclusion that their emotions were being juggled by a very neurotic guy.

"There was a time when Jerry would break up with me and he'd go back to her, and then he'd break up with her and come back to me," McNabb stated. "It was like back and forth."

Like McNabb, Effron, who wanted marriage and children just like her nemesis, realized Jerry had commitment problems. "In a lot of our conversations about marriage and kids," she would say later, "he'd say, 'Ugh, that's for normal people, and I'm not normal.' "

Sometimes Jerry mean-spiritedly threw Effron in McNabb's face, too, which infuriated her.

Jerry had taken Effron on a skiing weekend where they hooked up with the instrumentalist and composer, Kenny G, and his then girlfriend and future wife, Lyndie. Ironically, both were also friends with McNabb, and knew Jerry and McNabb were romantically involved. Nevertheless, the two couples spent the weekend hanging out.

McNabb found out and hit the ceiling.

"I was really, really jealous because Jerry and I didn't take vacations together, and he didn't take vacations much, anyway, and when he did he went off to Hawaii with George Wallace, not me," she declared. "On that skiing trip, he was with her. He wasn't with me. I was fuming about it because Kenny G and Lyndie were my friends, and the thought of him having fun with my friends and some girl really pissed me off."

Not long after, McNabb went to a Grammy Awards party and ran into Kenny G and Lyndie. "I remember Lyndie saying to me, 'Poor girl.' Because the whole weekend, to Jerry, they were like, 'Isn't Susan a great girl? Have you seen Susan lately? She looks great!' They were talking about me the whole time." Suddenly, as McNabb was hearing the inside story about the notorious weekend of schussing, Jerry made an appearance.

"We were all sort of standing in a circle talking and Lyndie mentioned something about the skiing trip and, of course, I immediately became uncomfortable," McNabb said. "And I remember Jerry saying, 'Yeah, that's the most fun I've ever had on a vacation.' Ouch! Thanks! He said that in front of my friends. It was a very hurtful thing to say, just thoughtless."

Jerry's ping-pong game with the emotions of McNabb and Effron continued into the early 90s when Jerry ended it with McNabb. At the same time, Effron, described as an independent kind of girl who does her own thing, gave Jerry the heave-ho. As she told a confidant later, "I may have been young, but I wasn't foolish. I knew what was going on, that there was another woman. I knew Susan McNabb's name. A woman can tell those things."

Several times during their rocky relationship, McNabb suggested that Jerry see a shrink, and she proposed it again as their relationship began to crash and burn for the last time right after her thirty-third birthday. Jerry's response was, "I've done it before. I don't really see that I need it now."

He was emphatic—he didn't want a relationship. And then in an ironic twist for a man who had dedicated his life to bringing joy to people by making them laugh, he declared, "I just don't want the responsibility for someone else's happiness."

McNabb said she didn't understand why Jerry couldn't have both a loving relationship and his career. "I said, 'Look at Bill Cosby, look at Bob Newhart. They have their own TV shows and they have wives and families. Why can't it be the same with you?' And he said, 'Well, I'm not as talented as they are.' I said, 'Of course you are.' But he said he didn't think so. He thought he had to work harder than everybody else."

Finally, she blurted out the question that she knew would end it all.

"We were talking on the phone and I said, 'Do you think you and I will ever get married?' And he said, 'No. I'm not going to marry you.'

"Oh, God, I lost it and started to cry. He said, 'I'm sorry. I don't want to hurt you.' "

After almost eight years, McNabb felt as if she had the word "fool" stamped on her forehead.

"I was in love with him, totally in love with him," she declared. "He was the man I wanted to marry for most of my adult life—and I thought I would marry him. Obviously, I wouldn't have hung around for all those years if I didn't think it was going to go anywhere.

"I was really devastated at that point and just basically tried hard to stay away from him, which I did. We did stay friends. I just didn't go to his house. I avoided spending time with him. For years I wouldn't see him. It was very hard and painful and then eventually enough time passed that we could have dinner together once in a while and talk on the phone.

"For years I didn't watch the show because it was too painful to watch. Everything about it bothered me after we broke up. I would watch half of one episode and I would see half a dozen little elements of things that we had said, or things that we had done, and I just found it very unnerving. It was so close to home. It wasn't that I objected to him using the material, it wasn't like I felt he had stolen anything from me, or owed me anything. It was just plain *weird*.

"Years later when I finally could watch it I thought, Elaine's perfect—he's written the perfect woman for himself: she's fun, she's one of the gang, he can hang out with her, and yet he has absolutely no responsibility for her."

Jerry, meanwhile, compared his fear of commitment to that suffered by his childhood comic-book hero, Superman, who as Clark Kent could never get it together with Lois Lane. "Why," Jerry once asked, half in jest, "is commitment the kryptonite of men, super and otherwise? If Superman isn't able to grow up and lead some sort of normal, adult life, what chance do I have?"

—31—

Birth of *Seinfeld*

One evening in the early 70s comic Tom Dreesen, relatively new to the stand-up game, is in New York, and naturally decides to explore the local comedy scene. New York comics fascinate him. Unlike those from Chicago, where he's from, or even L.A., where he's been working, they have a totally different approach to comedy: they are, he feels, edgier, more caustic.

At the Improv, he's knocked out by the work of a twisted stand-up, Larry David, a Sheepshead Bay, Brooklyn, boy who's been working the clubs for a couple of years. Completing his creative, imaginative set, David hurries by Dreesen, who stops him and says, "Gee, you're really funny. I really enjoy your material.' "

David who wasn't used to that kind of friendliness, or friendliness of any sort, pauses for a moment, says "Thanks," and moves on.

The next day Dreesen's at Catch a Rising Star and David's on, doing totally different material. "So he comes off the stage and I'm standing at the bottom and I say, 'Hey, you were really, really funny tonight. Stuff I didn't see last night. Absolutely great.'"

Once again David gives Dreesen the look as if he's thinking, is this guy queer?

The next night, Dreesen's coming out of the Stage Deli and he bumps into Larry and says, "Oh. Hey. Hi. How are you doing? I've got one thing to tell you, you're stuff is really funny. I really enjoy it."

David stops, peers darkly at Dreesen, and then hits him with the line: "What's with all the nice guy shit?"

And there you have him.

This Larry David anecdote has never been told before, unlike scores of others, some apocryphal, about how wacko and neurotic he is, the innumerable stories about how he would walk out on stage, check out the audience to see if they were up to *his* standards and, if not, say "Fuck you!" and walk off, leaving them perplexed, embarrassed, and nervous about this skinny, wired, edgy, angry guy.

Not so the in-crowd, not the Larry David cognoscenti, not the other hip stand-ups who are sitting at the bar *plotzing*, tears streaming down their faces. Because that was David: brilliant at times, off the wall at times, a pure comic with no gimmicks, who for years struggled along, not taking any shit, not kissing any ass, thoroughly devoted to his craft, a genius monologist and gifted writer known as the ultimate comic's comic, the one guy whom all the other comedians would run into the room to see. And totally insecure about everything from his nose to his weight to his hair—or lack thereof—which he worked into his act. To wit: "If Nazi war criminal Josef Mengele gave me a compliment, we could've been friends. 'Larry, your hair looks very good today.' 'Oh, really? Thank you, Dr. Mengele!'"

In his private life, he was equally strange and eccentric. He once threw a New Year's Eve party, and five months later his pal, Glenn Hirsch, went to visit David at his apartment and he still hadn't cleaned up. On another occasion, Hirsch recalled, "We were walking down the street and Larry stepped into a pile of dog shit. He stopped for a minute, took off the shoe that had been soiled, and threw it away, and continued walking with one shoe. He didn't think about cleaning it off."

David had graduated from Sheepshead Bay High School, and later said he had had "a wonderful childhood, which is tough, because it's hard to adjust to a miserable adulthood." After attending the University of Maryland as a history major, he entered the stand-up world in 1974, when Jerry was still a junior at Queens College. Like Jerry, David had had a number of odd jobs—cab driver, bra salesman, chauffeur.

The first time Jerry saw David work was in early 1977, at the Improv, and he wasn't impressed. In fact, according to Mike Costanza, who was with Jerry, "He thought, without question, Larry David was terrible. Jerry said to me, 'How the fuck do guys like him get up on stage and do this?' He said, 'I don't think I have too much to worry about after seeing this guy.' That night Larry David was being booed—*booed*—off the stage, and Jerry and I were goofin' on him. They didn't know each other at that point."

Bill Ervolino, once the comedy critic for the *New York Post*, thought David's material was good but needed work. "I remember sitting with him at the old Caroline's comedy club on Eighth Avenue where he was doing sketches and he was just like, you know, sitting at the bar holding his head, miserable, saying 'nothing's working for me.' He had gotten burned so many times that he was really kind of disgusted with everybody, at least that's the impression he gave me."

When Jerry began working, their paths crossed, and they became acquaintances, but no great friendship evolved. As David Sayh, who knew them both, observed, "I'm really surprised they ended up together, working on *Seinfeld* together, because they weren't in the same clique, didn't come up together. Larry was more edgy. Jerry's smart, but Jerry never came over as an intellectual, whereas Larry does."

Unlike Jerry who loved the road, David despised it and confined his stand-up mostly to New York. "I didn't like being away from my house," he asserted later, half in jest. "When I cried at night, I wasn't in my own bed."

Separately, Jerry and David, who was six years Jerry's senior, made the move west in 1980—Jerry to get his big break, David to take a writing and acting job with a short-lived late-night comedy show—a bad knock-off of *Saturday Night Live*—called *Fridays*, which also featured a zany actor named Michael Richards; that's where the future cocreator of *Seinfeld* and the future Cosmo Kramer met and bonded. *Fridays* aired from 1980 to 1982, before ABC buried it.

At that point, David returned to New York to take a writing job on *SNL* for the 1982–83 season, which was a disaster because he was unable to get more than one of his sketches on the air. Most were cut at dress rehearsals, and David came to despise the show's executive producer, Dick Ebersol.

One Saturday night, just before air time, David approached Ebersol and said, "That's it! I quit! The fuckin' show stinks! I've had it! I'm gone! I'm out of here!" But on his way home he confronted the fact that he'd be losing about fifty thousand dollars in salary, on which, he believed, he could live for another two years. That Monday morning, acting as if nothing had happened, he showed up at the office, and nobody, including Ebersol, said a word, and he worked for the remainder of the season. Later, for *Seinfeld*, he wrote, based on that experience, an episode involving George Costanza, one among many that mirrored his own real-life dramas and friends.

To add to his problems with Ebersol, no one on the show socialized with David, not even asking him to join them for a sandwich. "It was the only place I ever worked," he revealed later, "where I really, truly did not make a friend. I couldn't believe it." However, none of that was surprising to people who knew him well. They asserted that David had a "cynical bent" and didn't really like people, "but put up with them," which they blamed on his "insecurity." For example, when he was doing stand-up, if one person in the audience didn't laugh, David would lose the entire audience to get that one guy who pissed him off. He'd get into fights and screaming matches on stage, cursing at that one audience member, and then storm off stage.

As stand-up Nancy Parker, who had worked the clubs with both Jerry and David, remarked, "When you saw Larry on stage you said, this guy's never gonna open for Cher. Too edgy, not comfortable up there. Not really dirty, but a real outlaw. If the audience didn't like him he'd twirl his hair, say 'fuck you,' and walk off stage. Jerry couldn't deal with hecklers, either; he might walk off stage, but he'd never say 'fuck you.'"

After his horrible *SNL* experience, David had small roles in several films. He appeared in Henry Jaglom's *Can She Bake a Cherry Pie?* and *Second Thoughts*, both in 1983. In 1987, he played a "communist neighbor" in Woody Allen's *Radio Days*, and in 1989, a theater manager in the Allen-directed segment of *New York Stories*, an anthology.

He was doing stand-up again in New York, living in a high-rise in Manhattan's Hell's Kitchen area, at Tenth Avenue and West Forty-third Street, across the hall from a zany character named Kenny Kramer—the real-life Kramer—when Jerry began talks about doing a new sitcom.

It all started in early 1988, when Castle Rock Entertainment approached Jerry about auditioning for the lead in a pilot called *Past Imperfect*, a show that had been ordered by ABC but was on a last-minute hold because of a casting problem involving the current lead who wasn't up to snuff. Brimming with creative, comedic entrepreneurs, Castle Rock was the brainchild of Rob Reiner, son of Carl Reiner, cousin of George Shapiro, Jerry's manager. Others who formed the Castle Rock brain trust were from the world of successful sitcoms, writers and producers who had been directly or indirectly involved with "king" Lear—Norman Lear—and such ratings busters and breakthrough shows as *All in the Family* and *Maude*.

Still suspect of TV sitcoms after his *Benson* imbroglio, Jerry was hesitant, but he respected the credentials and goals of the Castle Rock people, and agreed to meet. And then agreed to take the part. In the end, however, ABC brass was wary of Jerry, despite his reputation as one of America's top stand-ups, because he didn't have a sitcom track record, and there was a mutual parting of the ways.

Not long after, NBC got wind of Jerry's TV availability. Rick Ludwin, a senior vice president in charge of late night, variety, and specials, whose mandate was to find the next Jay Leno, picked up the phone and called Shapiro to set up a meeting with Jerry. At the meeting, NBC made an offer Jerry couldn't refuse, which essentially was: whatever you want to do, we'll go along with it, within reason. Jerry couldn't go it alone and decided he needed a collaborator to develop a new-style sitcom in which he would star.

Jerry and Larry David were working Catch in New York shortly after Jerry's NBC meeting, and Jerry mentioned to David the opportunity that was being handed to him for a prime-time sitcom. After their sets, they went to a nearby deli. While wandering the aisles, kibitzing and joking about the foods, David suddenly stopped in his tracks. "This," he told Jerry, "is what the show should be about . . . comedians have a lot of funny conversations in their off-hours."

Not long after, the two—who had little in common—met at a Ninth Avenue eatery, the Westway Diner, not far from David's apartment and within walking distance of the Improv. Over coffee they conceived what would become the most successful sitcom of the 90s; a show about nothing and also about how comics use the situations around them to develop their material; the working title *The Seinfeld Chronicles*, with Jerry playing himself. "I would be the stand-up going around the city collecting material for my act," he said. "Then we'd show me doing the material in a club setting."

David was happily working as a comic in New York when Jerry initially approached him. "I was having fun doing that, and then everything took a right turn," he said. "It wasn't my ambition to go into television. The whole thing was a fluke."

Next stop, the office of NBC Entertainment president Warren Littlefield, where Jerry and David pitched the concept: the show would be about "conversation"—*almost* the same as nothing—and would not be "heavy" on story. To put it bluntly, Littlefield thought these two had lost it, and they were essentially laughed out of the building. But before they hit the street, David, the John McEnroe of stand-ups, lost it in front of Littlefield, acknowledging later that he conducted himself "in my usual unprofessional manner."

In the end, the idea they presented at their dog-and-pony show for Littlefield was similar to what eventually aired. While Littlefield was miffed, Rick Ludwin became Jerry's rabbi at the network and agreed to put up the money for a pilot. Jerry and David needed a producer—one who would be a partner rather than a boss. Shapiro called his friend Glenn Pandick, Castle Rock's honcho, who was hooked on the first draft of the first script. "It was so fabulous," he said. "Just everyday stuff, so recognizable." He gave Jerry the green light and a free hand. Castle Rock put up the money, ran interference with the network, took responsibility for the show's look—when Jerry wanted to shoot on film instead of videotape, a far more expensive and complex process, Castle Rock gave a thumbs-up.

The Seinfeld Chronicles pilot was shot in Los Angeles in April 1989. The next month the network conducted a viewer survey based on the pilot's airing to selected homes via cable. The results of the survey were, to say the least, ugly. All of the characters—Jerry, Michael Richards, who played Kramer, and George, played by Jason Alexander, were ripped apart by the survey participants: ". . . George is not a forceful character"; "Jerry is dense and indecisive . . ."; "Why are they interrupting the standup for these stupid stories?" Like that.

Kramer scored badly, George did the worst. The bottom line of the network's research was: "No segment of the audience was eager to watch the show again."

Littlefield felt vindicated. Brandon Tartikoff thought the show was too New York*y*, too Jewish, and wanted it shelved. And it was—until the middle of the summer of 1989, the day after Independence Day, when most TV viewers were at the beach, or at barbeques, or just about anywhere else but in front of the tube, and that's when Jerry and his gang made their debut.

Naturally, the first prime-time airing of *The Seinfeld Chronicles* was a disaster, ratings-wise. Littlefield was prepared to kill the project completely when Ludwin stepped in and agreed to finance four additional half-hours—twenty-two minutes per episode—of the show, he was that committed to the project. "How the hell," Jerry recalled asking David, "are we going to do four more of these things?"

The only caveat from the network: add a woman. They cast Julia Louis-Dreyfus. The rest, as they say, is television history.

There were still a number of obstacles to conquer, but Jerry Seinfeld was on his way to unimagined fame and fortune. The funny kid from Massapequa, as Caryn Trager observed, had won the lottery—big-time.

—32—

Press War

MAN: Waiter, there's a fly in my soup.

WAITER: I'll put it in Lost and Found, sir. If nobody claims it in seven days, it's yours.

Jerry grew up on cornball jokes like that, courtesy of Kal Seinfeld. Now, in the midst of the sensitive negotiations between Jerry's people and NBC to get *Seinfeld* up and running, in the soup appeared a fly that could have put the kibosh on the whole deal, or at least that's what was feared.

The fly, in this case, was more of a gadfly, and far more annoying and hazardous to Jerry's sense of well-being: an erudite, sophisticated, highly respected critic named Lawrence Christon who worked for the *Los Angeles Times*, and who didn't think much of Jerry's work. Unfortunately for the *Seinfeld* team and their supporters at NBC, he put his feelings into print at just the wrong time, when the network was about to finalize everything. But, as Jerry knew well after fourteen years in the business, shit happens.

After the comedy boom that Jerry rode in on beginning in the late 70s, a number of major metropolitan newspapers like the *Los Angeles Times* assigned critics to cover the genre—there were so many clubs, so many new faces. By the late 80s, Christon was a pioneer, the first full-time comedy critic. The editors at the *Times* saw L.A. as the comedy capital of the country; Christon's reviews were read by network decision- and tastemakers. So he had become a force to be reckoned with, a power broker who could make or break a comedy career; the first to take the tools of drama criticism and apply them to comedy.

Christon had high standards and soon found that most of the stand-ups he was covering didn't live up to those standards; weren't very good—no one on the par with, say, a Steve Martin who, he felt, found the secret to performing in screaming, foot-stomping, dope-smoking rock venues with his "King Tut" routine, with his "Well, EXCUUSSE me" retort, a comic Christon considered "a great showman." But the others, he felt, were so-so, and that group included Jerry Seinfeld.

"Seinfeld was amusing, but I wasn't ever very impressed," he asserted years later, after *Seinfeld* had gone into platinum syndication. He was speaking of Jerry as a stand-up, who, he thought, was "very low key—his frame of reference very small. He was very clever but within a very small area. To me, he was always bland, like the Bob Costas of comedy—pleasant and agreeable, but not terribly interesting."

To make his point, Christon remembered the time he'd taken Carol Channing, who was working on a project about comedy, to see Jerry perform at the L.A. Improv. As he watched Jerry work, he happened to look over at the blonde with the big mouth, pop eyes, and zany voice, and was shocked to find her snoring away. "Jerry Seinfeld put her to sleep."

So the last person in the world Jerry needed to review him when major decisions were being made at the Peacock Network about the future of *Seinfeld* was Lawrence Christon, comedy critic extraordinaire.

The venue for Christon's review was the enormous Bren Events Center at the University of California at Irvine, where more than seventeen hundred fans, most of them students, had come out to see Jerry in the middle of a three-day holiday weekend.

Also present at Irvine to write about Jerry's concert was *Playboy* magazine, represented by journalist Stephen Randall, there to do the T&A-

centerfold monthly's first profile of Jerry because of the "buzz around him," because of the sitcom that was starting, because he'd recently sold out New York's Town Hall, because he'd been the only legit guest for David Letterman's seventh-anniversary show.

As Randall, and therefore *Playboy*, saw him, Jerry was "on the cusp of being the Next Big Deal in comedy."

Christon had been hesitant about covering the concert because of his negative feelings about Jerry and probably would have passed on it, had it not been for the publicity material that landed on his desk. It billed Jerry as "America's Most Imitated Comedian . . . The Keenly Perceptive King of Observational Humor." When Christon read the release he thought, "That's an awful lot of hubris, so I went to see him."

"Here we are in the gym," said Jerry, opening his show. "We're in the gym and we're going to pretend it's a night club. We won't notice the scoreboard, we'll just pretend it's a little intimate cabaret somewhere on campus." He turned the fact that his West Hollywood condo neighbor had a pet monkey into material. "If you need a pet that roller-skates and smokes cigars, it's time to think about a family. You're so close." He riffed about TV commercials: "They say Tide cleans bloodstains. I say if you've got a T-shirt with bloodstains, then maybe laundry isn't your biggest problem."

Standard Seinfeld fare.

"Nothing shocking here," Randall would later write. "Not even a stray four-letter word, nor anything that could conceivably go over the head of anyone with even a moderate television education. That's one of the reasons the turnout is so large—Seinfeld has an appeal that crosses generational boundaries. It's the type of crowd you'd expect to see at a baseball game or a family reunion."

Which is precisely why the people at Castle Rock, and the suits like Rick Ludwin at NBC, knew *Seinfeld* had a shot at hitting big. TV had that same audience.

Because of the long lead time in the magazine world, Randall's essentially positive profile didn't run for months, not until August 1990. But a couple of days after the concert at Irvine, Lawrence Christon's review appeared in the *Times* under the headline "LAUGHING ON EMPTY."

> Jerry Seinfeld . . . bills himself with all due modesty as "America's most imitated comedian" as well as "the king of observational comedy." News to us . . . Seinfeld is a pleasant, effortless performer who works clean (no small feat in this time of howling, offal-heaving monkey) . . . He doesn't traffic in the mindless hate—or self-hate—that characterizes so many other stand-ups . . . He's expressive. He's clear. And he's completely empty . . . There isn't a single portion of his act that isn't funny—amusing might be a better word—but ten minutes or so into it, you begin wondering what this is all about, when is he going to say something or at least come up with something piquant. Seinfeld has no attention span . . . He has no frame of reference. Yeats had a line about "paying homage to unevent." Seinfeld pays homage to insignificance, and he does it impeccably.

The review hit Jerry particularly hard, not only because it might have an impact over at NBC, but because he felt it was mean-spirited. He wasn't used to that kind of criticism. Most mainstream articles and reviews, the kind found in *US* and *People*, raved about him, and he had the charm to veer writers away from writing anything negative or controversial about him.

The day the review hit, Jerry was hanging out at the Improv, as was Stephen Randall who, as a *Playboy* writer, had been given virtual carte blanche access to Jerry; despite his growing fame, he still wasn't *too* big to limit journalists who wanted to see him in action, so Randall had been a fly on the wall for weeks, touring with Jerry in Arizona, spending time with him at his condo, dining with him, seeing him in relatively intimate moments. And one such moment was when Jerry and his people were in a state of high dudgeon over Christon's observations.

Sitting at a table at the Improv, drinking Perrier, Jerry read and reread the review. "It's one guy, and that's what he thought," he told Randall. "There were eighteen hundred people there—what did they think? The audience makes the judgment and you can go right to them." Jerry was particularly upset by Christon's opening because, he claimed, neither he nor Shapiro ever used that language—*most imitated . . . The King*. He said it made him sound like "Siegfried and Roy . . . like a sideshow byline." He blamed the hyped billing on the school, and felt he should have said something about the language to the audience at some point, and not let it go.

Jerry was so upset that he did something entertainers rarely do because it usually opens up a bigger can of worms: he put in a call to Christon to explain what had happened. Christon was out and Jerry left a message.

As Jerry was having conniptions, his publicist, Paul Shefrin, was in a state of even higher anxiety, looking ill, pacing, worrying about his future, which he had every right to do since the publicist is the one who usually takes the blame when things like this go wrong, and gets the heave-ho. Jerry had hired Shefrin, and trusted him, because he handled a number of other comics in Jerry's circle—Garry Shandling, Larry Miller, Bill Maher—"a number of people Jerry was close with," Shefrin noted years later. "I had a lot of people knocking on the door, one guy would lead to another."

Shapiro, meanwhile, had gone into damage-control mode. Knowing a review like this could cause major problems for Jerry, he ordered that a stack of positive Seinfeld reviews be Xeroxed and hand-delivered to Littlefield, Ludwin, *et al.* at NBC headquarters. Too much was at stake. If Christon's diatribe was having any impact on the network suits' decision-making processes regarding Jerry's show, Shapiro wanted to offset it, any way he could, but without making it look like he was worried, which he was. Sending the positive reviews seemed subtle enough.

Jerry had more to think about, but he couldn't shake Christon from his mind. He was making another *Tonight Show* appearance and when he arrived at the Burbank studios, everyone was on edge about the review—Jerry's people and the *Tonight Show* people, among others—from producer Peter Lasally to Carl Reiner to Jay Leno, who telephoned Jerry and told him not to worry. Jerry wasn't up to snuff that night, so preoccupied was he with the Christon affair, but he introduced a new joke that seemed apropos of his mood. It had to do with happiness in life, or the lack thereof: "Nothing in life is fun for the whole family," he said. "There are no massage parlors with ice cream and free jewelry."

After the taping, Jerry had dinner with Shapiro, then went home to find a message on his machine from Christon. Stephen Randall was still buzzing around and, after listening to the message, Jerry suddenly played down the whole issue to the reporter. Now he described Christon "as gracious and charming a guy as you can find," and said that the critic "was going to print a retraction."

However, the paper never printed a "retraction" and never intended to print one; a correction, yes, but a retraction, no way. There's a big difference.

Years later, Christon acknowledged that he had, indeed, left a message for Jerry. "After it got back to me somehow that it wasn't Jerry who had done the billing at Irvine, that it was the publicist at Irvine who'd done it, I left him a message saying I'd see *if* we could run a *correction* about the billing issue. I felt I owed it to Jerry, who never called me back. But my editor had a heavy deadline, so the correction never ran. The correction would have only noted that Jerry wasn't responsible for the billing. My review of Jerry's performance and my thoughts about him, though, would have remained the same. It would not have been retracted."

But Jerry's people told the NBC people that Christon intended to print a retraction, which gave the impression he was going to take back everything bad he had said about Jerry. A week or so later, however, the *Times* did publish a letter from the Irvine official who took responsibility for the disputed wording in the promotional material. At the same time, though, the paper supported its critic by publishing a letter from a reader who praised Christon for exposing what the letter writer called "the most overrated comedian in America."

Meanwhile, as Paul Shefrin suspected, Jerry fired him, blaming him for not being on top of how the publicity was being handled at Irvine, and for the whole Christon mess. He was the first, but not the last, flack Jerry would dump over a brouhaha involving something that had been written about him.

Jerry also came after Stephen Randall, at *Playboy*. In his piece, Randall wrote about a conversation with Jerry—most of their interviews were on tape, some were in the form of handwritten notes—who talked about how he enjoyed the road and didn't mind flying so often; by the end of the 80s, he was logging three hundred thousand frequent-flyer miles in an average year. Randall quoted Jerry as saying, "I'm just a lucky fucker. I like what I do."

The word "fucker" is what Jerry had a problem with. He didn't want to be quoted using the F word in a high-profile magazine like *Playboy*, which might sully his Mr. Clean image, so he staunchly denied saying it. "He called me to complain," Randall said later. "He said I misquoted him, that I used the word 'fucker' when he said he did not use it. I don't know if I went back to the transcript and triple checked it, but I wouldn't have used it if it hadn't been there."

In the end, the Christon affair had no apparent negative impact on the negotiations with NBC, and if it did, the problems were smoothed over. Still later, Jerry made hay of the Christon affair, by tweaking and fictionalizing it a bit. Part of a *Seinfeld* episode involved a female heckler, a coworker of Elaine's, who destroys Jerry's pacing during a club performance while an important critic named "Lawrence Christon" is in the audience reviewing his act. Jerry then convinces Christon, who is not actually in the episode, to come back a second night to review him.

As with Jerry's stand-up act, *Seinfeld* didn't appeal to Christon, either. "In ten years people are going to say, 'What the hell was that about?' It wasn't like *Mary Tyler Moore*. It was darker. *Seinfeld* spoke to the young urban disaffected, people who weren't very pleasant, people screwing each other over all the time. I never thought he would become the most famous comic of the 90s."

For a long time the Christon review lingered with Jerry.

". . . There are things in that review that I aspire to," he said later. "Christon talked about revealing deeper truths and having social impact, and, yeah, I would love to say great things. Who wouldn't? I mean, these are my thoughts out there. So if they're shallow, I'm shallow. But I don't think they're shallow. They may be light, but I don't think they're shallow.

"I find that life is interesting on every level. I read a thing that Sam Kinison said the other day, about how it's tough to live his life so that he has something to talk about. The way I live, it's not a raw life. I have ambitions and disappointments, but I don't want to live a ragged, desperate existence just so I can talk about that. I don't want to necessarily make heavier observations, just better ones. I just want to be good."

—33—

Fame Is Free

Despite Lawrence Christon's criticism and Warren Littlefield's debunking, NBC was firmly committed to the show, but the all-important ratings would be closely watched.

In late spring of 1990 the first four episodes debuted, starring Jerry as Jerry, Michael Richards as Cosmo Kramer, and Jason Alexander, who had won a Tony Award for his role in *Jerome Robbins's Broadway*, as George Costanza. Under strict orders from NBC brass, a female *SNL* alumnus, Julia Louis-Dreyfus, was added to the cast as Elaine Benes, and the pilot title, *The Seinfeld Chronicles,* became *Seinfeld.*

Unlike the intrigues, feuds, diva behavior, flaring tempers, back-stabbing, sex scandals, drug and alcohol abuse, and whatever other nastiness that occurs behind-the-scenes on some TV sit-coms that result in supermarket tabloid headlines, the *Seinfeld* set would be a rarity in that it was sleaze-free, all-business, and mostly lots of fun for the cast and crew. If there were any differences at all, they were creative, and were mainly between Jerry and Larry David, who were both working to put out the best possible product.

Aired on May 31, the first episode was called "The Stake Out," in which Jerry goes to a party for a friend of Elaine's and meets a woman to whom he's attracted. On the advice of Jerry's father, Morty Seinfeld, played by the veteran character actor Phil Bruns, Jerry and George stake out the lobby of the building where the woman works.

Bruns—who found himself the target of Internal Revenue Service investigations between the mid-60s and the early 90s after he played Lyndon Baines Johnson in a biting satirical play called *McBird*—was summoned to audition at the last minute.

"My agent told me, 'They're looking for a father type; they're not excited with the man they have.'" Bruns went to the audition—"Jerry laughed like crazy"—and he got the part, or at least thought he did. The episode was shot before a live audience and "they didn't seem to catch the first scene," Bruns said. "Jerry and the rest of the cast didn't get many laughs and Jerry came off the set saying, 'Well, that's it. We're dead. We're down the tubes. It's over. We shouldn't have used this studio. It's bad luck.' Jerry was very angry and stormed into his dressing room. He was certainly upset with the audience response."

The next scene, however, went perfectly, with the audience getting it. Jerry's dark mood suddenly brightened. "He and everyone was extremely congratulatory and happy," Bruns said. "I thought, 'Well, golly, I might have a series going here.'"

But Bruns also sensed that Jerry wasn't pleased about something, or with someone. After most of the cast had left the studio, Jerry and Larry David told Bruns they wanted to reshoot part of a scene where Bruns is in bed with Jerry's mother, Helen Seinfeld, played by the actress, Liz Sheridan. As a veteran of TV—Bruns had played thousands of roles and had been Jackie Gleason's understudy—he knew all the pitfalls and the signs of danger ahead. After a couple of takes, Bruns noticed that the camera wasn't on him but on Jerry, and in a monitor he saw a dour expression on Jerry's face.

"I said to myself, 'Ouch, I may have gotten *too* many laughs.'"

As Bruns left the studio, he stopped to tell Jerry, "It was really fun. If this thing goes I would love sometime to meet your dad because he sounds a little bit like my dad." Jerry's response was "Yeah, yeah, yeah." A cold brush-off if there ever was one.

Bruns was never called back and never given a reason why. "No one ever told me anything." In subsequent episodes, he was replaced by another fine actor, Bernie Martin. Bruns figured he had angered or offended Jerry in some way. Could it have been his remark about wanting to meet his father? That Jerry was upset that Bruns didn't know Kal Seinfeld had passed away? It's doubtful. After all, his late father was being portrayed on the show, and was joked about often in Jerry's stand-up act, as if he were still alive. At the time, some writers who profiled Jerry thought his use of his father was somewhat curious since he was dead, and some actually believed he was alive and kicking in Florida until they checked the clips.

No, Bruns came up with two other possibilities for why he was booted. The first was that Jerry thought Bruns was getting too many laughs. "Your job as a character actor is to conceal the fact that the star can't act," he said. "The first rule is—don't get more laughs than the star."

His second thought was that religion played a part in the decision.

"If I ever run into Jerry again, the first question I intend to ask him—jokingly, of course—is: 'Why? What happened? I'm not Jewish? Is that the reason?' It's hard not to think that, although Bernie Martin is certainly not Jewish, but probably looks a little more Jewish than I do. Someday I'd like to get a straight answer."

While Bruns saw Jerry breathing fire on the set during the filming of the first episode, Castle Rock honcho Andy Scheinman said it could never happen. "Jerry never loses his temper—he doesn't have a temper to lose. I've never seen anyone more relaxed in a stressful work atmosphere."

But others discovered that normally mild-mannered Jerry possessed, as one veteran observer put it, "a wicked snideness that is much sharper than the comparatively harmless sarcasm associated with his act. This may come as a shock to fans who embrace the public image of puppy-dog innocence. Yes, there is venom bubbling beneath the nice-guy surface, ready to spew if not held in check."

Phil Bruns saw some of that in the filming of the very first *Seinfeld* episode.

Jerry himself acknowledges that he's not a people person. "The average person that I meet," he says, "I don't like. So I try to be polite, because if I didn't, I would really be a huge asshole."

The three remaining episodes in the first series of four aired as follows: June 7, "The Robbery"; June 14, "Male Unbonding"; and June 21, "The Stock Tip."

The reviews were so-so. *Entertainment Weekly*, for one, said, "*Seinfeld* isn't laugh-out-loud funny, but it's one of the most amiable shows on the air." The critic, Ken Tucker, pointed out that "the weakest aspect" of the show was Michael Richards's Kramer. "Richards is doing little more than an impersonation of Christopher Lloyd's Jim on *Taxi*, and he ought to cut it out . . ." But he added, ". . . when other series are drifting into reruns, *Seinfeld*'s worth catching." He gave the show a B+.

People, in its review, pointed out that Jerry "gets good mileage out of being himself. . . . The sitcom stuff, though fairly standard, is likeable and bouncy enough, but the snatches from Seinfeld's act bump this comedy up a notch."

The concept of mixing stand-up with sitcom action was not originated by Jerry. Going back in time, a number of comedy greats—some of them his idols—had done the same thing, quite successfully, either on early television or on radio—among them Abbott and Costello, Burns and Allen, Edgar Bergen and Charlie McCarthy, and Jack Benny, whose shows all sprung from their acts.

While the reviews for *Seinfeld* weren't boffo, the important ratings for the summer 1990 run were. *Seinfeld* won its time slot, averaging a very respectable 14.4 Neilsen rating and a 25 percent share of the audience.

"I thought we'd get a very small rating but a very good demographic," Jerry later told the *New York Daily News*' veteran TV writer, Kay Gardella. "Instead, we reached a much broader audience."

Thrilled, NBC ordered a full season of shows for 1991.

Jerry was head over heels with becoming a full-blown celebrity, a status to which the series had instantly propelled him. He was amazed every time he heard a movie or TV star complain about the price of fame. "I love it," he declared unabashedly. "I love everything about being famous. As I see it, there is no price. It's all free. You get special treatment everywhere. You can talk to people if you want, and if you don't want to you don't have to. Why are celebrities always whining? What's the problem?"

Besides a sudden burst of national media coverage, which would never, ever end, he was asked to cohost an NBC special titled, appropriately enough, *Spy Magazine Presents How to Be Famous*, which was a mock scientific look at the phenomenon of celebrity, a concept sold to the network by the country's then premiere mean-spirited satirical monthly. The idea looked good on paper, but didn't do well in ratings and reviews. "It begins funny . . . but before the hour is over, the jokes have become tiresome and cruel," the *People* critic observed. "The barbed, satirical tone does make this a unique proposition, a reminder of how bland and inoffensive most tube wit is."

Nevertheless, Jerry was riding high, feeling flush—the money floodgates opening still wider—and he bought his second property, much different from his first, the close-to-tacky Kings Road condo with the pet monkey on the other side of the wall. His latest acquisition was back east, in the city he loved—New York—with a glorious view of Central Park from wraparound windows, a far cry from the rent-controlled walk-up on West Eighty-first Street that he'd held onto for more than a decade, a place that looked out onto "tar beach"—what they call rooftops in New York*ese*. One of the first items Jerry bought was a powerful thousand-dollar pair of binoculars to scan the park.

Apartment 14-C in a handsome, awning-fronted co-op with a uniformed concierge on Central Park West set him back three million big ones, and another four hundred thousand for renovations that included glistening hardwood floors and a top-of-the line gourmet kitchen consisting of a Sub-Zero fridge—usually empty of food—and a Gaggenau stove, among other luxurious appointments. The colors were grays and blacks, and he called his style "clean tech, which is high tech without the attitude."

When Mike Costanza asked Jerry how he enjoyed his fancy new digs, Jerry offered a cynical chuckle. He said the place had tapped him out, which was an exaggeration—probably voiced to make his schlepper friends from the old days feels less discomfort when they visited him in his new, chic abode.

They had no idea how much mind-boggling revenue he was generating for the same kind of work he was doing when he started in 1976.

"At his peak, before *Seinfeld* hit, Jerry was a $2.5-million-a-year player," stated Robert Williams. "This wasn't a poor kid getting a break in TV. I remember Jerry and I laughing that every week he worked on the series in the beginning he had to take a pay cut. By the time *Seinfeld* went on the air full-time, Jerry could average $100,000 a week doing stand-up in amphitheaters, concert halls, doing conventions and corporate gigs. In the beginning he sure wasn't getting paid that for *Seinfeld*."

The hectic schedule of writing, producing, and acting in *Seinfeld* was actually cutting into Jerry's stand-up revenue, as evidenced by the fact that by the time the program entered the top-ten at the end of 1991, Jerry had to cancel eight to ten lucrative dates because of conflicts with the taping schedule.

He was literally paying the price of success. Of course, he soon was making tens and then hundreds of millions from the show, and while stand-up was his first love, the money was peanuts compared to television.

But, in the beginning, the program both hurt and helped him. Often he had to cancel lucrative engagements, other times he got bookings in cities that had never showed an interest in him until *Seinfeld* hit the air; cities such as St. Louis, where he practically filled the 4,300-seat Fox Theater when *Seinfeld* went into its first full season of reruns. Because of the show, he was known everywhere by Mr. and Mrs. John Q. Public, and the demand for him to make personal appearances was exploding.

Like a sports figure, he had sponsors and endorsements, but didn't wear the logos for Trident gum and Certs on his Armani sports jacket. Those sponsors provided free gum and mints for the audiences at his concerts, and made tickets available through contests from Warner Lambert, makers of Trident and Certs, which didn't require him to actually endorse those products. Other sponsors included Pontiac and Bacardi Mixers, but by the time the show hit Jerry refused to associate himself with tobacco and alcohol products.

Jerry's stand-up contract rider called for only him to be on stage with a stool and a glass of water. He demanded first-class audio and lighting, with a follow spot, and a limo to take him from the airport. Not much more. By the time *Seinfeld* was up and running, tickets for his concerts were in the seventeen-to-twenty-five-dollar range, but were more costly if union stagehand crews were involved. Of course, scalpers were charging hundreds per ducat.

"Jerry was, and is, an astute businessman," Williams asserted. "He had a lot of input in the business end, but it was a reverse process. He trusted his advisors—Shapiro, West, me. But he questioned decisions. He always had control of saying 'yes' or 'no.' There was a sense with Jerry—Shapiro knew it and we knew it—of what arenas he liked to be in, what areas he liked to be in, and what he didn't like. We would have discussions when we weren't one hundred percent positive, and we always brought him all the offers for his consideration. We would increase the pricing for him in accordance with what the traffic could bear.

"George and Howard were the right kind of managers because they didn't let their short-term greed interfere with their long-term greed—meaning when other things could have come from NBC for Jerry to start his own sitcom but not the show the way Jerry saw it, they didn't try to oversell him by saying 'Look, this is a perfect opportunity.' They were much more into long-term planning, of having a sense of what Jerry needed and wanted, and saying to him, 'You're on the right path,' and believing in that path.

"Jerry was persistent. *Seinfeld* showed his persistence. First he got the pilot, and it bombed. Then NBC worked on him for months trying to get him to headline in another sitcom, but he had the fortitude to say 'no.' He truly had a mission, and that was the businessman in him. Another artist might jump to do another sitcom. Not Jerry. I remember him sitting in my office one day talking about the show. He was a consensus-gatherer. He said, 'What are your thoughts?' I looked at his schedule and I said, 'You've got two million dollars on the books for this year, so the question is how bad do you want the show.' He said, 'If I can do what I want, I'm fine. If I can't, it's not that bad.'

"He was willing to give up the *Seinfeld* show if he couldn't get it structured his way. In the old days it was called 'stand by your guns.' He had a belief. He truly had a sense of mission."

Though now he could live in splendor like a prince, he still maintained, in some ways, a Joe Average existence. Jesse Michnick was dumbfounded when he visited Jerry at his palace on the park. "I looked at the shelves, there was nothing there. In one room he had a desk set up with a Mets cap hanging on a hook, and that was it. I said, 'What is this? The Holiday Inn?' He's like, 'No, I just don't have a lot of possessions.'"

Brushing aside Jerry's assertions that he lived that way because he was a Zen minimalist, Michnick had another view. "I think that he thought that there were no roots for a guy like him, that this was just a way station to the next venue," he said. "When I was at his place in California, he had a picture of George Burns Scotch-taped to the wall. I said, 'Buy a frame, Jerry!' And he had one of those ozone machines like you get at Sharper Image for $29.95 and he said, 'It gives the air the fresh smell that you get after a storm.' I said, 'You're coming home with me. You've been in L.A. too long.'"

Still involved with Jerry when *Seinfeld* went on the air, Susan McNabb said his public philosophical claims of being a minimalist might have been authentic, but she never heard them articulated by Jerry in the years they were together.

"He could have had anything, anywhere," she observed. "The way I saw it, he just didn't like things out on the counter, nothing on his coffee table. At one point we were talking on the phone and he said, 'Well, I gotta go. I'm going to make some lunch.' I asked him what he was having and he said, 'Macaroni and cheese.' I'm like, 'Jerry, you could hire a cook, you know.' He said, 'I don't need one . . . I'm fine. I still make my own food.'"

Other than the money and the fame, Jerry, in many ways, hadn't changed much from when he was poor, living in a New York hovel, just starting out in the business. LuAnn Kondziela recalled visiting him at his old apartment with Caryn Trager when Jerry suggested doing a frozen-macaroni-and-cheese taste test. He loved the stuff—Betty Seinfeld had fed it to him all the time at home—and he now wanted to determine which was the best on the market. "We bought all different kinds and cooked them and then tasted them," she said. "We decided Stouffer's was the best." She said she wasn't surprised that he was still eating the comfort-food combination years later when he was rich and famous.

That year, 1991, was supposed to be *the* year—Jerry, NBC, and Castle Rock were betting on it—that *Seinfeld* would land on its feet running and never look back at the pack. But the show's first full season started inauspiciously.

The season opener, scheduled for 9:30 P.M. on Wednesday, January 16, was episode 6—"The Ex-Girlfriend"—in which Jerry starts dating a sexy woman with whom George had broken up, and then loses her after she sees his act, which she doesn't find funny—even though, as Jerry points out, "but you're a cashier." The new season had gotten major promotion by the network, and critics were eager to review it. There was buzz. The time spot wasn't bad either—after *Night Court*—but it could have been better because Jerry was up against the well-entrenched, popular drama *Jake and the Fat Man*, with William Conrad, that had been on CBS since 1987, and an on again–off again ABC laggard starring one of Larry David's best friends, Richard Lewis, in a romantic sitcom called *Anything but Love,* costarring Jamie Lee Curtis.

Jerry arrived early at the home of his publicist, Lori Jonas, for a little celebratory get-together before sitting down to watch the season opener. But a then relatively unknown military man by the name of General Norman Schwartzkopf had other plans for the TV schedule that night—a blazing show with a winning title called "Operation Desert Storm." The Gulf War had exploded and *Seinfeld* became the first sitcom in TV history to be preempted by live pictures from a combat zone. The media dubbed Schwartzkopf "Stormin' Norman," but had any reporters been on hand that night to see Jerry, they would have witnessed a comic on the warpath, too. He was fuming that he'd been preempted by a fucking war. He was kicking things. He had to be taken outside and calmed down.

A week later, though, on January 23, the show aired. Jerry's competition were the casualties of war this time; *Seinfeld* turned in a solid 15.1 rating and a 25 share in the Nielsens—a share is the percentage of TV sets tuned to a program, and each rating point equaled 921,000 homes. The season carried through February 13, 1991, when episode 9—"The Phone Message"—aired.

The show picked up again on April 4 with a great new time slot, immediately following the blockbuster *Cheers*. That episode was called "The Apartment." The season ended with episode 15—"The Busboy"—on June 26.

The reviews were good, overall.

"*Seinfeld* is sort of like *The Dick Van Dyke Show* crossed with *My Dinner with Andre*," observed Steven Rea, writing for *Entertainment Weekly*, "antic angst and engaging ennui in a contemplative comedy about more-or-less real people."

Writing in the *New York Times*, Glenn Collins called *Seinfeld* a "decidedly off-center sitcom . . . the most idiosyncratic comedy on network television. So far, the comedian's efforts have been applauded not only by audiences and NBC, but also by critics . . . . The epicenter of the show's punch is Mr. Seinfeld's keenly observed, carefully timed, contemplative humor about life's minutiae, people's foibles, and mankind's quotidian moments of angst . . . . Though it is taped in Los Angeles, much of its energy comes from Mr. Seinfeld's New York experience."

Near the close of the show's winter-spring 1991 season, Jerry appeared on *Larry King Live* to plug the new time slot. "That ain't a bad lead-in," King growled, speaking of *Cheers*. "I can screw it up," Jerry responded. He caught a lot of people by surprise—especially those at the network and Castle Rock—when he called *Seinfeld* "a sideline." He said, "I love doing it as long as it lasts. But I'm a comedian. I don't want to get away from that. I got into show business to stand in front of a crowd of people and sweat."

Stand-up was his first love and he didn't let the show interfere with it if he was able to get away. During a break in editing one week, he flew to New York to do a one-night concert and stopped in to see his pals at Catch. Eddie Murphy was there that night, too, and he was asked if he wanted to go up on stage, and was just about to when Jerry strode in. The manager walked over to Murphy who, at that point, was five times the star that Jerry was, and told him he'd go on *after* Jerry. Murphy exploded. "Fuck that shit," he fumed. "I'm leaving," and stormed out. The feeling was, Jerry at that point in time had earned the right to be number one there. Jerry returned to L.A. the next day, and drove to San Diego for another concert, showed up for work the following day in Burbank, then flew to Sacramento and Boston for gigs before returning to L.A. to continue working on the show. With a ton of frequent-flyer mileage, he had sent Betty Seinfeld to Europe for a vacation.

September 18, 1991, was the official start of the third season, but it was the first time the show was on the very important fall schedule. Jerry now felt completely confident it had a good shot at sticking around.

Still, it was a tense time. When Warren Littlefield stopped by to give Jerry and Larry David his thoughts about the direction of the show, they listened and then David told him, "Okay, now get out of my office." Jerry winced. David was still walking on the edge, and Jerry was concerned because Littlefield had the power to pull the plug on the whole shebang.

There were also problems with the NBC standards-and-practices people who wouldn't permit Jerry and David to use the word "schmuck" in a script. A heated argument ensued. Jerry made the case that an episode of *L.A. Law* had recently used the very graphic line, "As long as I have a face, you have a place to sit." But the censors would hear none of it, and the cocreators caved in, deleting "schmuck." Jerry saw it as not being able to eat at the adult table because they were doing comedy, not drama. "Deep down," he said later, "we love it. We get to bitch and scream and act all indignant. It's fun. For fifteen minutes, it's the Sixties."

Seinfeld, now a solid favorite with the coveted eighteen-to-thirty-four-year-old crowd, had been nominated for three Emmy Awards—Jerry and Larry David for cowriting, and Tom Cherones for directing episode 7, which aired on January 30, 1991, "The Pony Remark," in which Jerry inadvertently offends and possibly kills off a distant relative by saying immigrants shouldn't have ponies; in Poland, she had one. David, on his own, was also nominated for writing episode 14, "The Deal," May 2, 1991, in which Jerry and Elaine deal with the issue of whether having sex would jeopardize their friendship.

Now big-time, Jerry had been selected as one of the cohosts for the August 25 Emmy Awards show, a couple of weeks before the new *Seinfeld* season started. Jerry performed well, but the patient died. The coveted trophys went to *Murphy Brown.* At a party afterwards, the *Murphy Brown* producers effusively complimented the *Seinfeld* team for their creative, funny work.

"It made me feel better," Jerry said later. "But not really."

Because of Jerry's growing fame and popularity, which the media helped propel with a jet stream of hyped feature stories and profiles, many rehashing the same tired quotes, misconceptions, and myths, Jerry was now dubbed as something of a "sex symbol" by *People* magazine and officially proclaimed "one of America's most eligible bachelors" by *Ladies' Home Journal.*

At the tail end of her roller-coaster ride with Jerry, Susan McNabb read the magazine accolades with him, and said he found them amusing but far off the mark. "He laughed about them but didn't put too much stock into things that he read about himself in the press," she stated. "I don't think he distrusted the media, he just saw it as a necessary tool for his career."

Along with the never-ending issue of Jerry's lack of commitment, his show had also put a cloud over the relationship because "dozens, hundreds, of things from our life together" began appearing on *Seinfeld*. "We talked about the show all the time. He was always taking notes."

Before the first episode after the pilot, Jerry told McNabb that they were adding a female to the cast to play his ex-girlfriend, and that the character was going to be named Susan, in homage to her. McNabb's reaction was, "Oh, cool. Great. I was flattered and thrilled that he would use me as a model." But later Jerry told McNabb he decided to change the character's name to Elaine because during script-writing, or rehearsals, "somebody would come up with something that she would say and Jerry would say, 'Oh, no. Susan'—meaning me—'wouldn't say that. This isn't really the way Susan is.'" Over the years the media had repeatedly, and erroneously, reported that Carol Leifer was the model for Elaine, and Jerry had perpetuated the myth by not straightening out the facts, mainly because he wanted his relationship with McNabb kept private.

For a time, McNabb had worked as a hand model, a job that Jerry found intriguing, and he asked her detailed questions about the business. Later, in one classic episode, George Costanza is discovered to have elegant hands, landing a job as a hand model, a concept taken directly from McNabb's life. Subsequently, whenever McNabb went on an assignment, the first thing she was asked was whether she'd ever seen the hand-model episode. "I'd say, 'No. I never watch the show,' and they'd say, 'Oh, it's so funny. You should watch it.' They didn't know it was based on me."

Around this time in Jerry's career a number of friends who were trying for a shot in show business were in contact with him—some he helped, others he didn't. While Jerry was rejecting or ignoring some pleas for assistance in getting an agent, or securing work, he approached his pal Jay Leno to ask for an important favor for himself. In the summer of 1992, Jerry was in the process of writing the episode in which he and George Costanza pitch "a show about nothing" to NBC. Jerry asked Leno to appear in the segment, which, he proposed, would be filmed on the *Tonight* set. Leno loved the idea and ran it by his longtime manager and *Tonight* executive producer Helen Kushnick, one tough cookie.

Even though Jerry was now a hot property and a close friend of Jay's, she gave an emphatic 'no'; she didn't want it to look like Jay was scattered and not just focused on *Tonight*. Moreover, she thought Jay's audience was the same one that watched Tim Allen on *Home Improvement*, which was *Seinfeld*'s main competitor that season, and she feared Jerry would drag Jay in the wrong direction. She also said she didn't want *Tonight* to become part of "Jerry's personal playground."

Leno told Jerry, who instantly put in a call to Kushnick, thinking he could charm her. Instead, he was met with an angry, off-the-wall barrage of four-letter words and insults. He immediately placed a call to Warren Littlefield, told him what had happened, how insulted he felt, and then said he'd never make another appearance on *Tonight*. Littlefield, who'd had some horrible moments in the past with Kushnick, called her. "I have to tell you something. Jerry Seinfeld may not have a 30 share yet. But we think this guy is a huge, major asset. We care about him tremendously and we can't have this happen. We cannot have this happen." As a result, Jerry agreed to appear on Leno's show again, the two patched up their differences, and Leno eventually canned Kushnick. But Jerry never got to use the *Tonight* set and was forced to drop his concept from the episode.

—34—

Love in All the Wrong Places

By the fourth season of *Seinfeld*, advertisers were happily paying a whopping two hundred thousand per thirty-second spot because the show's demographics were spectacular. While *Seinfeld* had ranked thirty-eighth among regularly scheduled prime-time shows, it finished an amazing second in its time slot, not far behind *Home Improvement*, on ABC. *Seinfeld* also was ranked eleventh in popularity among eighteen-to-thirty-four-year-old males, the most desired demographic—a network ad salesman's dream. And the critics were mesmerized. Tom Shales, the highly respected TV observer at the *Washington Post*, called Jerry's show "painfully amusing and amusingly painful."

Jerry, who once hustled schlock jewelry in front of Bloomies and conned the patriotic poor into buying light bulbs by posing as a damaged Vietnam veteran, had now started to respectably hawk Gold Cards, appearing as a spokesman in lucrative American Express commercials.

David Apicella, the creative director at Ogilvy & Mather, in New York, which handled the AmEx account, saw Jerry as the perfect pitchman for his product. "He had kind of a cult following," Apicella said. "I thought he was a nice combination of being Everyman—and appealing to every man—and being insightful in his observations."

With all the money rolling in, Jerry had added a classic 1958 Porsche Speedster and a new midnight-blue Carrera to his stable, along with a couple of vintage VWs; his closet brimmed with gleaming white Nike Air Mowabs and Air Jordans; he had furnished his ivory tower above Central Park with sleek brown-leather chairs and couches, and had the wood floors bleached. During the show's summer hiatus, he worked constantly, performing concert dates in twenty-five cities, raking in still more money and adulation from his fans.

When asked, Jerry never said he actually lived in L.A.—he'd say it was a temporary layover on his career path, that his real home was New York. When *Seinfeld* reached its finale sometime in the future, he'd be gone, he asserted. Jerry was one TV star who had no plans of ever going into the movies, except to see a film.

He considered his West Hollywood condo merely a stopover, not a home. But in the spring of 1992, with his show looking like it was going to stay around for a while, and with a pile of dough to invest, he decided to put down firmer roots by buying a beautiful place in the hills overlooking the glittering lights of the City of Angels. At the time of the purchase, his publicist, Lori Jonas, was due to have a baby, and he surprised her with the news by declaring, "This'll put you into labor!" Like others in Jerry's tight circle, Jonas never thought he'd fully commit to living in L.A. and was both shocked and delighted by the news.

The sixty-five hundred-square foot, multimillion dollar property surrounded by palm trees on Sierra Mar Place, in the Hollywood Hills above Sunset Boulevard and Doheny Drive, was a genuine movie star's mansion, in the tradition of old Hollywood, though it was a contemporary. It had been owned by the veteran actor George Montgomery, a brawny leading man who started acting in low-budget western "horse operas"—*The Cisco Kid and the Lady*, *Riders of the Purple Sage* back in the 40s—and later starred in musicals, romantic comedies, and the TV series

Cimarron City. For nineteen years he had been married to Dinah Shore, and had counted Ronald Reagan and Ann Lindberg among his closest friends.

Jerry immediately sank big bucks into a renovation but one of his major mistakes was removing the fireplace because he believed it didn't get cold enough in L.A. to warrant having one. When the winter chill and rains came, he wished he had kept it. At the same time, he didn't like the cozy "look" of an old-fashioned hearth, preferring cold, unfriendly surfaces—a house made of stainless steel would have pleased him most. He furnished it as he had his other places—starkly.

"It was a gorgeous place," said Susan McNabb, who was still in the picture but fading, when Jerry was house-hunting. "It was pretty spectacular. *Huge*. But anybody who obtained the kind of wealth he obtained at that point in time would have been insane not to buy a big house. When he bought the place, it was still a tiny fraction of what he could afford."

Not long after Jerry dumped McNabb he began a short but intense relationship with another TV star—more his style now. She was quite a dish, a far cry from any woman he'd ever dated.

With a body to die for and a thick mane of red hair, Julie Kitaen—known as Tawny—was playing Mona Loveland, the nighttime DJ and designated sexpot, on the popular sitcom *WKRP in Cincinnati*, when she and Jerry hooked up. While Jerry was seven years Kitaen's senior, she had lived what seemed like three lifetimes compared to his highly structured, focused, and insulated world.

While Susan McNabb may have *resembled* a Robert Palmer girl, Kitaen had been an *actual* player in the down-and-dirty heavy-metal rock 'n' roll scene of the mid-80s. Kitaen had gotten her break as Tom Hanks's costar in the 1984 film *Bachelor Party*, a sophomoric romp billed as "Shocking. Shameless. Sinful. Wicked. And the Party Hasn't Even Started." She then appeared baring much skin in a series of low-budget but popular films—the *Hercules* series, *California Girls*, *White Hot*, and *Happy Hour*. While Jerry was still mending his emotional wounds from the *Benson* fiasco, and putting salve on his Jewish guilt for getting himself involved in the mess, Kitaen was partying hardy, having a hot-and-heavy romance with sexy Robbin Crosby, the guitar player for a popular group called Ratt, and had been the babe on the cover of Ratt's album *Out of The Cellar*.

Somewhere in the 80s, she became involved with O.J. Simpson, reportedly during the first six months of his marriage to Nicole Simpson, which was revealed during his murder trial. She'd also dated wild rocker Tommy Lee, who later married and divorced buxom *Baywatch* bombshell Pamela Anderson. The two later appeared in a widely circulated videotape in which they displayed their bedroom hijinks.

Kitaen's biggest conquest, though, was David Coverdale, lead singer for the group graphically known as Whitesnake—and for good reason; Coverdale was said among the groupie cognoscenti to have possessed one. Whitesnake's videos were erotic, and much of that had to do with the fact that Kitaen, with her big red hair and skimpy outfits, had a major role in several of them—"Here I Go Again," "Is This Love?" and "The Deeper the Love."

When Coverdale was thirty-nine and Tawny twenty-seven, they married. The year was 1989. Before they got divorced two years later, he called her "my whore and my inspiration."

Jerry had a different description of her, but it meant much the same. He told Mike Costanza that being with her "was like being in the presence of greatness."

By the time Jerry became involved with Kitaen, she had shed her "bimbo" image by starring in a number of PG-rated TV movies and hosting a family-oriented TV show called *America's Funniest People*.

The two met cute—on the Studio City lot where their shows were shot. "Our sets, Jerry's office, our little dressing-room trailers, were all within sixty feet of each other. How fortunate is that—to have your office right next door to your boyfriend's? That was pretty cool. We'd see each other every day, we started talking, and he invited me to appear on the show."

Jerry thought Kitaen would be perfect for episode 26, called "The Nose Job." He cast her as a beautiful but vapid model named Isabel, with whom he became infatuated and had wild sex. In a breakthrough scene for prime-time TV, Jerry's brain and penis are depicted playing a game of chess, with the winner deciding whether he should continue sleeping with her. The title of the episode had to do with the girl George was dating, who possessed a big honker.

"After we filmed that episode, we started seeing each other," Kitaen explained years later, the first time she'd ever publicly discussed her romance with Jerry. "At first we'd be very low-key about it. Knowing people were always watching, we'd be on the lot and say, 'Hello, how are you doing today?' That sort of thing. And then we finally said, 'What the hell are we doing?' And we'd hug and hold hands."

Their relationship soon became intimate, she acknowledged. "We definitely had an enormous fondness for each other."

Not only had they stopped hiding their relationship on the lot, but Jerry, without Kitaen's consent, revealed during an appearance on Howard Stern's radio show that they were dating, a piece of news that quickly got picked up by the gossip columns like the one in the *Washington Post*, and by the supermarket tabloids, such as the *National Enquirer*. "He wasn't afraid of telling anybody," Kitaen recalled. "There was an article in the *Enquirer* about him for the first time—about us. He was so giddy about that. He called his mother and said, 'Look at me and Tawny, we're in the *Enquirer*.' It was the first of what was to be many times for him. But he called his mother to tell her. He felt he had made it."

Jerry started sending her cute cards with little love notes, and bouquets of flowers. She'd lighten his day by walking by his glass-walled corner office at the studio, catching his eye, and doing something silly. Surrounded by writers, Jerry would look up from his serious work and see her smile, and she'd see the sparkle in his eye. They took weekend drives up the coast when they had the time, and stayed at the Biltmore in Santa Barbara; he'd be so tired on the return trip that she usually drove. He bought her a Coach-brand leather purse, and a teddy bear, but no big-ticket items.

At the same time, he was driving to the lot each day in a different Porsche. "He had nothing else to really spend his money on," she observed. "He was just amazed that he could buy so many tennis shoes and Porsches. He was like, 'Wow, I can do this? Cool!' But he wasn't the kind of guy who wore a gold Rolex with his left arm out the window of his Porsche going, 'Hey, baby.' If he was, I wouldn't have spent a second with him."

Kitaen wasn't what she appeared to be on screen and in the sexy rock videos that aroused men's libidos from coast to coast. In fact, Jerry quickly discovered that she was a very bright and creative woman who was the brains behind the wildly selling Whitesnake albums—she was picking the singles, the album covers, the photographers. Moreover, her parents—a Jewish father and a gentile mother—raised her as a Jew, which appealed to Jerry. "My family came right from Ellis Island to San Diego, that was important to him," she noted. That and the fact that she was intelligent and articulate, and they both were in control of their own careers, masters of their domain. It was, for a time, a match made in heaven. She wasn't someone for whom he thought he needed to be responsible, à la Susan McNabb; Tawny Kitaen was a woman who could take care of herself.

"Jerry was definitely a control freak," she noted, "and I've never been the kind to be pushed around or played anybody's fool. I probably could understand Jerry at that time better than anyone. First of all, I was working right next door to him. Second, I was also in the same business, doing the acting thing, so I could understand his hours. I wasn't sitting there going, 'How come you haven't called me?' There was none of that. My show took a lot of my time. I had to jump on a plane every two weeks and fly to Florida and shoot four episodes down there. I was constantly moving all over the place. So, we were not like a regular couple who works nine to five and gets to come home and have dinner with each other."

Following the same pattern that he had with McNabb, Jerry never lived with Kitaen, and he rarely stayed the night with her. "He'd go back to his place, and we'd end up at ten o'clock in the morning driving into the studio guard-gate at the same time."

The aspect of her life that really appealed to Jerry—along with the fact that she was gorgeous, sexy, and smart—were those wild rock 'n' roll years of hers, and he couldn't get enough of her reminiscences and gossip about those days.

"I had done things that Jerry had never done," she said. "He was intrigued. I had been married to a rock star and traveled the world twenty-five times, had the Lear jets, all the amenities. I was living in the fast lane, living out a rock 'n' roll life, at the top of the heap, doing just about anything I wanted to do, and no one was going to say anything. That's the lifestyle he wanted to know about."

For Kitaen, who had been involved with some heavy-duty lovers—Tommy Lee, for one, shocked the women and men who saw his video lovefest with Pam Anderson, and Howard Stern often made comments on the air about Lee's equipment—Jerry had something else going for him that won her heart before and after their relationship became intimate.

"After I had dinner the first time with Jerry he fit in with that group because he's an extremely attractive man," she stated. "I judge someone's intelligence by their humor, so Jerry just becomes so gorgeous. He becomes an extremely attractive man. But I definitely didn't go out with Jerry because he's a Tommy Lee. And the reason I didn't stay with Tommy was because he wasn't exactly a Jerry Seinfeld."

As with McNabb and others, Jerry wasn't hesitant about using material in his show that came from private conversations with Kitaen. "I would just laugh," she remarked. "I'd watch his show and something that we had talked about weeks earlier was on, even if it wasn't a direct quote, and I'd go, 'Jerry, I just told you that!' I mean he really used *everything*. He wouldn't violate the sanctity of our relationship, but I heard material that we had talked about."

Jerry and Kitaen were involved for just under a year when she became pregnant, and for a time—a very brief time—she and he thought it might be his. The father, however, was her old boyfriend, the professional baseball player Chuck Finley, whom she had begun seeing again without telling Jerry. She had fallen for Chuck after her marriage to Coverdale had ended years earlier, but she and Finley weren't going together when she became involved with Jerry.

"I had always loved Chuck and Chuck called out of the blue while Jerry and I were up at one of our weekends in Santa Barbara and left a message. The following weekend I flew up to see him in Seattle and the rest of what was meant to be my life started that weekend. A month later I was pregnant."

Kitaen now had to tell Jerry, and she was concerned the news of the pregnancy and the fact that she was back with Finley would be devastating to him.

"I didn't want to hurt him," she said. "I didn't want to stop our friendship. I went into his trailer and I sat down and I said, 'Jerry, there's something I have to tell you.' And I told him and I just remember him coming over and giving me a big hug and being happy that I was happy."

Afterwards they remained friends.

"I'd be walking on the studio lot and he'd come up to me and I'd feel these arms go around me and he'd have his cheek on my shoulder and he'd say, 'You are the sexiest pregnant woman I've ever seen.'

"Our relationship ended at exactly where it was meant to be. There are never going to be days that Jerry and I resent each other, or have a bad feeling about each other. We left each other almost like somebody who has died; the way you preserve them on your mind. And maybe that's how we were meant to end it." In 2002, Finley filed for divorce.

Jerry had boasted to friends like Mike Costanza about his relationship with Kitaen, and those friends began to notice a change in his attitude regarding women as he became more famous.

"At one time Jerry was more into a woman's personality, her inner beauty, but when he became big he became more shallow, and so did his relationships," observed Costanza. "It was very funny for me to hear Jerry say, 'Mike, I just hang out at the bar at the China Club and they're standing on line to give me their phone numbers.' I said, 'Jerry, you've come a long way for someone who couldn't even get a date.' But now he was bragging. Until he hit the jackpot, Jerry was always low-key—low-key about financial wealth, low-key about acquisitions—but suddenly he was apt to brag about acquisitions of women. When the show hit, it was like there was no talking to him. I don't know what it's like to have people constantly telling you how great you are, and to get anything you want in life, but he did, and once he did, he couldn't resist it. Fame changed him."

Besides his long involvement with McNabb and relatively brief romance with Kitaen, Jerry also developed another close bond with a man, Mario Joyner, a friendship that members of Jerry's circle, and later the media, thought was odd couple-*ish*.

Like Wallace, Joyner was black, a stand-up comic and budding actor—only Joyner was younger and far more handsome. Where Wallace was considered an old fogey—insiders saw him as a father figure–type for

Jerry—Joyner was just the opposite, a guy who liked to get down and party, a brother who knew all the hip late-night and after-hours clubs in New York and L.A.

A close associate of Jerry's, who was very aware of the friendship, noted, "There came a time in the mid- to late 90s when Jerry was with Mario 24-7. In L.A., they'd go to parties together and they'd talk about these models who were there, and talk about how beautiful they were. But those girls were completely unattainable to them. Jerry and Mario wouldn't hit on them—they'd just stand there and *talk* about them."

Jesse Michnick, for one, thought of Joyner as "Jerry's cruise director. When Jerry came to New York, Jerry and I and maybe Michael [Costanza] would go out and when it got late, Jerry'd say he was going to meet Mario. Mario knew all the back doors of all the places, and Jerry went along to see what the next door opened up to. Jerry liked to stand a couple of steps behind Mario, and Mario probably used Jerry's fame."

Jerry's glib response to all the speculation was, "Most of my friends are outgoing. I'm not, so they draw me out. Take my friend Mario Joyner, for instance. He loves going out to dance clubs. He knows all the doormen, where the VIP room is, and who *not* to talk to. I would never go by myself. *Ever*. But I'll go along with Mario. I'm like the sidecar. I climb in the sidecar, they ride the motorcycle, and we both get there at the same time."

Jerry had been invited to be on NBC's ninetieth-birthday tribute to Bob Hope on May 14, 1993, two weeks before the great entertainer's actual date of birth. Among the all-star celebrants were Johnny Carson, Jay Leno, Roseanne Arnold, Paul Reiser, and George Burns, a number of them Jerry's friends or idols, and one would have thought that he would have been honored to be among them. But Jerry turned down the invitation. Speaking for him at the time, George Shapiro said his client hadn't had a vacation in fourteen years and was going to Europe, and planned to tour the Porsche plant in Stuttgart; Jerry now owned six of the cars and was planning to buy his seventh. Jerry's traveling companion was Joyner.

"We have fun with this new toy, my celebrity," Jerry said of the good times he had with Joyner. In Paris, the two were ensconced at the ritzy Plaza Athénée, and they roamed the city together late into the night, partying after midnight at a little café near Notre Dame where the owner broke out a bottle of champagne for the two young men. Jerry went into hysterics when a waiter yanked on the neck of a stuffed pelican over the bar and, as Jerry later described it, "an egg dropped out of its butt." When the long night ended, "We went to bed around 4 A.M. and got up at eight. We were up all night, the next night, too, but who cared? We were having a ball walking around and popping café au laits and rich French food."

From Paris, the two funsters continued their romp to the Porsche plant, which was the highlight of the trip for Jerry. The Porsche people arranged for one of their drivers to take Jerry and his handsome friend for a ride on the test track, doing 150 miles an hour on the straightaways and 100 on the curves. To Jerry, using an odd analogy, the ride "was like being shrunk down and strapped to one of Arthur Rubinstein's fingers as he's playing a piano concerto."

The Germans were intrigued by the two giddy young men, especially when they started doing Rodney Dangerfield impressions. Like teenagers, Jerry and Joyner decided as a joke that everything they said had to end with the words "Hey, baby, all right?" As in Dangerfield saying, "Let's have some ice cream—hey, baby, all right?" The Germans then requested that their guests imitate the way Dangerfield wiggled his knee when he delivered a punchline. "Mario did za knee," Jerry recalled, "and they went wild. I can still see all those beefy Germans standing in a chorus line and hoisting beer steins as they did za knee."

After Stuttgart, the two were on the autostrada north of Florence in a rented Beemer going about 130 miles an hour when Jerry had his first hissy fit of the trip. Jerry was annoyed because Mario, who was acting as chauffeur, kept sucking air through his teeth. He asked Mario to cool it and demanded to drive. Indignant, Mario told him to chill. "I'm just sucking my teeth." But Jerry became adamant and took over the driving. "Jerry's a show-runner," a huffy Mario stated. "He likes control. Otherwise he gets cranky."

As his fame grew, talk about Jerry's sexual preference had started to bubble up among journalists assigned to write about him. Lori Jonas, Jerry's publicist, was constantly being asked by reporters whether Jerry was gay. Her answer was "no."

Entertainment Weekly subtly made reference to sexual preference when it described Jerry in a story as being ". . . single, thin and neat, not that there's anything wrong with that."

The friendship between Jerry and Joyner, while known to key people in Jerry's life, was unknown to the public. But it became an issue in Jerry's first *Playboy* interview, which appeared in the October 1993 issue.

Their relationship was first written about when the magazine asked Jerry what he thought the high concept one-liner would be for the interview itself. "We already did that," Jerry responded. "It's 'The Outing.' A reporter overhears Jerry and George joking around and then writes that they're gay. Not that there's anything wrong with that . . ."

Jerry then acknowledged that he'd been thinking of "The Outing" on the morning the writer from *Playboy* arrived at his home. "My friend Mario was in the kitchen, because he's staying here," Jerry offered. "I was thinking, Gee, I wonder if this guy from *Playboy* thinks there's something going on."

PLAYBOY: Frankly, that did immediately come to mind.

SEINFELD: Oh, really? That's so funny!

PLAYBOY: Also, hey, he's a damn good-looking guy.

SEINFELD: Oh, that's funny. Too bad we've already done the show. No. Mario's just a comedian friend of mine.

PLAYBOY: But you've heard the rumors?

SEINFELD: Many . . . And obviously I've heard that I'm gay. I've been told that everybody at NBC in New York thinks I'm gay.

PLAYBOY: Now's your opportunity to state unequivocally that—

SEINFELD: I'm not gay.

PLAYBOY: When you're in the heat of passion—

SEINFELD: Do I think of men? No.

PLAYBOY: Has a guy ever approached you?

SEINFELD: Once, in Rome. I decided to take a trip to Europe on my own, to see if I could meet people. This was a complete miscalculation of my personality. I don't talk to anyone. I spent ten days there without having one conversation, except with this guy. I was so thrilled to have someone to talk to, and I didn't realize that he was hitting on me. But then it became obvious.

Playboy: Is this misperception part of the reason that you did "The Outing"?

Seinfeld: No . . . It's something I'm not comfortable with. I used to be. When you're younger, if someone thinks you're gay you get really upset about it. But now I just laugh. It means people are thinking too much about you.

—35—

L'Affaire Lonstein

Despite the incredible growing popularity of *Seinfeld*, despite a mind-boggling seven-quickly-going-on-eight-figures personal income, despite the fashionable bicoastal real estate holdings, despite an airport hangar full of mint, German cars that were worth a small fortune, despite the adulation of critics and the fawning of fans, despite the fact that he had been recognized by the *New York Times* as well as the *National Enquirer*, Jerry knew he had really, *really* made it when Barbara Walters came a-courtin'.

The Queen of the Celebrity Interview had now decided that Jerry was a "get," and her stamp of approval documented for him that he was, indeed, a star in the big-time entertainment firmament.

The location: Carnegie Hall, SRO for an evening of comedy with the Man. The setting: A small service elevator going up to a VIP area reserved for Jerry and his entourage. Among those on the lift were Jerry's Queens College pals—Mike Costanza, Chris Misiano, Jesse Michnick, and Joe Bacino—whom Jerry had generously supplied with backstage passes.

Bacino was standing near the front of the elevator pressed against the back of a tiny woman who was closest to the door, like a young filly ready to sprint out of the gate at Belmont. When the door slid open the woman quickly stepped out, followed by Bacino, and, once in the hallway, the two—Bacino and the woman—were standing fairly parallel to one another and there, about thirty feet away, was Bacino's old friend waiting to greet him.

"Jerry's coming down the hall, he has a big smile on his face, he has his hands out and he's looking at me and I go, 'Jerry!' I'm ready for this big hug from him and to tell him, 'Wow, you're great! You made it!' But he walks right past me and says, 'Barbara! How wonderful to see you! I'm such a big fan!' And he hugs her, and I look over and the woman is Barbara Walters.

"She told him, 'You are *so* great. You are *so* wonderful. *Wonderful*! I want to have you on my show! Can we use some of the footage from this show?' And George Shapiro was standing next to Jerry and Jerry looked at George and looked at Barbara and said, 'Well, I don't know if we want you using footage of this show, but you can certainly use footage of other stuff I've done.' And Barbara says, 'Well, that's great, Jerry! That's *so* great! And you know what, I'll call you and we'll set it up.'

"I'm standing right there and he never acknowledged me," Bacino continued, recalling the classic moment in graphic detail years later. "And now I feel like I'm holding my dick with both hands. At that point, with Jerry and Barbara Walters chatting like old friends, I felt like I should get out of there and I walked into another room where they were passing out hors d'oeuvres and champagne and people were waiting for Jerry to make his entrance. And then he came in and that's when he kind of acknowledged me and Mike. But it wasn't as friends anymore. He was gracious—'How are you guys?'—but it wasn't, 'Hey, these are my friends.' He wasn't the same guy we thought we knew."

Jerry wasn't, for example, the same person Bacino remembered being with about a year earlier when Jerry appeared at a club called Bananas, in Fort Lee, New Jersey. During that visit, Jerry had given his two pals—Bacino and Costanza—the royal treatment, food, drink, whatever. At one point, as they laughed and joked and recalled old times in Jerry's dressing room, the phone rang and Jerry had one his lackeys answer it. "The guy put his hand over the receiver," Bacino recalled, "and said to Jerry, 'There's this guy who says he went to temple with you on Long Island and he wants to know if you can get him a couple of tickets to the show.' And Jerry shook his head like, 'No, I'm not here.' And the guy said, 'I really can't find Mr. Seinfeld right now. Could I take your number and we'll have someone get back to you?'

"I remember saying, 'Jesus, Jerry, I can't imagine someone calling like that and asking for a favor. And he looked at me and said, '*You* could do it, and I would give it to you.' It was one of the few times that he really made me feel important to him, and I felt, wow, that's cool. It's cool that we have that, and he feels strongly enough about me that I could call him and he'd give me something if I wanted it. I felt he was sincere. The one thing I can detect is bad acting."

Through the summer, Barbara Walters's field producers and Walters herself conducted interviews with Jerry, his sister, his mother, and he permitted Walters to use certain footage of his past performances. Her crews shot him in his Central Park apartment and California estate, and walking with her near the Seinfeld homestead in Massapequa—he said he hadn't been back in twenty years, which wasn't the truth. When the heavily promoted, highly rated, thoroughly controlled by Jerry *Barbara Walters Special* aired, there was nothing new or revelatory in it. It was precisely what Jerry had wanted, a cathode-ray tube puff piece in living color from the most famous interviewer in the business.

As Robert Williams noted, "Jerry's a master at handling the media. He won't talk about his personal life, his family life, his business life. That's the deal."

But in spring 1993, when the birds and the bees were out and about and love was in the air, Jerry met a child-woman in glorious, green Central Park. And when the media got wind of it, his long successful privacy containment was punctured as fast as a *New York Post* headline writer could punch out the word T-E-E-N-A-G-E-R.

Shoshanna Lonstein, age seventeen, was to the thirty-nine-year-old King of Comedy what Priscilla Beaulieu, age fourteen, was to the thirty-five-year-old King of Rock 'n' Roll in the mid-60s, at Graceland, in Memphis, and what a thirteen-year-old nymphette named Myra Brown was to the twenty-two-year-old "Great Balls of Fire" piano-boogie shit-kicker Jerry Lee Lewis in the late 50s. All three icons had fallen for Lolitas, touching off public and media firestorms and cries of cradle-robbing, but more so for Presley and Lewis—to make matters worse, Myra was Lewis's second cousin—than for Jerry, because times had changed, somewhat.

Elvis and Jerry Lee Lewis with teenage girls, yes, no problem; they were badass southern boys from regions where dating and marrying young 'uns was as common as cow dung on your shoes—as long as you didn't get caught or weren't famous. And they did make those gals honest by eventually marrying them. But Jerry Seinfeld? The very controlled, extremely focused, legendarily commitment-phobic Jerry Seinfeld with a teen on his arm? The media-savvy, public relations–versed, clean-cut Jerry Seinfeld who carefully considered every word he spoke and every move he made, with a teen on his lap? Not the same Jerry Seinfeld who was never considered by his circle as a womanizer, a dude desperate for some young lovin', kissing, and hugging, and who knows what else with a teen?

Nevertheless, he'd fallen fast and hard for a high school senior, albeit a very sophisticated, very mature, very beautiful, very buxom, very sexy, very bright, very upper-middle class, very Upper East Side, very exclusive all-girl Nightingale-Bamford School jock who played varsity soccer and basketball and was president of the film club. She also happened to be twenty-two years Jerry's junior. When Shoshanna was just an infant, a babe in the arms of her successful parents, computer businessman Zach and one-time ballet dancer Betty Lonstein, Jerry was already schlepping out to Sheepshead Bay to hustle gigs at Pip's. The only things Jerry and Shoshanna seemed to have in common besides out-of-control hormones were their mothers' first names and a shared religion.

There are a couple of different versions of how they met in either late April or early May of 1993. The precise date, however, is probably in Shoshanna's private diary, and if she ever decides to auction it off on eBay, the world will finally know the answer. But no one else can accurately pinpoint it. If Jerry met her prior to his birthday on April 29, he would have been thirty-eight years old; post–April 29, he would have been the same age as the venerable Jack Benny. But Shoshanna was definitely seventeen; she didn't turn eighteen until May 29, and the news of their romance broke before that date.

The one fact that isn't in doubt is where they met, which was near Central Park's Sheep Meadow, a short walk from Jerry's apartment. Jerry's well-polished version is that he and a few friends had gone into the park to take pictures of all the gorgeous girls they could find, the idea being that later Jerry would show the snapshots to Mario Joyner to make him jealous. "It was," Jerry says, "a prank."

In any case, one of the groups of attractive females hanging out in the park that warm, sunny Sunday was Shoshanna's. Jerry zeroed in on her like an Iraqi Scud missile on an Israeli warplane: she fit his type—dark-haired, dark-eyed, confident, and she was wearing a snug, cleavage-baring top—*major* cleavage—and tight Levis. The photos, he says, were taken, and the girls and Jerry and his pals moved in different directions. But Jerry says he couldn't get her out of his mind.

And she was probably having a high schooler's tantrum because the famous comedian hadn't hit on her harder.

"We went our separate ways," he says. "But then later, I was talking to my friends and I realized I really liked that girl and would like to talk to her again." He searched the park, found her, introduced himself to her—as if she didn't know who he was; Jerry, the most popular comedian in the land, had become the dream catch of every Jewish girl from Bensonhurst on the East Coast to Burlingame on the West.

She coquettishly gave him her name and number and he promised to call.

Shoshanna couldn't believe her good fortune, and her friends were envious, too. It was like, "Oh, my God, Jerry Seinfeld . . . Oh, my God . . . Shoshanna you are *so* lucky. Oh, my God!"

To Jerry, the magic of his celebrity and boyish charm had worked once again. It was the China Club over and over: they just line up waiting to give him their number.

Mike Costanza was supposed to be part of Jerry's crowd in the park that day—he says, to play some ball, not to take photos of nubile young girls. But he and Jerry got their signals crossed and they didn't hook up. That night Jerry telephoned. "I met this incredible girl, but she's a little young . . . very young." When Costanza asked, "Under thirty?" Jerry's response was, "*Way* under thirty." Costanza remembers asking, "Is she legal? Fifteen will get you twenty." Jerry said, "She's just out of high school. I think she has a birthday coming up." If, in fact, Shoshanna had told Jerry that, she had fudged the truth, as women do about their ages. Costanza suggested that Jerry take her on their first date to Serendipity, a classy ice cream parlor where trendy East Side matrons throw parties for their brats. "You can take her there for a banana split," Costanza joked. "You never know what may happen."

The other version of how their meeting came about has a somewhat less mythical ring to it, and doesn't involve a practical joke on Mario Joyner. This version has Jerry sitting for a magazine cover-photo shoot, the theme of which was "America Loves Jerry Seinfeld," which would have him standing in front of Old Glory with a big red lipstick kiss on his cheek. But the woman who was supposed to do the kiss, Lori Jonas, had become ill, so Jerry decided to take a walk into Central Park to find another female to plant the kiss. If Jonas hadn't come down with food poisoning—if this version is the true one—Jerry and Shoshanna might have never met and there might not be a line of sexy swimwear and lingerie called "Shoshanna" in chic shops across the land today.

Be that as it may, Jerry called Shoshanna and she agreed to go out with him. Within a few days or a week of their meeting, he took her via limo to dinner at Mezzaluna, a trendy Upper East Side restaurant, and to see a show. Whether she told her parents at that early stage of their budding relationship is not known. Later, like the mother and father of Priscilla Beaulieu, the Lonsteins gave their consent to the May-to-December romance, with Betty Lonstein declaring, "We love Jerry . . . he's the nicest person in the world. We are not worried . . . and we love his show . . . we watch it every week."

Jesse Michnick, who had a number of conversations with Jerry about Shoshanna during and after the affair, said Jerry probably seduced the Lonsteins like no one else could. "Jerry's one of those guys who can charm the husk right off the corn. He was able to communicate beautifully with them and made them feel totally at home. He was successful. Famous. Rich. He has his feet on the ground. And it's a chance in a lifetime for her. They were probably in turbo-*kvel* mode; they're calling *everybody*."

Somehow, the press had missed the restaurant and the show dates, but, if Jerry was trying to keep Shoshanna a secret, he pressed his luck when, on the evening of Tuesday, May 25, he strode into Madison Square Garden with his teen angel on his arm to see a Knicks' playoff game. Just as luck would have it, a sports commentator spotted the famous comic and did a brief interview with him on camera, with Shoshanna in the shot.

Jerry sensed problems immediately. As soon as he got home he placed an emergency phone call to his publicist, who was aware of the relationship, and told her he'd been spotted with Shoshanna. Jerry's people were seriously concerned. Was dating a seventeen-year-old illegal? Was he committing a crime? The worst-case scenario in their mind was that Shoshanna's father might take legal action against Jerry. But later, Jerry said, "We all get along great. I feel the age issue is forgotten." That, however, was still to come.

As Jerry figured, the media was all over him like white on a lamb ready for slaughter. And the King of All Media was the first to arrive at the kill. Howard Stern, who had had a good relationship with Jerry, called him on the air to rib and mock him about his newly revealed cradle-robbing proclivities.

"So, you sit in Central Park and have a candy bar on a string and pull it when young girls come by?" he asked a barely controlled Jerry. "I am not, repeat, not a cradle snatcher," Jerry declared, fire detectable in his normally calm voice. And then he really blew it by telling an untruth. "Shoshanna," he asserted firmly, "is not seventeen. That is all I'm going to say about it. Case closed."

The intrepid gossip column people from the *New York Daily News* jumped in, tracked down Shoshanna and got her to confirm that she was his "mystery date" at the Garden. Moreover, the talkative teen contradicted her boyfriend by confirming she was seventeen, but was turning eighteen on May 29, the day after the *Daily News* item—headlined "HIGH SCHOOL SWEETHEART FOR TV'S JERRY SEINFELD"—ran in the giant circulation tabloid. Mentioning the fact that she was president of the film club at Nightingale-Bamford, the columnist wrote, "So it's not like they don't have anything in common." And rubbing more salt on the wound, the item stated: "Sad as it may be, if Jerry wants to see her in the fall, he'll have to travel to D.C. or return to the campus comedy circuit—she's going to George Washington University. As far as we know, they won't be going to the senior prom together."

The bottom-feeder of the supermarket tabloids, *The Globe*, claimed an interview with Jerry in its June 15, 1993, issue, and quoted him as saying, "I met this girl; she was very sweet and very pretty. I knew she wasn't forty, but I didn't know how old she was. It wasn't until the article came out that she told me. By then, I was in a tabloid rocket to the moon. I guess I hadn't adjusted to celebrity yet because it was hard for me to believe that anyone gave a damn who the hell I went out with or what I did."

Hard for him to believe? Had he forgotten that he had called his mother some months before to proudly boast that he and Tawny Kitaen were an item in the *Enquirer*?

Trying to put distance between himself and Shoshanna, at least publicly, and trying to shed the taint of scandal that appeared to be enveloping him, he wrote off the relationship, telling another reporter, "I didn't realize she was so young. This is the only girl I ever went out with who was that young. I wasn't dating her. We went to a basketball game and to a restaurant and that was the whole thing. She's a very sweet girl, very bright, but I have to go back to L.A."

But before she began her freshman year of college, Jerry flew her to Los Angeles where the couple spent two weeks together.

The public reaction to the news about Jerry's teen-queen wasn't as bad as he had expected. After seeing the positive tone, after testing the nation's pulse, George Shapiro told Jerry he was "bulletproof—even in the tabloids."

Shapiro was on the mark. Not only did *Seinfeld* skyrocket in the ratings after the new season began in September, but a secret project Jerry had been putting the finishing touches on when *l'affaire Shoshanna* hit the fan also became golden, and that was a 180-page compilation of his stand-up observations called *SeinLanguage*, which became an instant best-seller, with a first printing of 450,000 copies, published by Bantam Books.

The prepublication publicity blitz was enormous. In New York, Boston, Los Angeles, San Francisco, and Seattle, full-page ads supposedly written by Jerry appeared, stating, "It's mine. But you can read it." The ads, which Bantam president and publisher Irwyn Applebaum described as "buzz builders," offered readers the opportunity to buy the book in advance of the pub date and get a button that declared, I READ SEINLANGUAGE. Jerry even gave a special performance for booksellers at Caroline's Comedy Club in New York, after which he shook hands and posed for pictures with the vendors. There were other promotions, too, and Applebaum was in literary heaven. "Would that all of our authors had twenty-plus million people interested in them every week," he said. "If only 5 percent of the people who watch the show buy the book, then it'll be a best-seller."

It was. It sold some two million copies, at $19.95 a copy. It soared to the top of the *New York Times* best-seller list. "I still can't believe this book is in a bookstore," Jerry's introduction said, in part. "A bookstore is one of the only pieces of physical evidence we have that people are still thinking. And I like the way it breaks down into fiction and nonfiction. In other words, these people are lying, and these people are telling the truth. That's the way the world should be."

In the book, Jerry dealt with a variety of subjects—personal maintenance, dating, dressing, childhood, relationships, and even homophobia. He used the book to once again stress:

> I am not gay. I am, however, thin, single, and neat. Sometimes when someone is thin, single, and neat, people assume they are gay because that is the stereotype. You normally don't think of gay people as fat, sloppy, and married. Although, I'm sure some are . . . They're probably discriminated against because of that. People say to them, "You know Joe, I enjoy being gay with you, but I think it's about time you got in shape, tucked your shirt in and lost the wife."

—36—

Ups and Downs in TV Land

Back in the hardball world of prime-time television, *Seinfeld* was like an army of Green Berets sweeping the Taliban from their caves. The show was unstoppable, a juggernaut, flattening everything in its wake. With Jerry and Larry David as *Seinfeld*'s Dick Cheney and Don Rumsfeld, the program had instantly shot from the high thirties in the ratings to tenth place in early 1993 after NBC permanently moved it—the sixth such time-slot change in four years—from Wednesday night at nine against *Home Improvement*, to Thursday at nine-thirty, right after *Cheers*. "*Home Improvement* was killing us," Jerry observed, acknowledging, "It has a broader appeal."

And a month later, *Seinfeld* beat *Cheers* in the ratings race and never looked back. When *Cheers* ended its run, *Seinfeld* took its golden 9 P.M. time slot. A decade earlier, NBC had looked to Bill Cosby, Jerry's childhood comedic idol, to hold down the fort on Thursday nights; now it was Jerry's turn.

Seinfeld had become "the hip sitcom to watch . . . the cool club to join," with a fan base of "TV literate urbanites," declared *Entertainment Weekly*. By the fall of 1993, when *Seinfeld* had exploded to number two in the ratings—this time without the help of the *Cheers* lead-in—*Time* magazine observed, "The show has expanded its appeal beyond a core audience of yuppie tastemakers. [It] is funny enough for everybody."

Jerry and *Seinfeld* were the hottest commodity in TV—creatively, aesthetically, humorously, sociologically, even pathologically—and it had taken only a few years. It was a cultural phenomenon like no other sitcom ever. It had even added a new lexicon to the language—quirky words, terms, and phrases: Mulva, Bubble Boy, Vandelay, my boys, double-dipping, Prognosis Negative, Buck Naked, master of your domain, yada-yada-yada; the show and all of its parts were ruminated on, discussed, and analyzed around water coolers, dorm rooms, executive conference tables, over picket fences, and on the World Wide Web.

Glenn Pandick, one of the Castle Rock honchos, could all but contain himself. "Giddy" was the only way he could describe the team back at the home office who watched the books and saw the ratings soaring and the revenues rolling in. Right after the program drubbed *Cheers*, he said he was finally convinced Jerry's baby could not be stopped. "The company has been energized," he shouted aloud. "Did I ever think *Seinfeld* was in danger of being pulled? A million times."

But not now.

All those involved with the show have acknowledged that its success—in 1993 alone, it had been nominated for eleven Emmys; Jerry kept the statuettes he won over his TV set—was due to the credo Larry David had set forth from the start, which was four simple words: "No hugging. No learning." Simply put, he meant this was to be a sitcom without moral lessons and syrupy family love and compassion, à la *Father Knows Best*. "It is a bit presumptuous to think that you can really teach people a meaningful lesson after twenty-one minutes of bad one-liners and try to get philosophical in that last minute and have it hit home," Jerry stated.

His rule for writers—and everyone from Hollywood professionals to New York cabbies were trying their hand at *Seinfeld*—was that no idea must sound like it came from another sitcom. "If the script is not written by Larry and me," Jerry pointed out, "changes can be fairly drastic."

David's philosophy was underscored by the infamous Emmy-winning masturbation episode, "The Contest," which aired on November 18, 1992. David said the show "probably gave me more pleasure than any other show I've ever done."

It was an episode of *Seinfeld* that would live in infamy and would haunt Jerry because most everyone who watched the show firmly believed that TV Jerry was real-life Jerry, so his onanism was of great interest. Interviewers, for instance, constantly wanted to know whether he was or wasn't "the master of his domain," the never-to-be-forgotten line from the show.

Masturbation, he acknowledged in the most appropriate of publications, *Playboy*, "is probably my biggest secret, the biggest skeleton in my closet. I didn't discover masturbation until after I lost my virginity," he asserted, presumably with tongue in cheek. "I don't understand how everybody else knew about it and I didn't. Nobody told me about it. I don't know how they found out about it. I didn't know this technique was available to me. I don't know how it happened, but somehow I was absent that day. And when I discovered it, I thought, 'Well, that's the end of that. I'm never going to get upset about a woman ever again!'"

Asked who told him about self-gratification, he said his college roommate, whom he did not identify, but who most assuredly was Larry Watson, with whom Jerry bunked at Oswego. "We were talking one day, and he told me."

Was Jerry embarrassed?

"Are you kidding? I would love to tell people about this. What a tremendous gift to give another human being, to tell them, 'You know, here's what you can do . . .' That's a funny version of it."

Besides *SeinLanguage*, which was going through the roof—Jerry eventually acknowledged that his role in putting together the book was "numbering the pages"—and besides his American Express spots and his stand-up when the show was on hiatus, Jerry also had cut a deal with Kellogg's, in Battle Creek, Michigan, so that his grinning face suddenly appeared on boxes of the cereal company's new Low Fat Granola—a savvy corporate move since the world knew Jerry was a cereal junkie. And then there was a series of revenue- and publicity-generating greeting cards that Jerry and the cast posed for, all of which caused caustic critic Tom Shales, who watched Jerry's popularity and momentum as closely as a Central Intelligence Agency Middle East analyst keeps his eye on Yasser Arafat's movements, to warn, "He risks overexposure. He ought to cut back on those things. For starters, how about the granola?"

Later, Jerry acknowledged that he wasn't happy with his decision to hawk the greeting cards. He said it seemed like "a good idea" at the time because the show "needed every bit of promo" it could get. "It's not really right for us anymore."

In early 1994, *Seinfeld* grabbed more awards, winning three Golden Globes—the best comedy series, Jerry as the best lead actor, and Julia Louis-Dreyfus as best supporting actress in a TV comedy series. That year the Golden Globe for the best drama film went to Steven Spielberg's *Schindler's List*, which the *Seinfeld* team later included in a hilarious but controversial episode in which Jerry makes out passionately with a girlfriend during a showing of the Holocaust tearjerker. In terms of Emmy Awards, *Seinfeld* got a dozen nominations for the '94 season, one more than the year before.

America celebrated Thanksgiving night 1994 with Jerry narrating an NBC special *Abbott and Costello Meet Jerry Seinfeld*. The show was his idea, a commercial way of paying homage to his idols, Bud and Lou, whom he watched as a kid. "Watching Abbott and Costello was one of the things that really got me interested in humor," he said. "It was the purest humor. Sitcoms and stuff weren't really pure. This was burlesque- and vaudeville-based style." He said that in helping put the special together he had watched the comedy team's famous "Who's on First?" routine more than two dozen times and laughed every time. "I went through a lot of effort, and a lot of money was spent to get that particular version of the routine," Jerry noted. "I think there are five or six different filmed versions. This one, from the movie *Naughty Nineties,* was the museum piece. It was an amazing performance of a great routine." Heavily promoted, the special got enormous ratings but few positive reviews. "[It's] from the MTV school of editing," the Associated Press TV critic, Bob Thomas, wrote.

By now Jerry was being paid $200,000 an episode, and a hefty piece of the advertising action, which was nearing $350,000 per thirty seconds—not to mention the revenues from all of his other gigs.

"Crazy money," his confidant, George Wallace, called it. "Sometimes he tells me how much he's making. I just hang up on him."

That summer, during hiatus—Jerry once said that the word "hiatus" sounded like something the Kingfish would say on the old, black-stereotype Amos 'n' Andy TV show—he did thirty-eight stand-up shows in thirteen cities. "That's what keeps him going," said George Shapiro. "I will have longevity," Jerry offered up at the time. "I'd like to play the London Palladium when I'm a hundred, just like George Burns."

And when Jerry wasn't figuratively on the road, he was literally on the road: a car nut, he participated in a road-racing driving school course, wheeling Formula Fords around a circuit at a hundred-plus miles an hour. "You could kill yourself," Jerry said afterwards. "Because the car doesn't know you're only learning."

In his office at the studio, Jerry kept a blown-up print of the original, very negative, audience-testing report on the *Seinfeld* pilot; the ironic piece of décor was a gift from Castle Rock, which was responsible for the show's look, commemorating the hundredth *Seinfeld* episode, a program of highlights from the first ninety-nine episodes, which aired on February 2, 1995. The show included Jerry's five favorites, among them "The Outing" and "The Contest."

In the 1995 season, *Seinfeld* had finally tied with *Home Improvement* as the number one show, and TV's most expensive buy; it now cost an advertiser $390,000 for a thirty-second spot

Once a detractor, Warren Littlefield, the president of NBC Entertainment, started acting like a bosomy Dallas Cowboy cheerleader at the Super Bowl. With *Seinfeld* now getting an average 35 share of the audience, he was calling NBC "the *Seinfeld* network—and that's a strong drawing card," he declared joyously. "Talented people look at that and say, 'We want to be a part of that.' It extends from on-camera talent to writers and producers who can assemble smart, funny adult comedy. *Seinfeld* really is the flagship for NBC."

Jerry has always attributed a good part of the show's success to the fact that it was always kept under the umbrella of the network's specials and late-night division—not the prime-time series department—something its first big backer at NBC, Rick Ludwin, had decided. As a result, Jerry said the show wasn't "meddled with at all . . . Even when they didn't like what we were doing or didn't understand it, they put it on the air."

But critics were starting to get cranky about the show and its cast. There was talk it was losing its edginess, even rumors that Jerry might close down shop—talk that he didn't go out of his way to deny, leaving open the possibility that anything could happen. "My antennae," he stated, "are always up." Magazine writer Ron Rosenbaum, for one, zeroed in on what he called the show's "smugness . . . What emanates unmistakably from the whole Seinfeld group is how enormously pleased they are with themselves." *TV Guide*'s Steve Pond pointed out that while the show "still expands the quirky minutiae of life into unlikely television fodder . . . computer-literate fans . . . [are] chiming in with harsh assessments."

The denigrating remarks infuriated Jerry, especially the comments about the show getting soft.

"We're on top of every page of every episode," he said emphatically. "There's no way in the world that the show is slipping. It's just the nature of the cycle. People get used to it—they've seen Kramer, they've seen George and me. If you went back to see *Schindler's List* every week for six months you'd say, 'I think this movie is slipping.'

"Critics generally, they want to be the first one to say it's bad. So go ahead. But obviously, we would know. I know—we're not hitting the wall."

Jerry compared *Seinfeld* to the gritty and swordlike black humor in Quentin Tarantino's film *Pulp Fiction*, saying it was "much in the tone of a lot of things we do. Some of that coffee shop stuff between John Travolta and Sam Jackson—I thought, 'That's like a me-and-George scene.'" As it turned out, Tarantino was a big fan of *Seinfeld* and watched it regularly, which might, in fact, have sparked the brilliant coffee shop scene when he was scripting the movie.

In interview after interview in the mid-90s, Jerry defended the level of the show, and in the end his blunt feeling was, as he told *TV Guide* in an interview that read like a combination tirade–hissy fit: "It doesn't matter what the people think . . . They're neither relevant nor irrelevant. It does matter what they think, but I don't concern myself with it because I can't change human nature. There's a point when people go, 'OK, that's enough,' and I'm very comfortable with people saying that. They never say that about any other show, you know. Nobody ever talks about other shows slipping, because they were never anywhere to begin with. So when they say we're slipping, I take it as a backhanded compliment."

But off and on, even Larry David suggested privately to Jerry and others at the top that the show might be losing steam; that he was running out of ideas; that he was losing his punch. Jerry had to act as a cheerleader and analyst, assuring his cowriter and coproducer that he was still as sharp and as dangerous on paper as a box cutter. The ego build-up was usually followed by "a Brink's truck, backing up to his house," Rick Ludwin quipped. But David's feelings would intensify—already Burbank was buzzing with insider reports of a rift between the two—causing major problems for Jerry and the rest of the crew down the road.

Meanwhile, Jerry exuded confidence. To prove the show hadn't lost its edge he boasted about an episode called "The Beard," in which Elaine acts as a girlfriend for a gay man, and tries to win him over to the heterosexual team. Besides "The Outing" and "The Beard," the issue of homosexuality appeared in a number of memorable episodes.

In 1995, Jerry could have chucked it all and, as they say in Hollywood, gone to the pool, at age forty-one. That year *Seinfeld* went into what's considered sitcom heaven—syndication through Columbia Tristar Television Distribution—at a fee of between $2.5 million to $3 million an episode: that's as much as $300 million for the first hundred episodes, with a big slice of the pie going to Jerry. In syndication, the show generated a record-breaking $100,000 for a 30-second national barter spot in the first season and escalated from there. Jerry also struck a deal for a new contract with a stupendous salary of a reported $500,000 per episode, beating Ted Danson's old record on *Cheers* of $450,000 a half-hour.

Even if some viewers and critics were starting to pan *Seinfeld*, Jerry clearly was laughing all the way to the bank.

—37—

Designing Woman

The summer before eighteen-year-old Shoshanna Lonstein started freshman year in the nation's capital—just two years before another sexy, young girl named Monica came to town to begin her May-to-December romance with another older, famous man—she went on the road with Jerry.

Zach and Betty Lonstein apparently had no problem with the tabloid "fast fact" that their little girl was traveling with a man twice her age—they thought of Jerry as a good, thoughtful person, and respected the relationship. In fact, for a time the whole Lonstein *mishpacha* took to the road with their bodacious daughter and her new favorite boychick. The fashionable, well-to-do Manhattanites had abandoned their posh Fifth Avenue digs to meet up with Jerry and their baby doll in Vegas, to catch his show and hang with him.

Seemingly loving every moment of the glitz and glamour of big-time show biz, the Lonsteins got backstage and hung out with Jerry in his dressing room, chatting amiably and respectfully with their daughter's Prince Charming who, according to a trained observer present, looked more like a broker after a hard day at Goldman Sachs—starched white shirt open, Armani jacket hanging neatly on a hanger—than America's most beloved comedian and sitcom star, now entangled with a teenage Jewish American Princess.

Later in the summer, Shoshanna—virtually overnight she had become a tabloid first name–only celeb, like Madonna, Cher, and, well, Monica—went off with Jerry, George Shapiro, Lori Jonas, and the ever-present Mario Joyner, sans *mère* and *père*, to New Orleans, where Jerry had a gig at the Saenger Performing Arts Center. Jerry and Shoshanna and their entourage—"Mario was *always* with them," an insider asserted—stayed at the Ponchartrain Hotel. Looking around, Jerry dryly observed, "There's not enough wrought iron here."

To Jerry, Shoshanna "is a person, not an age," he declared. "I'm not an idiot. She's funny, sharp, very alert. We just get along. You can hear the click . . . My interest in her is very proper."

Naturally, Jerry's pals, business associates, and former girlfriends were shocked by the relationship.

"It was like ugh, ugh, ugh! Talk about icky," declared Susan McNabb, recalling her feelings years later. "Because Jerry was a celebrity I was forced to hear about him constantly. Probably a friend told me it was on the news. People on elevators talked about it. People at parties talked about it, not even realizing who I was. I was sort of bombarded with it and, honestly, I didn't think it was true at first. I couldn't imagine that he would really date someone that young. There was nothing during the years that we were together to suggest he had an interest in teenage girls, so I was *shocked*. At first I would defend him to people when they'd say, 'Did you hear? Jerry's dating this young girl.' I'd say, 'Oh, it's gossip-magazine stuff.'"

At some point after he started dating Shoshanna, Jerry called McNabb to chat; they were on talking terms though she was in a new relationship.

"I said, 'Oh, God, please tell me it's not true.' He said, 'No. It's true. She's great, really nice, I like her a lot.' I was just like, 'Wow!' When I saw photos of them together I thought she was very beautiful and, of course, she's got huge boobs and I said to myself, 'Yeah, that'd be the girl he'd like.' The whole thing was bizarre."

Others in Jerry's orbit were equally as appalled and thought that he had reached a point—with all the money, power, and adulation—where he felt he could do no wrong, professionally or personally; that he might have even felt above the law, that he was the Teflon comic.

"Jerry was just entering middle-age-hood and someone like Jerry can develop a feeling when they're immensely successful that they can do *anything*, make their own rules, and I'm sure that contributed," observed Lucien Hold, who was introduced by Jerry to Shoshanna on the *Seinfeld* set. "I don't know a man around who isn't attracted to young women, but they say, 'She's way too young, it's inappropriate.' So they don't pursue her. But with Jerry, *immensely* powerful, there isn't as much of a hindrance. He felt his success protected him.

"His relationship with Shoshanna was very public and I don't think he was displeased about that; he liked the idea. I was on the set with them and she was there with two friends from high school, and they were off to one side talking like teenagers do, and here's Jerry, at forty or whatever, in the middle of producing and directing this major TV show. It was strange, but she was *quite* well put-together and her physical beauty and the fact that she was young obviously captured him."

To Jerry's agent, Robert Williams—who by this point had amicably severed their business ties but had remained good friends—"the Shoshanna relationship was Jerry's worst nightmare because he's a *very* private man. He sort of had the ability to keep his private life private, as Cosby had that ability, as Leno had that ability. When he got involved with her, the bottom line of it was Jerry was at a point in life where he felt less susceptible to the heat. He absolutely felt above the law because he was so successful. He felt above the hit and it was as if he were saying, 'Good, bad, or in between, I'm in love. I wish she was a couple of years older, but she isn't—so sue me and Elvis!'

"I actually didn't see it working against him because at that point he was loved by America. The times had changed. We were dealing with a society in this country where 50 percent of all marriages ended in divorce. There was a different viewpoint. More importantly, it's the first time Jerry got head over heels—I mean blown away! The analytical businessman I knew went out the window. This is the same guy who NBC couldn't get to star in a sitcom he didn't want to do. He *never* could be dictated to. So part of me said he's really got to love her because he's a guy who carefully watches every move he makes."

Jerry's college friend and confidant, LuAnn Kondziela, watched the scandal unfold and, in the end, "I figured he dated a really young woman so the relationship would stay very surface-oriented. He knew it couldn't get really deep, and therefore it wouldn't lead to anything serious."

It's true, Jerry was smitten. But Shoshanna also represented something else, something important to Jerry's career, and that was an image enhancer. As a trusted member of his tight-knit organization acknowledged years later, "I used to tease him that everyone thought he was gay, like the way he dressed, like the tight black jeans and the hideous pink oxford shirts and the white shoes—a total geek, very gay-looking.

"Part of the whole Shoshanna thing changed that because Jerry, finally, publicly, came out with a girl. He had never done that before; had never allowed himself to be involved with a woman in the public eye, and it made Jerry look sexy, look macho, going out with this beautiful young thing. He never was sexy before that; he tried to be sexy and he wasn't, and he realized that, and he joked about it all the time. He'd say things like, 'Do you think people think I'm gay?' And after he started seeing Shoshanna, he'd say, 'No one's gonna think I'm gay now, right?'"

At the same time, Jerry teased the press by not permitting too many sightings and photographs to be taken of the two of them. "He was thrilled that being with her was scotching the gay rumors," the well-placed source said, "but Jerry didn't want to have to talk about the relationship and tried to be very private about it."

When he appeared at a public star-studded event, like an Emmy Awards ceremony, he'd have publicist Lori Jonas drop him off at the front entrance and let out Shoshanna, who was hiding in the back seat, at the rear entrance. Then, Jonas or someone would have to sneak her through the kitchen. "Jerry," the source said, "would never walk in with her." His edict was, "Shoshanna's never allowed to be seen with me."

For the celebrity press it was a coup to get them together on film or videotape. Along with the tabloid paparazzi, freelance video cameramen were now on the lovers' heels, ready to offer their tape to the highest bidder, be it *Entertainment Tonight*, *Hard Copy*, or similar shows.

In one instance, in L.A., Shoshanna and Jerry, with their arms entwined, were leaving the Source, a popular health food restaurant on Sunset, when E. L. Woody, a video paparazzo, spotted them. The lovers hopped into Jerry's Porsche. Woody expected them to peel rubber, but instead they got out and inexplicably pretended to have a fight as his tape rolled. "When he's with her, he's more like a teenager," the stunned cameraman concluded.

"Jerry dangled the relationship," said the insider, "and it hurt Shoshanna's feelings tremendously. He sometimes wanted to be seen with her and sometimes not. That was the half-committed Jerry Seinfeld, always—24-7. At the same time, he was very caring with her."

After Shoshanna started classes at George Washington University—her required courses included Spanish and chemistry—Jerry visited her on campus when he could and the two were spotted by fellow students walking hand-in-hand. He publicly called her "the most wonderful girl in the world," and noted that "the age issue" had been "forgotten." George Wallace, who knew Jerry intimately, said at the time that he thought "it's serious between them. She's beautiful and mature. She's good for him." And George Shapiro declared, "I've never seen him happier."

On weekends, she'd jet to Los Angeles for the taping of the show and they'd have brunch with friends and members of the cast at one of Jerry's hangouts, Jerry's Famous Deli, in Studio City, and then they'd go to Beverly Hills to shop at Armani. Back on campus, she started going Hollywood, wearing sunglasses even on dreary Washington afternoons. Despite her fairy-tale existence, Shoshanna wanted to have it both ways. "I would like my life to be normal and just go about being a student," she was quoted as saying at the time.

Shoshanna spent a year at GW and then transferred to UCLA to be closer to Jerry. In L.A. she got an apartment and was with him exclusively, mostly at his Hollywood Hills manse—not that he was there often; he spent most of his time at the studio. At Jerry's request, Lori Jonas took Shoshanna into her publicity firm as an intern so that she could get credits for school, and she was listed as an intern for two years. Jonas and Shoshanna became close friends and some thought they even were sisters because they had a similar look. When Jonas and Shoshanna went out together, Jerry made sure they were escorted if he wasn't around. He usually chose George Wallace and Mario Joyner because to him they looked like they could be bodyguards.

To celebrate New Year's Eve 1994, Jerry and Shoshanna combined business with pleasure. Unlike Jerry's one-time fiancée, Shoshanna had no qualms about sitting in a club on New Year's Eve watching him work. She accompanied him to Florida where he was giving a black-tie New Year's Eve performance at the Sunrise Musical Theater near Fort Lauderdale. Afterwards, he took her to Delray Beach where he proudly introduced her to his mother. Betty Seinfeld was quoted at the time as saying she was "thrilled" with Shoshanna, but friends revealed years later she was far from impressed with the girl. "She didn't think Shoshanna was for Jerry," recalled a friend, Gloria Pell. "Betty said, 'She's too immature. It doesn't matter the age, it matters the maturity.'"

By the spring of 1995, his or her lack of maturity caught up with them, and their relationship appeared to be on the rocks. The celebrity gossips had put out the word that Jerry and Shoshanna had called it quits; others said they were taking some time off. One report had it that she was even seeing Guy Starkman, proprietor of the deli where they ate. Shoshanna's pals claimed for public consumption that she was miffed because Jerry spent most of his waking hours on the job, and if he had free time he was with his male buddies. There also was a questionable tabloid report that Jerry had proposed marriage—"SEINFELD TO WED TEENAGE LOVER . . . THEY'RE ALREADY MOVING IN TOGETHER—WITH HER DAD'S BLESSING." Another report said he demanded a prenuptial agreement, that Shoshanna hit the ceiling and proclaimed she didn't need his money, that her daddy had his own big bucks.

One classic tabloid tale during the whole affair had Jerry blowing his stack when Shoshanna allegedly introduced him at an L.A. restaurant to long-maned, muscle-bound, 6'3" Fabio, the romance-novel coverboy whom she allegedly claimed she had known for years. According to the report, Shoshanna was fifteen and Fabio twenty-nine when they met at a Club Med in the Caribbean where she was vacationing with her parents. They later connected back in New York, it was said.

Years later, a close associate of Jerry's said, "He and Shoshana got back together and broke up and got together and broke up *so* many times. There were times when they wouldn't even tell anybody they were seeing each other again. It was a whole long, huge ordeal—like months and months and years. For four years they were together on and off."

In August 1996 they were still definitely together. Shoshanna had come to a taping of *Seinfeld* and afterwards, as he often did, Jerry took questions from the audience. A female fan stood up and asked, "What size cups does your girlfriend wear?" Her question was met with an icy silence. "Well," he finally responded, "we've suddenly taken a nasty turn here, haven't we? I don't think that's any of my business, much less yours. By the way, what size jock do you wear? Looks like a large to me."

If Jerry seemed cranky, he had a right to be. The year 1996 was not a vintage one for the *Seinfeld* team, so the last thing Jerry needed was to have to defend himself to an audience member regarding some young stuff in his life.

Over the last few seasons, as the ratings soared, Larry David became more of a problem, always threatening to quit, asserting his creative juices had dried up, that he was fried, with nothing left to offer the show. Privately, Jerry told at least one friend, "He's a fucking pain in my ass." All efforts by Jerry, Castle Rock, and NBC to keep him aboard failed and, in 1996, he quit.

In an interview at the time, he admitted, "I've wanted to leave since this show began . . . I should be doing something else." He claimed that he and Jerry had agreed to end the show, but that Jerry had decided to move forward with at least one and possibly two more seasons. "I can't stop them from doing the show," David said, adding, "I probably won't watch it." Jerry said he knew how David felt, that he was under "tremendous" pressure. "Larry is sensitive about the press we get. For some of us, this is a glorious struggle. For others, it's just a struggle."

With Larry David gone, the full burden fell on Jerry's shoulders. One of his first decisions was to kill the show's stand-up act because he had neither the time nor the energy. "It's a one-parent family now," he said. "I wish that he would have stayed, but the show was very difficult for him. He labored over every little detail. Sometimes that's a personal asset, and sometimes it's a liability, when it becomes so difficult you can't work anymore. He just was worn out, and the show was kind of running on its own momentum, in terms of the writers. I wish him nothing but success and happiness."

There were other issues, too, involving Jerry's first and only true love—the show—that were causing growing friction in Jerry's relationship with Shoshanna, things that were keeping his mind off her. After all, Jerry was the job, always, and Shoshanna, despite her youth and beauty, couldn't change that.

Warren Littlefield had shocked Jerry when he proclaimed that if NBC was to hold on to its prime-time leadership among the networks, he could not cede control of the schedule to anyone—even someone with a hit like *Seinfeld*. "There's nothing in the contract that would keep us from moving [*Seinfeld*] if we determine that's what we should do," he said, and added that he was willing to take heat from Jerry if that's what it took to keep the network from losing its lead. "We have to remain 'masters of our domain,'" he declared.

Money had become an issue, too, for Jerry and the cast, who demanded more. To keep *Seinfeld* in the fold, NBC was being forced by Jerry to pay more than they ever had paid for a half-hour comedy—as much as $5 million an episode. Jerry had become a free agent in the spring of 1997, and was using his power to get the most buck for his bang. After all, he intimately knew the numbers: there were nine thirty-second spots in each episode, and NBC was now peddling each one of them to advertisers for about $475,000 a pop, soon to be upped to $600,000. Including reruns, NBC was making an estimated $7 million an episode, most of it profit because the network paid Castle Rock Entertainment, which owned the show, a little over $2 million an episode. Jerry didn't even have to do the math, and neither did Jack Welch, the honcho of NBC's parent company, General Electric. So valuable was Jerry and *Seinfeld* to the network that Welch, in a rare move, visited personally with Jerry in New York and L.A., reportedly suggesting that GE stock options might sweeten any deal that was made with him.

After a well-publicized negotiation, Jerry's costars quadrupled their salaries in an eleventh-hour deal, signing a two-year contract that gave them an estimated $600,000 per episode each, while Jerry got a million an episode.

Meanwhile, on the home front, Jerry and Shoshanna were on rocky ground. One of the more curious episodes involving their relationship occurred when Jerry agreed to be profiled on *60 Minutes*, a segment that aired on February 9, 1997. Interviewed by Steve Kroft, Jerry went through the usual—breaking into the business, working his way up to a hit show, and then there was footage of Shoshanna—"a UCLA student half his age," Kroft said in the voice-over. Visibly nervous, looking pained, Jerry said, "It's just a nice relationship going on. We're—we're—we're very happy. We have a good time. I think maybe it's because I'm so immature and she's so mature that we—we meet in the middle."

Showing surprise, Kroft asked Jerry if he was immature. Jerry said, "Yes." Kroft asked in what ways, and Jerry said, "Name one." Kroft asked, "Sexually?"

JERRY: Yeah.
KROFT: Why?
JERRY: I—I don't know.
KROFT: What do you mean?
JERRY: I'm embarrassed.
KROFT: Huh?
JERRY: I am not going to talk about being sexually immature on *60 Minutes*.
KROFT: But you think you are?
JERRY: It's not *60 Swinging Minutes,* you know?
KROFT: Right. But you said, "Name one," and I named one.
JERRY: OK. Well—well, I'm forty-two years old. I'm not married. That's not real mature. That's nothing to be proud of. You know, I should have settled down by now.
KROFT: Are you going to get married?
JERRY: Yes.
KROFT: To Shoshanna?
JERRY: I hope. I don't know.
KROFT: Have you asked her?
JERRY: No.
KROFT: Are you going to ask her today?

JERRY: I mean, I'll get married. You know when I'll get married? I'll get married when somebody like John F. Kennedy Jr. gets married.

KROFT: He got married.

JERRY: Really?

KROFT: Yeah. Are you married?

JERRY: No . . . You're used to talking to too many liars . . . Can we get some powder on my face? I'm sweating . . . God!

A month or so later, Jerry's relationship with Shoshanna was kaput.

One of the first in his circle to know was Jesse Michnick, who had had several phone conversations with Jerry just before and right after the final split.

"Shoshanna was one of these people for who, after a while, the Hollywood thing sort of wore off and she wanted a little more of Jerry," Michnick said. "But Jerry was keeping those long hours at the studio. His whole life centered around that show. And when I spoke to him about Shoshanna, he said, 'All she does is yell at me. I emptied out half of Tiffany's for her and she's still yelling at me.' I think that's what he thought would put out that fire—was giving her gifts.

"He wanted peace with her, and there was no way that he was going to have a peaceful time because he was so dedicated to micromanaging all the details of the show," Michnick continued. "Jerry was definitely emotionally impacted by the breakup, but he was okay mentally about the whole thing. He felt he wanted to get on with life, but didn't know whom he could trust in the world of women. At the same time he was so pulled and drawn with the show that he didn't really have time to spend a lot of time reflecting about the split. He referred to Larry David and Shoshanna in the same sentence. He said, 'They're both off my desk—my emotional desk.'"

Mike Costanza said Jerry told him that he had actually proposed to Shoshanna and that she had asked him to leave the show. "She was apparently more ready to think about things like building a family and having a home than he was," Costanza said. "I heard a lot of emotion in his voice, even when he tried to joke about the breakup and rationalize it, saying, 'The problem with her is she's the kind of girl who thinks that when I say I'll be right back that I really will be right back.'"

Jerry also confided in Susan McNabb.

"It was right after they had broken up and he said something to the effect that he had bought her a bunch of jewelry at Tiffany's, and I remember feeling a dab of jealousy, thinking I had never gotten anything like that, not that I minded. I was just as thrilled with the VCR he bought me. He did say that he would have married her. I asked why they broke up and he said, 'She really wanted commitment, and the thing is, I would have married her.' I remember going, 'Ouch,' because he never asked me to marry him."

The summer after their split, Shoshanna and a friend went to Europe where she tried to figure out what to do next with her life. She now had a degree in history and art history, and name recognition that money couldn't buy. There were a couple of job offers waiting for her back in New York—a corporate PR job, a job in investment banking. But with financial help from her father she decided to go into the *shmata* business, albeit upscale *shmatas*—halter dresses, bustiers, and lingerie for women who had the same outstanding physical assets she possessed. Like Jerry, who used his last name for business, she called her label "Shoshanna," and it took off. Among the first stores to carry "Shoshanna" were Bloomingdale's and Lester's, on Coney Island Avenue in Brooklyn. She also started dating a handsome and wealthy banker closer to her age, Sherrell J. (Jay) Aston Jr.

And she finally decided to tell her story, choosing the enormous circulation of *TV Guide* to get it across, with a guaranteed cover, sexy shots of herself, and enormous advance billing. Even *Time* thought the fact that *TV Guide* was devoting a four-page article to her was newsworthy in itself. The news weekly quoted a *TV Guide* senior editor as saying, "America first got to know Shoshanna through her relationship with one of America's biggest TV stars of the past 30 years. Since then, she has become a pop-culture figure . . ." And that was the reason for the piece—headlined "SHOSHANNA"—which, in the end, revealed little about her and Jerry.

"We did have a great relationship, and he's an amazing man," she declared. "Circumstances for us were difficult. I was really in an embryonic state of adulthood. Marriage wasn't presenting an issue for me. But it was an incredible relationship that was very special to me. We were in love. I think that it was my first grown-up relationship, and it was wonderful. I think we made each other laugh—maybe he made me laugh a little more."

She said the age issue isn't a factor "in whether someone's a good person," and acknowledged, "It bothers people . . . I am sure why, but I'm not sure I care why."

She suggested she wasn't happy in L.A. Besides being with Jerry, she felt there was no reason to be there. "I wanted to hang out a little and play, and there wasn't a lot of that. I ended up going to the same few restaurants and the gym." Going to the Emmys, she noted, was just ". . . one day and then you go back to school and you have five papers to write . . . Probably the greatest day of my life is the day I graduated college. I can't believe I did it, 'cause I'd been waiting since, like, second grade to be done with school." She acknowledged her fame was due to having been involved with Jerry, and that because of it she had gotten offers ranging from selling vitamins on QVC to hosting wrestling. "Those things," she declared, "weird me out."

By the year 2000, still cashing in on the relationship with Jerry, Shoshanna was named a contributing editor at *Cosmopolitan*, where, in one article, she shared the secrets of her success with men. One of her tips was to be "a realistic romantic . . . The second you make eye contact with a cute guy, the night suddenly seems more interesting." Her business boomed and she was generating millions in revenues, she became a regular boldface name in the gossip columns, "canoodling" with this one and that one. On Jerry's network, NBC, she was given a regular segment on style on the *Today Show*. In late January 2002, she became engaged to one Joshua Gruss, the twenty-eight-year-old investment banker son of a couple described as Manhattan "social lions"—his mother was on the board of Lincoln Center.

The *New York Post* declared, "JERRY SEINFELD, EAT YOUR HEART OUT."

The 1997 Lonstein breakup may have hurt Jerry, but by the end of the year he was over it enough to visit a former girlfriend, Karen Greene, the one-time flight attendant whom he had dated in New York before he made the big move to L.A. Greene had long ago left the friendly skies and enrolled in law school at Southern Methodist University, where she got her degree in 1991. For a time she worked as an assistant district attorney for Dallas County, and as an appeals court briefing attorney. She then became a staunch Republican conservative associate of George W. Bush in the final years of his Texas governorship before he became president.

While Jerry is publicly considered to be apolitical, he quietly showed up in Dallas in December 1997 for a $250-a-person cocktail party fund-raiser to support Greene's reelection as a Republican Dallas County criminal court judge—she had earned a reputation as extremely conservative in her rulings. It was a post to which Greene had initially been appointed a year earlier by Bush, after she served as executive director of the Texas Criminal Justice Division of the Office of the Governor, a job that involved developing policy and overseeing legislation for Bush, who was pro–capital punishment.

On her desk, in a frame, she often kept an old photograph of herself and Jerry when they were dating.

"I don't know that you'd call [Jerry's attendance] an endorsement," said Greene, who raised $35,000 at the event and subsequently won the election. "It's important to raise money, and it's important to have name recognition, and Jerry can help with both."

Jerry had arrived at Dallas Love Field in a private Gulfstream IV jet—he had recently arrived back in the country from Switzerland, where he had been looking at new cars to buy, and was whisked to the party in a limousine hired by Greene, where he rubbed shoulders with conservative businessmen such as Texas oilman Al G. Hill Jr., and wealthy financier Clint Murchison, owner of the Dallas Cowboys. Seinfeld's attendance shocked some Texas political observers who felt it was rare for a judge seeking election to have such a glitzy fund-raiser. And Greene's only opponent in the GOP primary, Dallas attorney Teresa Hawthorne, said Seinfeld's appearance was inappropriate, and urged voters to elect a judge based on experience, "not because we're in cahoots with stars who have nothing to do with Dallas."

Greene said Jerry was not a Republican, and described him as "really pretty apolitical—nothing at all. I'm the only person on this planet he would do this for . . . This was more because of our personal relationship, not any political ideology on his part." But she added that they did discuss judicial philosophy, and noted, "He's very proud of my conservative philosophy and my approach, which is to strictly interpret the law and work with integrity. He agrees with all of that."

With Greene at his side, Jerry spent about two hours at the event and spoke for less than a minute after the judge introduced him as being "largely responsible for my being where I am today." What Greene meant, according to a close friend, was that Jerry gave Greene "the confidence to think she could do anything she needed to do." Without elaboration, Jerry said, "That's true . . . Karen is a great person and the reason I came here tonight is because I really believe in her, and I'm really proud of her. And what was that flyer thing you hand out, 'Keep Karen Greene,' which is really true. It's terrible to throw out a perfectly good person. You should keep Karen Greene judge."

Todd Gilman, a reporter for the *Dallas Morning News*, had been invited to the fund-raiser by Greene, but was not permitted to bring a photographer. Gilman was mainly interested in telling Jerry that his brother had been friends with Jason Alexander when they were growing up in New Jersey. But when he approached Jerry, "Seinfeld just totally blew me off. Wouldn't talk to me. Turned away, like I was a bug. He was just a total asshole."

—38—

The End of Life as We Know It

In late 1997, the rumors started flying that *Seinfeld* had reached the end of its run. Jerry was about to make *the* decision, and everyone from Warren Littlefield on down to Joe Cabbie was asking the same burning question: would he, or wouldn't he, sign for another season—the tenth? Jerry called it "a thorny issue." He said, "Ultimately I'll have to decide." He said, "It's my deal. It's my show." He said, "I'm the only holdout." He said he would make up his mind right after the start of the new year.

With Larry David out of the picture, Jerry had been working twelve- and sometimes eighteen-hour days, seven days a week, overseeing all of the writing, the editing, *everything* along with performing. The actor Larry Thomas, who played the Soup Nazi, one of the series' most memorable characters, in an episode that aired on November 2, 1995, in season seven, recalled watching Jerry closely on the set right after a rehearsal near the end of the show's run, and being shocked by the amount of work, the number of decisions, he was involved in.

"It was just amazing what was thrown at him," said Thomas who, like most character actors, needed a day job. He worked as a bail agent and court investigator when he wasn't on a set. For his classic role on *Seinfeld*, he received the upper end of scale—$2,610. "I remember watching Jerry. He was in a chair, and a line of people that wound out of the studio and into the hall formed in front of him. It was amazing. There were assistants, writers, production people, and all of them needed an answer from him, an okay from him, a signature from him, on some matter, on some decision. He was the one. It was like opening the hatch of a submarine and letting the water run in.

"I wanted him to sign my original 'Soup Nazi' script," Thomas continued, recalling the moment several years later, "but he was so busy and I was waiting for the right time. At one point he was sitting in the dark in the coffee shop set and all the action was happening on the other side of the stage, and he was there by himself trying to memorize his lines. I sat down quickly and said, 'I'd like you to sign my script as soon as you get a second.' We finished shooting at 3 A.M., so I finally just walked up to him, and he had food in his hand and a cigar, and I said, 'Can you put everything down and sign this, Jerry.' And he said, 'Yes, of course.' He was very gracious. But the guy was exhausted."

Jerry loved what he was doing. Nothing else meant anything, or took priority, in his life. But nearing forty-four, after almost a decade of the prime-time grind, of the pressure, he was drained and realized he had no life outside of the show. More important, he felt it was imperative for him to get out at the top of his game. Critics were already crawling all over the show like a swarm of lady bugs, pointing out growing weaknesses, which Jerry adamantly refused to publicly acknowledge. At all costs, he wanted to avoid overkill, to leave a winner, with everyone wanting more, to be among the sitcom elite—*I Love Lucy*, *The Dick Van Dyke Show*, *The Mary Tyler Moore Show*—that closed shop at exactly the right time. All Jerry really, truly cared about was the quality of the show, and he felt if he stayed around any longer the shine would fade.

Certainly money wasn't an issue, as it was with an actor like Larry Thomas. Jerry didn't have to work another minute. *Forbes* estimated that he had earned $94 million just from the '96 and '97 seasons—$66 million in '97 alone—and noted that he could live off the interest on his interest forever. "It ain't quite Bill Gates territory," the magazine observed, "but it ain't bad." By now Jerry had two Santa Monica Airport hangars filled with some sixty cars, among them two dozen—count 'em, twenty-four—vintage Porsches, and in preparation for his final move to New York he had sold his Hollywood Hills place and bought a $4.5 million 3,400-square foot duplex at the very tony Beresford, on Central Park West, where his neighbors included the likes of Peter Jennings, Tony Randall, and John McEnroe. He paid cash. Money? No problem.

In a top-secret meeting the Sunday before Christmas 1997, Jerry met in New York with NBC's CEO and president Robert Wright, and Jack Welch. Also on hand were George Shapiro and Howard West. The scene was Wright's apartment on Central Park West, not far from Jerry's. The object of the meeting, of course, was for the corporate honchos to try to persuade Jerry to remain for at least one more year. They put on a dog-and-pony show formally titled "Seinfeld: A Broadcast Phenomenon." It involved charts and stats; it showed data that the audience was still growing, at least among the eighteen-to-forty-nine age group—the key one; it showed that while *Seinfeld* gained during the previous five seasons, *Home Improvement*, ABC's big winner, suffered audience losses. Welch even told Jerry that GE was prouder of *Seinfeld* than even any of its other products—the washers or dryers or toasters or refrigerators or light bulbs or parts to make deadly missiles.

It was a convivial and in some ways emotional meeting. After two hours, Jerry and his managers left—still echoing was talk of an unprecedented $5 million an episode if he reupped. They spent the rest of the day strolling the Upper West Side, discussing which way to go—to leave or not to leave. No final decision had been made by the time they flew back to L.A.

As Robert Williams noted, Jerry was the ultimate businessman. He knew the game inside out and knew how to play it. "I know the agent's game, the network game, the producing game . . . I'm in every camp when there's a negotiation," Jerry has acknowledged, "I know what everybody's agenda is. I've had a seat at every table in the whole world of television."

It was Jerry's decision and no one else's. Soon after the big meeting in New York, he said, "I felt . . . the moment. That's the only way I can describe it. I just know from being on stage for years and years, there's one moment where you have to feel the audience is still having a great time, and if you get off right there, they walk out of the theater excited. And yet, if you wait a little bit longer and try to give them more for their money, they walk out feeling not as good. If I get off now I have a chance at a standing ovation. That's what you go for."

And that's precisely what he went for.

Two days after the meeting he telephoned Welch and told him of his decision.

Friday, December 26, 1997. Front-page headline, *New York Times*: "SEINFELD SAYS IT'S ALL OVER, AND IT'S NO JOKE FOR NBC."

> *Seinfeld*, the most popular comedy of the 1990s and the centerpiece of the most profitable night in television history, will stop production at the end of the season, Jerry Seinfeld, the show's creator and star, said yesterday.
>
> "I wanted to end the show on the same kind of peak we've been doing it on for years," said Mr. Seinfeld. "I wanted the end to be from a point of strength. I wanted the end to be graceful."
>
> NBC issued a statement, saying, "To keep a show of this caliber at its peak has been a great undertaking. We respect Jerry's decision that at the end of this season it's time to move on."

Just before the public announcement, Jerry met with Jason Alexander, Julia Louis-Dreyfus, and Michael Richards—the cast who made it all happen, with Jerry as kind of a ringmaster. They had expected the ax to fall, but had been holding out hope that they'd get in another year; after all, they'd just gotten whopping big raises. "They just started making good money . . ." Jerry noted later, "but they were generous enough to respect the timing . . . not that they could have talked me out of it."

The meeting took place in Jerry's dressing room. "It was," Louis-Dreyfus recalled later, "pretty heavy, pretty wild. There were no tears shed, but there was a lot of heart thumping." Richards later said the fun had gone out of doing the show, replaced by exhaustion and what he called "work, real work, and we were losing our sense of play." As Alexander noted, "It was time to go."

> House Impeaches President. Seinfeld Steps Down. Newt Gingrich Resigns. Seinfeld Steps Down. Paula Jones Suit Dismissed. Seinfeld Steps Down. U.S. Bombs Iraq Again. Seinfeld Steps Down. The Starr Report Released. Seinfeld Steps Down. Viagra Introduced. Seinfeld Steps Down.

Despite domestic political scandal and upheaval, Middle East conflict, and a new wonder drug that gave men erections—could an episode involving Viagra have been more of a *Seinfeld* bit?—Jerry's decision to pull the plug became the really big story of 1998, or at least the one that millions of rabid fans really and truly cared anything about.

It was, literally, a shot heard 'round the world. The *Daily Mail* of London, for example, ran a huge headline over a picture of Jerry and Elaine that read: "A NATION OF FANS MOURNS AS SEINFELD SAYS NO TO £3 MILLION PER SHOW." Said the paper, "Distraught viewers across America expressed their sorrow in TV interviews and radio phone-ins. Executives at the NBC network were devastated at the loss of the show, shown in Britain on BBC2, which helped boost the station's profits by £650 million last year."

Back at home, even rival networks like ABC reported the demise of *Seinfeld.* On *World News Tonight with Peter Jennings,* Forrest Sawyer noted, "Millions of TV viewers are likely wondering what they're going to be doing on Thursday nights next fall."

Virtually every magazine in America, virtually every newspaper—small and large, dailies and weeklies—printed odes to *Seinfeld* and laments about *Seinfeld*'s passing. All media was jumping on the *Seinfeld* farewell bandwagon. From *People* to *Time* to *Newsweek* to *Entertainment Weekly* to the supermarket tabloids to the trade publications, even to the *National Review*—". . . the success of *Seinfeld* was an implicit rebuke to PC pieties . . ."—there were stories that, in a nutshell, stated that *Seinfeld* wasn't just any old TV show but "a barometer of our times, a balm for the afflicted. A breaded veal cutlet for the hungry," as *MEDIAWEEK* so appetizingly put it.

Within a few days after Jerry gave his emancipation proclamation, writer Lynn Hirschberg approached *Vanity Fair* with the promise that she could spend time with Jerry leading up to the last show. Her pitch was "If you want him, you get me," an in-the-know *Vanity Fair* source revealed later. "Lynn came to the magazine with the idea of full access to Seinfeld, which, I think, in some ways she had already worked out with him." At the time, Hirschberg, who profiled celebrities, had been in and out of most of the best slick, glossy magazines over the years, was a contract writer for the *New York Times Magazine*, and had previously worked for *Vanity Fair* in the early 90s, when Tina Brown was in charge.

According to the *Vanity Fair* insider, "there had been some hard feelings between the magazine and Lynn about the conditions under which she had originally left."

But *Vanity Fair*'s editor-in-chief Graydon Carter green-lighted the story and assigned a relatively new editor, Ned Zeman, to handle it—a May cover piece that would be on the newsstands and in subscribers' mailboxes by mid-April to coincide with the much ballyhooed final episode. Hirschberg was infamous in celebrity and journalistic circles because she could be tough on her subjects—when she vilified Courtney Love in a 1992 profile, fans released a song called "Bring Me the Head of Lynn Hirschberg." She also could be very soft on her subjects, which would be the case with Jerry.

Hirschberg, who was contracted to do the story for what was said to be a sweet $40,000 for six thousand words, got the access she promised because she had good friends in Jerry's orbit, friends who hoped their pal could go out in a blaze of Hirschberg puff piece glory. One such friend who might have helped with access, who might have put in a good word for Hirschberg with Jerry and with Lori Jonas, was Gavin Polone, a former International Creative Management agent turned producer, and manager of Larry David, who had returned to cowrite the final episode as a favor to Jerry. Later, Polone became the executive producer of David's post-*Seinfeld* HBO hit, *Curb Your Enthusiasm*.

On and off for about a month in March '98, Hirschberg got to hang out with Jerry on the set; ate and drank with him at the Hamburger Hamlet on Sunset; toured his Porsche storage hangars in Santa Monica; heard the revelation from his own lips that his lucky number was nine and that was one of the reasons why he had decided to quit with the ninth season. A more than competent writer and reporter, Hirschberg turned in a good read, but nothing earth-shattering or truly revelatory. Her piece moved through several galleys and was in the copy-reading and fact-checking stage when problems suddenly developed.

One involved her request to tone down a phrase in the story that dealt with the consequences to the show after Larry David quit. The paragraph in question reportedly said, "For the last two seasons, much has been made of former producer Larry David's absence from *Seinfeld*. 'I know the critics say the show is not as good,' says Seinfeld, showing a rare sign of vulnerability." Reportedly, the portion Hirschberg wanted to change was a sentence stating that after David left the show, "a certain broadness set in" to the characters and plot. Hirschberg wanted to make it read "the characters shifted" after David left. Zeman refused to approve the change.

As the editorial process was nearing its end, a fact checker routinely called Lori Jonas to double check on the number of cars Jerry owned. He didn't tell Jonas how many cars were mentioned in the story, but Jonas already knew, and she corrected the number.

First thing the next morning the fact checker raced into Ned Zeman's office. "I've got a feeling," he is said to have told Hirschberg's editor, "that they know some things about the story, and I don't know how they know them."

One of the cardinal rules of journalism, and a definite no-no at *Vanity Fair*, is never to allow a subject of a story to see the finished version in advance of publication.

Zeman instructed the researcher to immediately call Jonas. "How did you know this?" he asked the publicist. "How did you know what was in the story before I even asked?"

"Because," Jonas responded, "I have the story sitting right here on my desk."

Another account had it that *Vanity Fair* managing editor Chris Garrett was notified and asked Jane Sarkin, the features editor, to call Jonas, who confirmed she had the story and stated that it had come from Hirschberg. That scenario said that Jonas also gave *Vanity Fair* an envelope with Hirschberg's name on it, which probably contained the article. Jonas also faxed to Sarkin a copy of the story that showed it had come from Zeman's office; the only copy Zeman had given out was to Hirschberg.

And then, as one *Vanity Fair* insider revealed later, "All hell broke loose." Immediately, Graydon Carter attempted—unsuccessfully—to get Lynn Hirschberg on the phone, to get her to come into the office, to get her to deal with the situation. The *Vanity Fair* editor kept leaving messages on Hirschberg's answering machine: You've got to come in. We have a problem. Did you give him the story? Did they steal the story? Did you leave it somewhere?

The magazine was willing to give Hirshberg the benefit of the doubt, because it was all so puzzling. But when she refused to answer their calls they became convinced she was involved in leaking the story to Jerry.

Lloyd Grove, a journalist who had done work for *Vanity Fair*, and also was doing a piece on the out-going *Seinfeld* for his paper, the *Washington Post*, telephoned Zeman, who also was Grove's editor on freelance work. According to the *Vanity Fair* insider, Grove, who had been given access to the set, told Zeman that he had seen Jerry reading a galley of the story, and thought it was unusual that he would have an advance copy.

Zeman had a friend who was a writer on *Seinfeld*, and he called that friend to get further confirmation. Yes, the writer said, he had read it, too. According to people familiar with the situation, Jerry had expressed pleasure with Hirschberg's profile. "It's pretty good. It looks okay."

Zeman was dumbstruck. He immediately put in another call to Hirschberg who never responded. A message was left on her answering machine, which essentially said, "Don't bother coming in anymore. We don't want anything to do with you."

The impression in the upper echelon at *Vanity Fair* was that Hirschberg "had fallen in love with Jerry, as she's done in certain cases," the insider asserted. "And in order to get him to go with her to the glitzy

Vanity Fair Oscar party, she'd given Jerry the story. We didn't blame Jerry for that. If it were me, I'd want to read it, too. We were humiliated. There were consequences to us. Whether Jerry was promised an advanced look at the story from the get-go, or whether he asked Lynn to make the copy changes, was never determined."

The imbroglio between the magazine and the writer quickly became public, and *Vanity Fair*'s publicist called the nasty affair, "a very serious breach of the editorial process."

James Ledbetter of the *Village Voice* wrote a column headlined "SEINED, SEALED, DELIVERED." In it, Hirschberg denied the conclusion reached by the internal investigation at *Vanity Fair*. "I did not give the article to Jerry Seinfeld," she stated. The *Voice* quoted her as saying that she had a theory as to how Jerry acquired the advanced copy, but declined to say what it was. A couple of years later, in a profile on her in *Brill's Content*, she once again claimed innocence. "I did not give Jerry the piece, and I feel no remorse about anything that took place on my end regarding that situation. I feel much remorse about other things in my life but not about that, because I did nothing wrong."

Hirschberg never wrote again for *Vanity Fair*, which ran her much disputed story.

Another player in the ring who went down for the count was Lori Jonas, a longtime loyalist of Jerry's. She learned he had fired her when she drove up to the studio gate on April Fool's Day 1998—a week before the ill-fated *Vanity Fair* piece hit the stands; a month and a half before the last episode aired—and was barred from entry. She was devastated, in tears.

Some people familiar with the situation believed that Jonas was out to nail Hirschberg because the writer was too aggressive and had done end runs around Jonas, had become too friendly with Jerry, and had cut Jonas out of the loop. Jerry had asked Jonas to "take the fall" for the scandal and she refused. Jerry became infuriated, and that's why he pink-slipped her, according to one scenario.

"Page Six," the *New York Post*'s gossip column, led with Jonas's firing and quoted a "top-level television industry insider" as saying, "People think she's an ineffective, destructive and mean-spirited excuse for a publicist." Another "detractor at a major network" was quoted as saying,

"If you were to poll everyone here, they'd say she's a big bitch and a huge pain in the ass." At the same time the column quoted another source who said Jerry's decision to fire Jonas so close to the show's final episode was "a big fuck you to her."

While Jonas was not the most liked publicist in Hollywood, there was enormous sympathy for her and little for Jerry because of his dismissal of her. As one player put it, "The revealing personality trait about Jerry in this whole episode is that he shit-canned the publicist who had worked for him loyally for nine years."

The *Voice*'s Ledbetter wrote that the "Seingate scandal . . . masks a serious ethical question: how far is celebrity-profile journalism willing to go in exchange for access to the entertainment industry's most glittering sources?"

A couple of years later, though, Ledbetter said the whole scandal "underscored for me the fact that Lynn, Graydon, *Vanity Fair,* and the whole celeb-media apparatus are sucking off the same bloated tit as the people they cover. Under those circumstances, any insistence on journalistic niceties seems almost quaint, though probably still worth preserving."

Jerry, who knew how to cover his tracks, stayed out of the fray. His new publicist, Elizabeth Clark, told the press, "He's stepping back from this, and not commenting."

Another journalistic endeavor in the final year of *Seinfeld* received little or no media attention, but caused great problems and immense emotional pain for Jerry and for Jerry's close friend of a quarter-century, Mike Costanza.

It all started when Costanza read an article in which Larry David asserted he was the genius and creative force behind the show's George Costanza character. David's claim infuriated the real Costanza who felt the TV character—and many of the story lines and premises involving him—were based on his life and on incidents that Jerry had eyewitnessed over the years of their tight friendship.

After reading David's claims, Costanza telephoned Jerry. "I asked him, 'How come you don't put any rest to this story Larry's telling? Why don't you acknowledge the real truth of our friendship, and how it became part of the show?' Jerry said, 'I just don't want to make any waves.' He said, 'It seems to be going good the way it is.'"

The two old friends then had their first big argument. Costanza asserted that things might be going good for Jerry, but not for him—that as a result of the TV character's antics, the real Costanza felt he had "zero credibility, and people don't even believe that you and I are friends."

Costanza then told Jerry he wanted to write a book memorializing their long friendship, and moreover, "get the whole thing squared away" regarding the real versus the TV Costanza. Jerry became incensed, and the conversation turned into a "screaming match."

"There's nothing to write about," Jerry declared, and then he floored Costanza by adding, "We hardly knew each other!"

Disbelieving what he was hearing, infuriated by Jerry's Kafkaesque stance, Costanza made a threat of sorts. "Listen, Jerry," he declared, "I'm going to write this fucking book, and I'm going to refresh your fucking memory."

It was a curious statement on Jerry's part, because there was a close circle of friends who could easily document his long friendship with Costanza. Moreover, in 1992, Jerry even threw Costanza, then a struggling actor, a bone, asking him to appear on the show playing "a loudmouth truck driver"—the first and last time, even though Jerry left hope that there would be more opportunities. On the set, Jerry introduced Costanza to the cast and crew as the real deal. In addition, Costanza had kept an answering-machine tape in which Jerry had left a message saying he was going to use the name "Costanza" for the character who was his best friend on the show. Now—with the possibility of a book in the works—Jerry was denying those close ties, and Costanza was deeply wounded. Years later, in recounting the whole affair, he broke into tears. "I always loved this guy," he said, his voice cracking.

Apparently to get things under control, Jerry made a promise to Costanza he never kept.

"Look, when the show's over we'll write a book *together*," he said in a subsequent talk. "Just leave it to me. We'll do it within a year or two."

Jerry then instructed Costanza to begin preliminary work. He asked him to put together notes on their relationship and the funny incidents that had occurred, and write some sample chapters. When Costanza mentioned he'd have to take time out from his growing Long Island real estate business, Jerry sent him $20,000 to cover the expenses of starting up the project. Costanza said he didn't want or need any upfront money from Jerry, but rather hoped for his creative and editorial input. Later, he came to think of the money as a "bribe" for *not* writing the book.

Mollified for the moment, though, Costanza went to work, wrote about some of the off-the-wall things they had done together, and sent it off to Jerry who called him and told him he liked the material.

But a few weeks later, there was another, more urgent call from Jerry, this time to say, "We have to stop the book." Jerry told Costanza that on the advice of his attorneys he didn't want it published. Jerry told Costanza that admitting to knowing him and using incidents in the show based on their friendship might cause problems down the road for Jerry, Castle Rock, NBC, and others affiliated with the show. "Jerry had never signed a release with me to use any of the stuff involving me," Costanza explained, "and I think the attorneys were concerned that the use of the material could bite Jerry in the ass."

Shocked and surprised, Costanza told Jerry he intended to write the book anyway.

"No you're not," Jerry shouted. "Yes I am," Costanza responded. "You're a fucking brat," Jerry yelled. "I'm not a fucking brat, and I'm telling the truth and there's no reason for people not to read this. There's nothing disparaging. It's a loving reminiscence."

Costanza said Jerry had always been a "control freak and didn't want anybody doing anything about him that he didn't control."

At that point, Costanza teamed up with professional writer Gregg Lawrence, who had written several well-received books. Costanza was hoping to cash in on the last *Seinfeld* episode, as were others who were writing quickie books.

Near the book's completion, a copy of the manuscript was sent to Jerry along with a letter from Costanza: "Change anything that you don't feel is appropriate." "It was sent back to me unopened," he says. "Jerry was acting like a lawyer, and listening to his attorneys."

However, there was one subsequent response—from one of Jerry's lawyers. "It was just pure hostility with the threat of legal action," Lawrence said. "Jerry wasn't pleased that his old friend was doing this book, even though it was a positive book. There were areas that Mike steered clear of, like the issues of drugs and sexuality. Mike didn't write the book with the intent of hurting Jerry at all. He really cherished that friendship even though it cost him in personal terms." Lawrence wrote back to the attorney saying much the same, but there was never a reply.

Lawrence approached a number of editors he knew at major publishing houses. "There was definite interest; everyone took it very seriously," he said, "but none bit, particularly because the book had to be in the stores by May—it had to be a crash book and that time frame really made it impossible, so I just thought it would be fun to self-publish it."

The book, titled *The Real Seinfeld, as Told by the Real Costanza*, was published in June 1998. "We had high expectations," Costanza said later, "and it turned out to be pretty bad. It didn't make any money."

Still angry that his life was "usurped" by the show, Costanza filed a $100 million suit—the huge amount of money was Costanza's lawyer's call, he said—in New York State Supreme Court against Jerry and other principals in the show. The suit was eventually thrown out on the grounds that it was "frivolous," and the judge ordered Costanza and his attorney to pay $2,500 each in sanctions. Costanza appealed, and lost. As he later saw it, "It's really amazing that I didn't get my day in court, but nobody wants to be the one to put Jerry Seinfeld on trial. I did what I had to do. I told the truth. I told my story, and that's the end of it. This was a Costanza ending."

May 14, 1998, is a date that will go down in TV history. That night an estimated 79 million viewers tuned in to see the final episode of *Seinfeld*—some 42 percent of all TV households. NBC charged a record-breaking $2 million for each thirty-second spot—$700,000 more per spot than what was charged during the Super Bowl—and earned a whopping $32 million from just the final episode. Viewers around the world where the show was aired awakened as early as six in the morning to watch—in Israel, in France, in Belgium, in Canada, and in Latin America.

The final show has Jerry, George, Elaine, and Kramer sentenced to a year in prison for breaking a recently passed "Good Samaritan Law" in a Massachusetts town because they stood by while a fat man was mugged and carjacked in front of them. They were laughing and making jokes about him rather than coming to his aid. As *Washington Post* TV critic Tom Shales observed, ". . . the characters were at last held accountable for what a prosecutor called their 'selfishness, self-absorption, immaturity, and greed.'"

As *Variety* concluded: "The New York Four, as they are dubbed in the final episode . . . in effect are without peer. No one, we know deep down, can be truly like them. Yet everyone knows part of themselves to be just like them: They are the flesh-and-blood neurosis and the mismanaged moments and bad habits of all our lives. America laughed for nine years at *Seinfeld* because in its characters we see ourselves."

Indeed, *Seinfeld* left in a blaze of glory. There would never be another new episode of the sitcom about nothing, a show that transformed America's expectations of what TV comedy could and should be all about.

—39—

The Marrying Kind

A month after first-run *Seinfeld* signed off, a momentous performance of another kind took place far from the now eerily silent Sound Stage 9 in Studio City—an episode whose aftermath would have a profound impact on Jerry's life, public and private.

The setting was a chic country-house hotel called Blantyre, in the bucolic Berkshire mountain town of Lenox, Massachusetts. Between June 12 and 14, 1998, Blantyre, an edifice of feudal architectural features replete with towers, turrets, and gargoyles, modeled after an ancestral home in Scotland, had been taken over for what was billed as *A Wedding Weekend in the Country: A New Musical.*

During the memorable three-day weekend the bride and groom and their 175 or so guests were housed, fed, and pampered in splendorous English rooms with fireplaces, period furniture, Oriental rugs, parquet and marble floors, and dressing rooms, all situated on a hundred acres of tranquil flower gardens, majestic trees, and manicured lawns, where one could play tennis or croquet. Indoors was a heated pool, sauna, Jacuzzi, exercise equipment, and in-room massages.

The lavishness of the setting harked back to the grandeur and service of a bygone era, of a time known as the Gilded Age. Not far from Blantyre was Arrowhead, where Herman Melville wrote *Moby-Dick*, and also nearby was the home of novelist Edith Wharton. Guests could go hot-air ballooning, visit Hancock Shaker Village, or the Norman Rockwell Museum.

It was a very special weekend—at least on the surface. Underneath, however, was a whole different, very dark and very gothic story, one in which Jerry would soon figure prominently.

The two stars of the weekend—which was ominously stormy, with drenching rainfall—were Jessica Sklar—the young bride, a relative commoner, and Eric Nederlander—the young groom, a scion of one of America's great theater dynasties.

Apropos of his background, the guests were each given a faux *Playbill*, which described each "act" of the weekend—from the Friday, June 12, wedding party rehearsal on the croquet lawn at 6:30 P.M. sharp, to the Saturday June 13, 6:30 P.M. wedding ceremony, to the 10 A.M. Sunday, June 14, "Farewell Brunch."

Inside the *Playbill* cover were photos of the bride and groom—simple headshots that looked much like high school or college yearbook pictures, but each had a thoughtful caption expressing their love for one another and respect for this most joyous of events.

"Jessica Sklar (Bride) is thrilled to be making her Broadway debut. Jessica is looking forward to spending the rest of her life with Eric Nederlander. She thanks him for the wonderful five years they have spent together and can't she wait [*sic*] for the next 50.

"Eric Nederlander, the groom, noted that, *A Wedding Weekend in the Country* marks [his] Broadway debut. He is so excited to have landed a starring role in this production, especially opposite such a fantastic leading lady. In finding the role of a lifetime, he anticipates a long-running smash hit!"

But this was to be no *Fantasticks*, no long-running smash hit like *Cats*, no boffo at the box office like *The Producers*, and the reviews would be terrible. In fact, the merger of Nederlander-Sklar would turn out to be among the shortest—and most scandalous—marriages in the history of contemporary gossip columns, thanks in part to Jerry's role in the drama.

Those guests in the know, those in both the bride's and the groom's inner circles, chortled when they saw the romantic setting of this play and read the *Playbill* because they knew the marriage was more analogous to a magic show—all smoke and mirrors, deception, and a disappearing act as a finale that was still to come—than a musical, as billed.

Before long, Jerry would become *the* featured player in the Nederlander-Sklar production of *A Wedding Weekend in the Country*—the villain, some observers would say—who would strike the set and steal the male lead's bride.

Who was this woman, Jessica Sklar, who caused such a highly publicized and scandalous tug-of-war for her affections between these two men of power, fame, and wealth?

She was born Nina Danielle Sklar on September 12, 1971—the year before Jerry went off to Oswego—on the poor side of the tony town of Oyster Bay, New York, to Karl and Ellen Sklar. She was the second of the Sklars' three daughters; the first, Rebecca, had come into the world six years earlier, and the third, Elsbeth, two years after Nina.

Somewhere along the line Nina grew to despise her name and started calling herself Jessica. So emphatic was she about the name issue that in 1995, a couple of years after she became involved with Nederlander, she filed an application in New York's State Supreme Court to formally change her first name from Nina to Jessica, asserting, "I have always gone by the assumed name as a matter of preference."

There were close friends of hers who never knew Jessica's given name had been Nina—even her groom—and some thought it odd that she would hide such seemingly innocuous information about herself from them. "We were very shocked at all the lies and things that we found out about her after she left Eric for Jerry Seinfeld," a former longtime female friend said. "I mean, her real name isn't even Jessica. And as long as we knew her—five or six years—we did not know that, she never revealed that to us."

The precocious child also referred to herself as "the middlest" because she ranked between her two sisters, even though she was considered the favorite of her parents, sparking early and lifelong competition and resentment with the older Rebecca, who years later—when Jessica was already involved with Jerry—blurted out to a journalist, "My mother always liked Jessica best!"

When they were growing up they were mean to each other, tripping each other, making each other fall, according to Rebecca's first husband, James Meiskin, head of a successful New York commercial real estate partnership who, by coincidence, won a national Jerry Seinfeld look-alike contest as part of the hoopla leading up to the final episode. "Jessica definitely had it in for Rebecca and Rebecca definitely had it in for Jessica. Rebecca didn't like the fact that Jessica got more attention, was invited to more family outings and gatherings than Rebecca, and that incensed Rebecca," he stated. "As adults, the two sisters would attend Thanksgiving dinner at their grandmother's apartment in New York and wouldn't say a word to one another."

The Sklars, Karl and the former Ellen Fried, came from liberal, Jewish homes—he from Milwaukee, she from New York, where she attended the very progressive and private Little Red Schoolhouse in Greenwich Village from third grade through graduation in 1961. It was there that Ellen became lifetime friends with Kathy Boudin, daughter of the prominent civil liberties attorney, Leonard Boudin, who made his name as a lawyer defending the most clients summoned before the House Un-American Activities Committee, and had himself been accused of being a communist, which he denied.

An only child whose parents were said to have divorced, Ellen spent much time in the Boudin's elegant town house in Greenwich Village, later describing the scene there as "like a salon—a little intimidating." Speaking of Kathy Boudin and her brother, she said, "There was an understanding that, whatever they were going to do, they were going to contribute in a big way."

Like her parents, Kathy Boudin had radical leanings, which would surface in the headlines in the late 60s.

Boudin went to Bryn Mawr College where she graduated magna cum laude, while Ellen had gone off to the University of Wisconsin where she met her future husband in her first week as a freshman. After college, they married and moved back to New York, settling—in the mid to late 60s—into a three-story walk-up apartment on East Tenth Street in the Village, where they savored the leftist, radical, bohemian ambience. "They made no bones about those leanings," said a Sklar family observer. "They were interested in socialist culture; that was a commonality, and probably drew them together."

Kathy had also returned to the Village after college, and the two young women continued their friendship. In 1969, though, all that changed: the Boudin home was destroyed when a homemade bomb went off, killing three of Kathy's friends, members of the violent Weather Underground organization. Two young women survived—one of them Boudin, who wandered out of the rubble naked and disappeared into thin air—wanted by authorities. Before going underground, though, she coauthored a book on how to deal with the police when arrested; the cover showed Lady Liberty being gang-raped by policemen.

Boudin didn't surface again until a dozen years later. Using the alias Barbara Edson, she was arrested for her role in the October 20, 1981, $1.6 million holdup of a Brink's truck in Rockland County, New York, during which one of the armored car guards was shot and killed, and another wounded; the gunmen were linked to another militant radical group, the Black Liberation Army, an arm of the Black Panthers. When she was arrested, Boudin was the mother of fourteen-month-old son.

Boudin was sentenced in 1984 to twenty years to life after pleading guilty to one count of felony murder. During her years in the maximum security Bedford Hills Correctional Facility in New York, she and Ellen Sklar, who went on to become a Vermont corrections department official, stayed in contact. Involved in victim reconciliation work, Sklar urged Boudin to reach out to the family of the murdered Brink's guard. When the journalist Elizabeth Kolbert did the first interview with Boudin since her incarceration for the July 16, 2001, *New Yorker*, Boudin recommended that she interview her friend. "Ellen talked to me to help Kathy," Kolbert said.

While Boudin led a life that ended with her behind bars, Ellen Sklar took a more traditional course, as a middle-class wife and mother. After Rebecca came along, Ellen Sklar stayed home. Karl Sklar, who with his beard and bushy hair had a resemblance, some thought, to Karl Marx, held a number of jobs, among them teaching children with learning disabilities and emotional problems. Unable to afford to buy or store a car in the city, he got around on a motor scooter.

In the early 70s, the Sklars' apartment building was sold. Given the choice of buying their small, cramped flat, or moving, they decided on the latter. Now working for a company that developed audio-visual systems,

Karl Sklar had scraped together enough savings to make a down payment on the family's first house, said to have cost $18,500, in what Ellen Sklar later described as the "ordinary part" of Oyster Bay, on the affluent North Shore of Long Island, where Jessica and Elsbeth were born. The Sklars later said that the house had a tree growing through the roof when they bought it and they had to do extensive renovations.

At the beginning of the twentieth century, the town's most famous resident had been Theodore Roosevelt, the only president from Long Island, who built Sagamore Hill there as the summer White House. Meiskin had been told that the Sklars had actually bought Sagamore Hill, and "when the family moved in, there were rat droppings all over the place, it was a fixer-upper; they purchased it but didn't have enough money to renovate it, or hold on to it. It was like a shit pit." But the story he heard from a Sklar relative was fantasy because the twenty-three-room Sagamore Hill is a carefully maintained National Historic Site.

In any case, the Sklars settled into their new life in Oyster Bay. Jessica was bright and alert. When she was three years old she saw a male baby being diapered and she yelled, "Jason has something on his vagina."

When she was four her parents took her on a fishing trip and Jessica caught a big one. She was so happy she started dancing around, gleefully singing, "Aren't you proud of me?" Jessica always wanted to be part of her older sister's crowd, but when Rebecca's chums once ordered the five-year-old to leave a birthday party, she stood up to them, declaring, "I'm Rebecca's sister—and I'm staying! My feelings get hurt *very* easily."

By the time she entered junior high, Jessica was a jock of sorts, very outdoorsy—she played field hockey, she went sailing and boating on Long Island Sound with far more affluent friends. "A very enthusiastic child," was the way she was described.

When Jessica was fourteen and in ninth grade—not the greatest time in the life of an adolescent for a major upheaval—her parents decided to move. The reason the Sklars gave for the move was that they had grown bored with the world of Oyster Bay and wanted a more active cultural life for themselves. But where?

Karl Sklar had become a follower and supporter of the "Progressive Movement" of Bernie Sanders, the Socialist mayor of Burlington, Vermont, who had been in office since 1981, and later was voted the state's lone independent congressman in Washington. The Sklars took a trip to Burlington and fell head over heels in love with the New England town located on the eastern shore of Lake Champlain, between the Adirondack and Green Mountains. The first house the agent took them to—a small, tractlike ranch built in 1959 on a bus line; they had grown to hate "carpool culture"—they bought. By coincidence, Bernie Sanders lived across the street. The Sklars were seeking a Utopian existence, and were convinced they had found it. At first Jessica's father worked as a carpenter, and later developed a computer software language program, which became his business.

Utopia, shmutopia. Jessica was furious about the move, about leaving her well-to-do friends behind, and she despised Burlington, which seemed to her like Mars—far from the excitement of Manhattan, where she often visited, staying with her grandmother who lived on the lively Upper West Side, Jerry's real-life and TV stomping ground.

"She was tremendously unhappy," Ellen Sklar said later about the move. "She was filled with anxiety and she never got over it." The fact that the Sklars subscribed to the *New Yorker* and the *New York Times* in order to keep up with life in the city didn't make Jessica any happier. "But we had faith in the move," her father said. "Burlington's a great place. You can run into the governor on the street."

For Jessica, though, the anticipation of the move was worse than the actuality. Her mother met with a school guidance counselor and described her daughter's fears and anxieties. The counselor kept an eye on Jessica and before long reported back that there was nothing to worry about, that she had a new crop of friends and was popular with boys.

"Jessica didn't talk to us about boys," her mother said later. "There must have been some things we didn't know about. But she was very straight with us. There was nothing going on in her life that she kept from us. She's a direct, *honest* person."

A friend in high school, who was a year older than Jessica, remembered her wonderful sense of humor and dynamic personality. "She was witty, clever, generous, always great to be around," he said. "She was a pretty good athlete, played field hockey, skied. I would hate to use the cliché that she was the life of a party, but she was a treat to be around."

Others from that time remembered Jessica as fun, lively, and entertaining, with a talent for mimicking celebrities. Her imitation of comedian Steve Martin was said to be especially good. She quickly was able to make people laugh with her take on the Vermont country accent, and sometimes embarrassed her parents by imitating their friends in front of them.

While in high school, she held down part-time jobs as a waitress and as a checkout girl at the local Grand Union. She went to her prom with a group, not a date. At home she cooked the turkey every Thanksgiving, and though she was not one to follow recipes, her very special dish was pancakes. The family belonged to a local congregation, Ohavei Tzedek.

As a student, Jessica was "driven by ambition," her mother has boasted. "Anything she did, she did 150 percent."

Burlington is the home of the University of Vermont, and while Jessica wanted to go away to school—New York was her first choice—the family couldn't afford the tuition and her grades didn't warrant the kind of scholarship needed to foot the cost of a New York school. So she became a member of the U of V class of '93.

Situated on a glorious hillside overlooking the placid blue waters of Lake Champlain and the snow-capped Adirondacks to the west, the U of V had two disparate reputations when Jessica matriculated: it was dubbed "a public Ivy League" school, and a party place, and once tied with Brown University as the number-one social school with the best combination of academics, activities, and partying.

As a close male college friend of Jessica's said, "You had sixty-five bars within a mile of your dorm room. From a party point of view you could drink in class if you wanted to. You could buy beer on your meal card. You could have kegs in the dorm. You could party as much as you wanted to and still get very acceptable grades.

"It was also a druggie school, mostly pot. But later they had something called 'the bomb'—a mixture of Ecstasy, cocaine, and mescaline rolled into one big mess. It was more of a rave-type drug. I thought, 'Boy, that sounds like fun.' "

Rather than remain at home, Jessica eventually moved into a dorm on campus, and later lived in several off-campus apartments. Because of the university's reputation, it attracted rich kids who couldn't get into genuine Ivy League schools but wanted a decent education and a place to have fun. If Jessica harbored ambitions for a better lifestyle, the scene on campus opened her eyes to the possibilities. "A lot of the people were from New York and New Jersey—a lot of Jewish princesses whose parents bought them convertibles for the summer and four-wheel drives for the winter," a female classmate of Jessica's said. "That was Jessica's crowd and she was in the same dorm with them, Coolidge. We called them 'The Coolidge Girls'—all very girly-girly, all about clothes, all about Gucci, all about Prada. They were all looking for their MRS degree.

"There was a lot of money at UV and Jessica was very much a part of that group, even though she didn't have the money," her classmate continued. "There was a lot of that fashion culture there—'What car are you driving?' 'What island are you going to?' That's who she was hanging out with. I met her father and he was a nice guy, down to earth, so it was hard for me to fathom how she got that way."

Naturally, like her friends, Jessica wanted a car. "It wasn't automatic to us that we bought the children a car," said her father who turned down her request. But Jessica was persistent and wrote a letter stating her case. "The letter was eloquent in its persuasiveness and its logic," Jessica's mother recalled. In the end, Jessica was permitted to buy a secondhand light-blue Volkswagen Jetta.

Meiskin said that when he met Rebecca Sklar she led him to believe that her parents were "well off." As he got to know her, though, the story and image of the family changed.

Meisken said, "Even though they were good people, and hardworking people, and smart people, they don't have money, and that was always a sore subject for the Sklar sisters."

Incredibly, in the five or so years Nederlander was involved with Jessica before she left him for Jerry, and during the years Meiskin dated and was married to her sister before they divorced, neither had been invited by their respective Sklar women to the family home, but stayed in area hotels when they visited Burlington, which was rarely. Meiskin said, "When Eric finally saw the house on *Extra* after the Seinfeld thing broke he called me and I asked what it looked like. 'It's a shack,' he said."

Meiskin met Jessica for the first time when he and Rebecca visited her at college. "Her tiny dorm room was *full* of clothes and looked like a cyclone had hit it. She didn't even lift her head to greet me. She was engrossed in folding her stuff. It was a very weird and unpleasant experience."

A classmate remembered Jessica as "a little prissy—like the rich kids who came to UV to play with their parents' money. She was also a bit of an ass-kisser to the professors. She'd always be sitting in the first row in lecture classes and kind of flirt with the professors, try to be the apple of their eye. It just rubbed some people the wrong way."

Jessica had an intense relationship during college, which continued on and off after she graduated, and she reportedly went back to him at least once after she became involved with Nederlander.

For a time Jessica wanted to be a lawyer, and she served an internship at the state attorney's office. She also worked in the psychiatric ward at a local hospital.

After graduation in 1993 with a B.A., she decided to move to New York where Rebecca was married to Meiskin.

"Jessica had the most confidence," her father boasted. "Her genes were good to her." Added Ellen Sklar, "She didn't have the Barbie syndrome, but she was cute and she knew it. She wanted to do well, but she never thought she'd be in the national spotlight."

In New York, James Meiskin introduced Jessica to his friend, Eric Nederlander. The two young businessmen had met in true *Seinfeld*ian fashion. As bachelors, they had both lived in a high-rise apartment building on the Upper East Side, and one Sunday Meiskin threw chicken bones and a newspaper out of his window and the debris had blown into Nederlander's window. When Nederlander tracked down Meiskin to complain, the two bonded.

Meiskin, who met Rebecca in Central Park à la Jerry and Shoshanna, married her under what he described as intense pressure in a ritzy wedding at the Beekman Hotel in October 1991, and they went to Italy for a costly honeymoon that she arranged. Later, Meiskin told Nederlander, "Do I have a girl for you." The girl, of course, was Jessica.

To an outsider, Eric Nederlander's life growing up in a beautiful Tudor-style home on Private Lake Drive in the affluent community of Bloomfield Hills, Michigan, looked golden—like life in one of those *Father Knows Best*-style "learning-hugging" sitcoms of the 50s that Jerry so assiduously avoided emulating—but the kind of elegant-seeming setting

Jessica Sklar might have imagined being part of when she was growing up in a simple house in working-class Burlington.

The youngest of two boys of Robert and Caren Nederlander—he the multimillionaire president of the Nederlander Theatrical Organization, she an artist, photographer, and therapist specializing in sexual dysfunction, behavior modification, and marriage counseling—Eric was educated at the very exclusive Cranbrook School in Bloomfield Hills. The campus is considered one of the nation's great architectural treasures, replete with landmark buildings, formal gardens, woods, lakes, and America's largest collection of outdoor sculptures by the Swedish master Carl Milles. And its academic reputation is such that wealthy parents from all over the world send their children to board there.

Eric was more of a jock than a student. He loved baseball and tennis in the summer, attending an exclusive camp every year to sharpen his game; his father had been captain of the University of Michigan tennis team in his day, and Eric inherited his competitive genes. On summer nights, Eric and his best friend, Ricky Rosin, who lived next door and was like a brother to him—Rosin was Nederlander's best man at his wedding to Sklar—stayed out until all hours playing ball in the sprawling apple orchard behind their homes, land their entrepreneurial fathers had hoped to buy up from the owner and develop, but, to their sons' delight, were unsuccessful.

"We hung out every single day like brothers," said Rosin, who was two years younger. "I mean our parents had to scream at us to come in from playing ball, throwing the football, or whatever we would mess around and do."

In the frigid Michigan winters—Bloomfield Hills is a half hour and a world away from Detroit—the boys played Nerfball in the Nederlander basement and attended every home football game of Robert Nederlander's alma mater, the University of Michigan, where he was a Regent and extremely influential.

Eric and Rosin were like Jerry and George—inseparable—and there were family connections, too. Rosin's father, who was in residential and commercial development, and Nederlander were partners in some lucrative business ventures, and like their sons, the two fathers were close as youths.

The boys also partnered up on a venture of their own—obsessively collecting every baseball card they could get their hands on, spending on their investment every cent they could pinch from their parents.

"Our parents would always yell at us, 'Why are you guys spending all your money on that shit?'" Rosin fondly recalled. "Meanwhile, our 'shit' as they called it is worth a lot of money. We collected some unbelievable stuff—we bought Mickey Mantle rookie cards, Pete Rose rookie cards, Nolan Ryan rookie cards. We had *tons* of baseball cards, football cards, and Eric had all the New York Yankees memorabilia, autographed everything—every year, bats, balls, uniforms." His idols were Reggie "Mr. October" Jackson, amazing relief pitcher Goose Gossage, and Yankee captain Thurman Munson, who died tragically in 1979. "When Munson was killed in a plane crash," Rosin said, "Eric was devastated."

The New York Yankees were virtual family to Nederlander. Among his father's many business ventures and holdings, he was managing partner of the team. Babe Ruth. Mickey Mantle. What more could a kid ask for? When Nederlander married Sklar, one of the many prominent wedding guests at Blantyre was baseball great Willie Mays.

In an odd, six-degrees-of-separation coincidence involving *Seinfeld* and Nederlander, George Costanza worked for the Yankees organization for a time, and George Steinbrenner, a close friend of Robert Nederlander's in real life, was poked fun at. The face of the *Seinfeld* Steinbrenner was never shown; only his voice, which was that of Larry David's, was heard. Like Nederlander, Jerry was a rabid baseball fan, but a follower of the Mets, and he created one *Seinfeld* episode in which he appeared to fall for hunky Mets star first baseman Keith Hernandez.

And like Jerry, Eric Nederlander had a compulsive personality that surfaced early.

"Eric would spend hours and hours and days and days organizing his baseball cards," Rosin recalled. "He could be so incredibly neat, organizing his clothing in his closet perfectly."

Despite the family's wealth, Robert Nederlander was frugal, and Eric wasn't spoiled. Unlike Sklar's nouveau riche friends at college, Nederlander didn't get a new car when he turned sixteen, but was given his older brother, Robert Nederlander Jr.'s beat-up Pontiac Sunbird, an inexpensive car to begin with. While Eric's mother drove a flashy

Corvette, "his father drove a Lincoln Town Car basically until the wheels fell off," noted Rosin, asserting, "Eric was raised *very* low-key."

When Eric was growing up, his father owned the enormous fifteen-thousand-seat Pine Knob Amphitheater in Clarkston, Michigan, setting for standing-room-only music concerts. "As kids, it was great," Rosin reminisced. "For Eric's birthday, or whatever, we'd go to the concerts and sit in the front row. And if Eric wanted, we'd go backstage and meet the bands. Bob Seeger always used to come to town around Eric's birthday and we would have the privilege of sitting in the front row. But Eric was *never* pushy, he was the last guy who would try to get in by using the Nederlander name, or try to get special parking, or special privileges. Even though he could, he never would. Despite everything they had, they were very humble people. Despite their wealth and power, they were no-show, all-dough—not your New York–L.A. flamboyant types."

Looking back fondly on those years, Rosin asserted, "We definitely had a good childhood."

But there was another side to Eric Nederlander's life in the big house on Private Lake Drive with the poetic apple orchard in back.

According to family insiders, Eric's mother was known to be cold, and was dubbed "Hazel the Witch" by those familiar with her personality. And while Robert Nederlander was any boy's dream father—how many sons can make the claim that their dad has a piece of the greatest baseball team in history?—he was a driven businessman who was rarely around during the week.

"Bob, the father, worked hard and wasn't home until late, except on weekends," Rosin said. "He was good to Eric and they were close, but he was busy with business and was around mostly on Saturdays and Sundays when he spent time with his sons, took them to the movies, the football games, the tennis courts."

In Eric's room, his mother painted a blue sky with clouds on the ceiling. Eric, the jock, tried to redecorate as best he could. "He put all of his Little League pictures in his room," Rick Rosin said, "and he hung his baseball hats on this little antique chair with a mirror. He put his baseball player posters in his closet. You'd open up the closet and it was all sports."

Meanwhile, many of the walls of the Nederlander home were covered with Caren Nederlander's abstract art and photography.

She apparently inherited her talent with a camera from her father, a self-employed photographer who was also a master bridge player. He and Caren's mother had earned the title of Life Masters, one of the highest honors awarded to bridge players.

In June 1991, Caren Nederlander's collection of "impressionist" photographs was exhibited at a gallery in Bridgehampton, New York. By that time, she and Eric's father had divorced after twenty-three years of marriage.

A press release from the gallery that presented her exhibit described her as a "world traveler and New York socialite" and went on to say that her collection "represents the intellectual creativity and emotional energy she redirected after her glamorous and privileged lifestyle was interrupted by divorce . . . . Although Nederlander also is a psychologist and marriage counselor, her foray into fine-art photography was her way of coping with the 'traumatic' adjustment to 'aloneness' after the divorce."

Nederlander was quoted as saying, "It was never my intention to be a photojournalist . . . I developed my own distinct camera technique. From a car or train, while walking or on horseback, I would take my photographs—joining motion with color to achieve a 'painterly' illusion."

The release noted that Caren Nederlander and her ex-husband "are the proud parents of two adult sons, Robert Jr. and Eric." The divorce, which received no publicity and was delicately handled in a typical low-key Nederlander style, was finalized when Eric was in his late teens.

After high school, Eric jumped from college to college—the University of Michigan, Michigan State, the University of Arizona, and finally Boston University.

Before he met Sklar, "Eric never had a lot of girlfriends," Rosin stated. "When I was at Arizona with Eric, there were lots of girls but he would never really get involved. He kept to himself. He was never a womanizer." He was, then, much like Jerry when it came to the opposite sex.

According to Rosin and others in Nederlander's circle, problems that existed at the Nederlander home when he was growing up had a profound impact on his later relationships with women, Jessica Sklar in particular.

People Sklar befriended after she arrived in New York described her as "outgoing . . . wanted to be the life of the party . . . had a hyper level of energy . . . was very positive and always willing to give a kazillion compliments, almost to the point of making you feel uncomfortable . . . she wanted everyone to like her. She wanted to be liked and she wanted to be the center of attention and the one who got the laugh."

Within the first year of their turbulent five-year relationship Nederlander and Sklar had a major fight and she left him for the first, but not the last, time, at that point, supposedly for her old college boyfriend. Because Sklar was some five years younger than Nederlander, problems developed due to their experiencing different things in their lives, as one friend of the couple observed. Jessica wanted to play, go clubbing, things that Nederlander, who was now trying to establish himself in the family business, had already done. When Sklar left him that first time, Nederlander was devastated and sought the help of his best friend's wife, Drita Rosin, who had not yet met his girlfriend.

"Eric called me nonstop, night and day, asking me how to get this woman back," she said. "It's funny because he literally wrote down exactly every word I told him to say, and he got on another line and left a word-for-word message—and she fell for it hook, line, and sinker. I told him to tell her—'Pick up your stuff. It's downstairs with the doorman. It was nice knowing you. Have a good life.' Sure enough the next phone call I got from Eric they were back together. It worked like a charm. Just knowing that aspect of her, but not having met her, I wasn't sure what to expect."

But the Rosins were pleasantly surprised.

They were introduced to Sklar for the first time at the opening of the first Nederlander play on Broadway that Eric was involved with, called *Play On.*

"She was extremely warm, fun, bubbly—and very enamored with the fact that Eric was putting on this play and there was all this limelight to it," Rick Rosin said. Afterwards, the couples went to Nobu, a chic Manhattan sushi restaurant, and had an overall great first weekend together.

"Jessica became like a little sister to me," Drita Rosin, a few years older, noted. "Early on Jessica was very friendly—*very* friendly. I really liked her and found her to be a nice person. Is she extraordinary, a super-special person? No. But she was very bright, had a lot of knowledge about a lot of things, fashionable and chic, wearing the right labels—labels and designers are very important to her, eating at the right restaurants is very important to her.

"They stayed with us in Aspen and she called me two months before the trip, to make reservations at the places to be seen. She liked to be in the right spots, with and seen by the right people. They traveled a lot, ate out every night. He introduced her to some really great people, some celebrities. Eric's dad, for instance, is very close friends with Mike Wallace, so these were people she wouldn't have exposure to otherwise. She definitely enjoyed the lifestyle that Eric was giving her."

Nederlander's best pal was thrilled for him, thrilled that he was thrilled. "Everything between them," said Rick Rosin, "seemed really healthy and happy. I thought it was going to be a really good thing. Jessica seemed extremely genuine."

—40—

Stop, Thief!

In 1996, after they were perceived by everyone in their circle to be a devoted couple deeply in love, Jessica Sklar went into hypermarriage mode. The pressure on Eric Nederlander was constant. Though he took her on first-class vacations and bought her expensive clothes at Barney's and chic boutiques in trendy SoHo—they shared good taste—Sklar's objective was the MRS, with a big ring.

"A year before they got engaged Jessica was *relentless* about trying to get Eric to propose to her," stated Drita Rosin, who admitted to having some culpability in the all-out assault on Nederlander's bachelorhood. "As Jessica saw it, all of Eric's friends were getting married and she was a bridesmaid umpteen times. So between her and I we were always scheming. We used to gang up on Eric together to get him to pop the question. I mean she was *ruthless* about it. She just didn't hold anything back. You know, 'Eric, let's go to Tiffany's,' 'Eric, what are you waiting for?' It was constant.

"But Eric just doesn't jump into anything. He's cautious about everything. He wanted to really make sure that she was the one. As much as he loved her and there was no doubt that he did, I just sensed that he didn't want to rush into anything. She was the one, but he was going to take his time and make sure."

Ricky Rosin, a businessman, had a Wharton School take on his best friend Eric's reluctance to take the next logical step in his relationship with Jessica.

"It was like Eric had an option on a piece of real estate and he kept extending the option and finally it was like you're either going to close, or you're out of business. You have no more due diligence time. It's time to put up," he said. "Jessica was making the marriage demand and she wasn't going to waste any more of her life.

"She wanted marriage and she wanted to be a mother, there's no question. But Eric had been on and off with her for almost five years. As sweet as Jessica was, she was definitely into being Mrs. Nederlander. She was not going to marry Mr. Stockbroker, even if he was comfortable. She wanted more than comfortable."

A former close friend of Jessica's in New York, a woman in media, noted that she never recalled Jessica saying she was in love with Nederlander.

"In fact, after Jerry came into the picture, she said she *didn't* love Eric. But Eric was mesmerized. Jessica can turn on the charm. She's fun and entertaining. She's very complimentary to people—*overly* so, in order to ingratiate herself—'You're *so* beautiful,' you're so this, you're so that. But from what I saw and heard they fought a lot; they both sort of got off on fighting."

The wife of a couple who had been extremely close to Jessica and Eric for the entire time of their relationship—and were shocked and devastated by the manner in which it ended—were surprised at how aggressively she pursued him to get married. "We would go out with several couples and she would say in front of Eric, 'How lucky you guys are that you are engaged,' or 'It's great that you two got married.' And then she'd turn to Eric and confront him with, 'Are you *ever* going to marry me? Why won't you marry me?' And she'd actually break down in tears in front of everyone, and it was embarrassing for Eric and for us—and that's a pretty intimate and emotional scene for her to play out in front of people, even close friends.

"But I certainly didn't see her wanting to get married to him for another reason, other than she cared about him and loved him. I never questioned her motives. I cried at their wedding. But I'm one of the people who she pulled the wool over."

Barraged, besieged, and harangued by Jessica and her cheerleaders—with the words MARRY ME!—ringing in his ears, as if he had taken a shot from Mike Tyson, Eric finally was knocked to the mat, and in spring 1997 asked for Jessica's hand in holy matrimony, to have and to hold and yada, yada, yada.

Now that she had him where she wanted him, the next step was the rock. On his own, he ordered it through a relative in the jewelry business. While Nederlander did shower Jessica with gifts to some extent, he also was very frugal and watched every penny he spent. As one pal put it, "If he bought a pack of gum, he'd count the change and if it was a nickel short he'd go back and get it." So he shopped for the ring, looking for the best price. And when he gave it to Jessica, she wasn't thrilled.

It was a unique diamond, estimated to have cost $25,000 wholesale, and weighing perhaps 2.5 carats, and cut rectangularly with most of the weight in its depth. When one was looking down upon it, the stone appeared smaller than it was because it was so deep.

A day or two after he slipped it on her finger and she went *yuck*, they celebrated their engagement with the Rosins.

"It was a great disappointment to her when she saw it," Drita Rosin said, "because it *looked* small, even though it wasn't small. But she wanted a *big* ring. She tried not to look like she was disappointed because it would have seemed like she was mostly interested in having a big ring on her finger. But she seemed very, very happy to be engaged. It was definitely a celebrated moment."

Nederlander had given the ring to Sklar in New York, where James Meiskin heard from Nederlander that he was so angry at all the pressure she had put on him that when he finally got the ring "he wound up throwing it at her, 'Here's your fucking ring.' "

In any case, the deed was done and according to friends, Nederlander and Sklar appeared happy. Right after the engagement, about a year before the wedding, Sklar became a woman of leisure, quitting her job at

Golden Books, where she had started as an assistant in the marketing department of the publishing company's entertainment division. Her job entailed working on some of the company's licenses, ranging from "Pat the Bunny" to "Rudolph the Red-Nosed Reindeer." By the time she quit she had reached a position of manager in the licensing department, but "her level of expertise in the marketing area was very limited," a Golden Books' executive said. "At the end of her time at Golden she was entrusted to try to license product out, but it was never with any great success."

In the months leading up to the wedding, Sklar's life was taken up with three leisurely activities: preparing for the big event, attending many, many, many wedding showers thrown for her by her friends, and getting herself in shape.

As a gift he will probably always regret giving, Nederlander presented his fiancée with an expensive membership in the tony Reebok Sports Club, on West Sixty-seventh Street and Columbus Avenue, on the Upper West Side, a short walk from Nederlander's doorman building at 45 West Sixty-seventh Street, and not far from Jerry's multimillion-dollar duplex on the park. The 140,000-square-foot colossus, which had everything from the highest-tech workout equipment to rooftop inline skating, a spa, beauty salon, swimming pool, and even a rock-climbing wall, catered to Manhattan's young, upwardly mobile, and affluent. There also were a number of celebrities who belonged, one of whom would soon include Jerry, who liked to keep fit and boyish-looking.

"If Jessica hadn't been engaged to Eric, she never would have met Seinfeld," Rick Rosin observed. "That's the reality of it because it was Eric who bought her the membership to the Reebok Club."

With very little else to do with her life, Sklar went to the gym on an almost daily basis, and when the weather got warm enough she'd walk to and from the facility in her tights, shorts, and sneakers, and while she didn't possess a *Baywatch* figure, she was cute and did attract the stares and sometimes the whistles of men. In fact, one of the reasons she wanted to go to the gym—at least initially—was to get in shape, so she'd look buff on her wedding day.

On a visit to the Rosins a week or two before the long-awaited nuptials the subject of Sklar's workouts at Reebok came up. The two couples were sitting chatting after dinner and Eric had been complimenting Jessica about how she was getting in shape, and even asked her to show her arm muscles to demonstrate how defined and toned they were.

"Look at Jessica," he boasted. "She's really getting buff. She's in great shape." When Ricky Rosin and Nederlander moved out of earshot, Sklar had a Cheshire grin on her face.

"Ricky and Eric had kind of moved away and Jessica told me, 'You know, I'm working with this trainer and he's really cute and he's gorgeous and we flirt with each other, and Eric's jealous, but don't say anything about this around him,'" Drita Rosen said.

Rosin saw Sklar's revelation as an invitation to rib Nederlander, thinking there wasn't anything to it. "I'm like, 'So Eric, tell me about this handsome trainer Jessica has.' And Jessica went, 'Shhhh, don't say anything, don't say anything!' And I said, 'Oh, is he cute? What's going on with those two? Jessica tells me he's really cute.' I was just kind of instigating, joking around and little did I realize there might be more to the story.

"Eric said, 'Why, what did she say? What did she just tell you?' And I said, 'I hear he's really hot and she's into him.' Jessica freaked out and Eric freaked out. At the time I never seriously thought that anything was going on between her and her trainer.

"All Jessica told me was that they flirted with each other."

A friend of Jessica's, who also worked out with the same trainer at Reebok, said Eric was convinced that Jessica had a relationship with the trainer, but the friend thought it was "probably" nothing more than a flirtation. "I don't know," she said. "He's a big flirt, very athletic, *real* cute, very jovial. Jessica's not great-looking, but she's also flirty. She can turn it on. She can be very sweet, very complimentary and anybody with an ego eats it up." The friend said she talked to the trainer about Jessica after the scandal with Jerry became public, and all he would say about her was, "She's crazy. They're *all* crazy."

Sometime in the month between the May 14 end of *Seinfeld* and the June 13 wedding of Nederlander and Sklar, Jerry and Jessica met at the Reebok gym. This was later confirmed by prominent New York publicist Howard Rubenstein, who represented Nederlander after the Seinfeld-Sklar relationship became public.

"They knew each other from the gym *before* the wedding," Rubenstein stated. Jessica, according to a friend at the time, confided that she had met Jerry and that they had hit it off and were getting along. The friend was dumbfounded, because she had been invited to Jessica's wedding.

For Jessica, meeting Jerry was incredible, because she was a rabid fan of *Seinfeld* and had often expressed that she loved him—found him so funny and sexy. And now he was showing interest in her in real life. If life imitates art, the meeting at the gym of the fan and the star was a textbook case. In "The Contest," the classic *Seinfeld* masturbation episode, Elaine meets her dreamboat John F. Kennedy Jr. at a Manhattan gym—he flirts with her, she flirts with him, yada, yada, yada. It was as if Jerry was turning his script into a real-life situation.

Yet it's difficult to imagine that Jerry, always business-first-pleasure-second, had time to flirt with a cute girl at the gym, because the period right after the final episode of *Seinfeld* was a hectic time. Immediately after the hullabaloo over the show had subsided—he was said to have driven back to New York in one of his Porsches—he took his stand-up act on the road again, playing Australia, Iceland, and England.

However, after all those years of commitment-phobia, the plan of settling down was definitely on Jerry's mind, at least according to his close childhood friend Cliff Singer, who had become a psychiatrist with a practice on the West Coast. Singer also had become something of an unofficial spokesman for Jerry and whenever the gossip columns called, he happily offered up his thoughts about his famous pal; he was later gagged by Jerry's publicist. When the *New York Daily News* came calling in the period leading up to the final episode and asked Singer what Jerry intended to do with the rest of his life, he told them Jerry's plan was to move back to New York and start a family, which, of course, is precisely what happened.

Jerry actually did appear to have a master life plan, with everything falling into place in his favor. Break into stand-up comedy, check. Move to L.A. and appear on Johnny Carson, check. Become one of the country's most beloved comedians, check. Develop a hit TV show, check. End the hit TV show at precisely the right time, check. Now, Jerry believed, with all the other passages out of the way, it was his time to get the wife and then the kid.

Outside Blantyre, the beautiful country estate, the sky opened and the rain fell in torrents on the weekend of June 12–14. Inside, it was emotionally dark and stormy, too, despite the fact that one of the most elaborate and incredible weddings ever held there was about to take place. Finally, Jessica Sklar was getting everything she so desperately wanted, everyone thought, but to those guests and principals who knew her best, she didn't appear happy, not like a woman in love who was about to take the sacred vows of holy matrimony.

One guest who watched her pose for a photographer with her groom said later that her smile was forced and her face looked dead.

Drita Rosin, who stayed at Jessica's side much of the time, vividly recalled, "Jessica just didn't really seem in love. She didn't have that in-love, newlywed feeling."

Knowing her better than anyone else present, or at least he thought he knew her best, the groom was extremely concerned. At one point he took Rosin aside. "Drita, go talk to Jessica. Go see what's wrong. Tell her I'm worried about her. Tell her I love her. Tell her everything's going to be okay."

Heading for Sklar's rooms, Rosin remembered thinking, "She's probably just suffering wedding-day jitters; it's so sweet that Eric's so concerned."

She also thought that some disturbing and anxiety-inducing incidents that had occurred during the wedding weekend had upset Jessica.

For instance, Eric had gotten into a fight with Gladys Nederlander, the woman his father had married after his divorce from Eric's mother. After the marriage, the talented new wife coproduced a number of well-received plays.

"They were having a rehearsal and that was when Gladys found out that she was not going to be part of the wedding ceremony," Drita Rosin recalled. "This happened during the day of the wedding. Gladys had hoped that she would walk down the aisle with Eric, with her husband on one side of him and maybe Caren Nederlander on the other. But Eric didn't want Gladys in any part of the actual ceremony, so she started to make a stink, and Eric yelled at her, 'Shut up, you fucking bitch.' This was probably the first time he was this frank with her, and he said it in front of the immediate family and close, close friends."

The meeting of the two families—the Sklars and the Nederlanders—did not come off as well as everyone had hoped it would, either. "They were like oil and water," one Nederlander family friend observed later. "The Sklars are like nice country folk, really not into the glitz and glamour, and here were the Nederlanders with all their money and all their glory. It was a strange mix and there was no real bonding that I could see." The Rosins liked the Sklars, but as Drita pointed out, "They weren't the kind of parents you would want to go around flaunting to all these wealthy New Yorkers, high society, phony kind of people. It was important for Jessica to be in the right crowd and they didn't mix into her world. They were more like Flo and Joe, the next-door neighbors."

The other issue that Drita Rosin thought might have put Jessica in a sour mood on this happiest of days was the behavior of her sister Rebecca, who was one of the bridesmaids.

"Jessica didn't really get along with her older sister," Rosin said. "Rebecca's prettier than Jessica, but here was Jessica marrying Eric Nederlander and having this incredible dream wedding and I could see a lot of jealousy from Rebecca."

By the time of Jessica's wedding, Rebecca and her husband, James Meiskin, had separated. Rebecca Sklar Meiskin had left public relations and now was working as a contributing producer in the New York bureau of the celebrity TV magazine *Entertainment Tonight*.

In any event, the day of Jessica Sklar's wedding had been "a very emotional day," according to Drita Rosin. "There were all these crazy things going on behind the scenes that I thought had distracted Jessica. As Eric had asked me, I went into Jessica's suite. I told her, 'Eric sent me here. He wants to know how you're doing. Is everything okay?' And she was just emotionally blank. She wasn't depressed. She wasn't down, but I didn't see a loving, happy about-to-be bride sitting there. I told her how lucky she was, how much Eric really loved her. I would have been touched. I expected her to say, 'Oh, I love him. I'm so excited.' That was not her response. Her response was just unemotional. There was not that loving, sweet, how-thoughtful feeling. She was just blank."

Later, Rosin looked back on that day and on how Sklar was acting and came to the conclusion that her lack of emotion was due to her having Jerry Seinfeld on her mind.

Nevertheless, the wedding ceremony came off as planned. There were flowers everywhere, thousands of dollars' worth, there were so many rose petals covering the long aisle leading to the *huppa* where Rabbi Stacy Bergman joined Jessica Sklar, 26, and Eric Nederlander, 32, in matrimony at 6:30 P.M. on June 13, 1998, that guests couldn't see the expensive carpet below. Afterwards, tears and celebratory Dom Perignon flowed through the night.

The next morning Jessica was up early to read the New York papers delivered to the bridal suite before the farewell brunch. The wedding made the *Times,* naturally. And reading through the other New York papers, Jessica Nederlander and a girlfriend were overjoyed to see that her nuptials got bigger play than those of the daughter of a noted designer who had also gotten hitched that weekend. "They were just loving it," Drita Rosin recalled, "and I rolled my eyes and thought, It's that fucking New York *thing*. But it made Jessica happy, having people hear or learn about her."

Not long after the wedding, the Nederlanders flew off to Italy for what was to be a month-long honeymoon at the spectacularly luxurious Villa D'Este on Lake Como, a playground of the rich and famous—from Napoleon to the Duchess of Windsor and now to Jessica Sklar Nederlander. Jessica's sister, Rebecca, had honeymooned there with Meiskin, so Jessica wasn't about to be outdone by her.

Upon the Nederlanders' arrival, Jessica ran into the Long Island–bred fashion designer Michael Kors, a gossip-column boldface name. Nederlander later complained to Meiskin that he was forced to stand there like a spot of caviar on the rug while his bride gossiped, networked, and talked fashion for a couple of hours. "Even Michael Kors was looking at Eric like, 'What the fuck's with her? What's going on here?'" Meiskin recounted.

In Italy, Eric turned over his Jerry Seinfeld–promoted American Express card to Jessica who went on a shopping spree to end all shopping sprees, racking up thousands of dollars of charges on clothing and accessories and Prada, Prada, Prada.

Then, three weeks into their honeymoon, she demanded they cut it short and come home early. Later, Nederlander told friends, "Jessica was suddenly antsy to get back home. She wanted to end the honeymoon as quickly as possible." Their arrival back in New York coincided with the one and only postcard Jessica sent to their friends, the Rosins. She said how great the shopping was and that she was able to *handle* 40 percent off Gucci, recalled Drita Rosin. "She also said the wedding was great, we had so much fun, we miss you. But there was nothing romantic in it."

About five days later, Rosin was rushing around doing errands when the phone rang and it was Jessica, who was breathless and talking a mile a minute. "I said, 'Wow, married life must be good. You just got back from your honeymoon and you sound like you just got out of bed with Eric.' And she said, 'Oh, my God, you're not going to believe what's going on.' She sounded hysterical. And I said, 'You caught me at the worst time. I'm running out the door, call me back in an hour or two,' and she said, 'Ugh! Okay.'

"I came home a couple of hours later and Eric called. 'Are you sitting down?' he asked. I asked why and he said, 'I think you need to sit down,' and I said, 'Well, what's up?' and he said, 'Jessica's leaving me for Jerry Seinfeld!' And I swear it took fifteen minutes for him to convince me. I thought it was a joke."

Eric told Rosin that out of the blue Jessica demanded that they end the honeymoon early. Back in New York he discovered that Jessica was seeing Seinfeld whom, he said, she met at the gym, and now she was making up excuses to argue with him and was threatening to pack up and move out. "We spent many, many hours on the phone," Rosin said.

At first Eric demanded that Rosin not talk to his wife, but Rosin convinced him that she should hear the whole story from Jessica, but they made a pact that Rosin wouldn't let Jessica know when she called that she was aware of the Jerry situation.

"Jessica called me that same day," Rosin recounted, and told her "how miserable she had been, how their sex life was nonexistent, how she couldn't take him anymore. She was telling me things you would hear from a couple who had been married for years, not a month. And she said these were the reasons why she wanted to pack her things and move out. She never once mentioned Jerry Seinfeld. My advice to her, without my being able to deal with the Jerry Seinfeld factor, was, 'You love each other. You just came back from your honeymoon. Try to work it out. This doesn't make any sense.' I told her that nothing had happened up until this point to warrant this kind of reaction from her. I don't know if I made much headway, because we were not dealing with the real issue, which was Jerry Seinfeld."

Rosin now thought back to the way Jessica had acted on her wedding day and it all began to make sense. "Yes, she did meet Jerry before the wedding," she remembered thinking, "and then she just wasn't into marrying Eric."

Rosin's next conversation was with Eric Nederlander. She told him what Jessica had said, "some of which was embarrassing for him to talk about" with her, and then he gave Rosin permission to bring up the Jerry Seinfeld matter with Jessica.

Like Bill with Hillary during Monicagate, Jessica stonewalled.

"I said, 'Let's talk about Jerry Seinfeld,' and she said, 'He has *nothing* to do with it.' She admitted she had met Seinfeld and said they were friends but that nothing was happening between them. She said she and Eric had been having problems long before they got married. I said, 'Jessica, you are not being honest with me. If you really are seeking my help and advice then let's be honest. I can understand how you would be flattered and impressed by Jerry Seinfeld, but you are a newlywed, you've got issues to deal with.'

"But she denied any relationship with Seinfeld. I couldn't get her to tell me she was interested in him. I said, 'If you ever want to talk to me *honestly* and discuss what's really going on call me, but other than that my loyalty is with Eric, and if you can't be honest with me, then we have nothing to talk about.' She said she'd call me back. We never talked again."

Some months later Drita Rosin saw Jessica Sklar Nederlander with Jerry at a restaurant in Aspen. "He had his arm around her and they looked very much in love," she recalled. "She avoided sitting near us."

James Meiskin had heard about Jerry and Jessica from her sister, Meiskin's ex, Rebecca, with whom he had remained on talking terms.

"Rebecca told me about the Seinfeld-Jessica affair before Eric even knew about it," he asserted. "Rebecca told me, 'Believe it or not Jessica's seeing Jerry Seinfeld!' We were having breakfast at a restaurant, a little diner on Third Avenue, and Rebecca was on her cell phone with Jessica and talking to Jessica about the situation, telling Jessica that she should go back to her husband, that Eric's a great guy—'What are you doing? You're making a mistake.' I said to Rebecca, 'Why is she doing this?' And she said, 'Because she knows what happened to the two of us and she doesn't want to be unhappily married.' According to Rebecca, Jessica said she didn't want to go through a divorce five or six years into the relationship with Eric, and that she'd rather get it over with now. I don't think Rebecca ever thought that Jerry was going to marry her."

Besides the Rosins, Jessica cut off all communications with other longtime friends of her and Eric's.

In July, with Jessica back in New York struggling with her demons, Jerry once again happily took to the road—his one and only true mistress, Sklar would soon learn—and began a national comedy tour: San Antonio, Omaha, Des Moines, and the Comedy and Magic Club in Hermosa Beach, south of L.A., where he did three sold-out nights at the 250-seat club as a favor to the owner who gave him a break as a headliner there two decades earlier.

He ended his tour on August 5 in New York—setting for his lionized sitcom, his new life, and his wife-snatching—with ten performances at the Broadhurst Theater (not part of the Nederlander chain).

Jerry said he was glad to be back in New York and noted that his favorite thing about the city—his affair with Sklar had not yet become public—was the Yellow Cabs with the messages recorded by celebrities. "I only wear my seat belt," he told the *New York Times*, "because Judd Hirsch tells me to."

His show *I'm Telling You for the Last Time* culminated in a live August 9 HBO eighty-minute special with Jerry's pledge that he would "never do any of these jokes again," and his vow to begin work on all-new material, that is if, some insiders thought, Eric Nederlander didn't come after him with a shotgun first.

Nederlander told Drita Rosin that Jerry had asked his wife to attend a performance of the show and she accepted.

"Eric told me she came home from the gym and she wasn't wearing her wedding ring and she told Eric that she met Jerry Seinfeld and he invited her to his show," Rosin said. "And Eric asked Jessica, 'Did you tell him you were married?' and she said, 'No.' And he said, 'Why not?' and she said, 'I don't know. I just didn't.' And Eric wasn't threatened by it. Why should he be? He had trust in his wife. I believe Jessica went to the show with a friend and the next day met Jerry again and didn't mention she was married, and had a date with him that night.

"Eric had left to go to the Hamptons for a stag party, and Jessica called him while he was away and started an argument that was, in Eric's mind, just out of left field. It didn't make any sense to him, but it scared him. And she threatened that she was going to move out. So he drove back from the Hamptons that night and they talked and he felt they had patched things up and he went back to the Hamptons. When he came home the next day, a Saturday, she got into another huge fight with him, and she packed her bags and she went to her grandmother's or a friend's. She hadn't officially moved out, but she staged an argument, made something up, and couldn't get out of there fast enough."

Not long after that scene, Nederlander was said to have learned that his wife and Jerry had met before the wedding.

Meiskin said he was told by Nederlander that he had returned to their apartment and "all of her stuff had been moved out by Jessica in big, green Hefty garbage bags, and she had moved in with her grandmother. Can you believe it? Coming home to find that your bride's moved all of her shit out?"

Meanwhile, with the scandal still not public, Jerry's show got mixed reviews. *Variety*'s Charles Isherwood called it, ". . . arguably the most overhyped media event since, well, the last episode of a certain sitcom about nothing . . . . Here's a philosophical question: Without *Seinfeld* would the hoopla about Monica Lewinsky's blue cocktail dress have been possible? I doubt it. One somehow feels Kenneth Starr was a 'must-see TV' fan." Isherwood gave Jerry's show a C.

Entertainment Weekly's critic, A. J. Jacobs, asked after watching Jerry's old tried-and-true bits, "When was the last time the lionized multi-gazillionaire had to fly coach and watch that little curtain separate him from first class? It makes you wonder what Jerry's future bits will sound like: 'Didja ever notice that when you're on your own private Learjet, they never have enough pâté de foie gras? What is the deal with that?' . . . No matter, if you suspended your disbelief, you were happier than Kramer puffing on a Cuban . . . [Jerry] was elegant, seamless, and as tidy as that fictional apartment 5A."

Luckily, Jerry's affair with Sklar Nederlander didn't affect his sold-out show because the story didn't become public until August 28, 1998.

Under a headline reading "JERRY & A MRS. GET PHYSICAL," the lead item of the *New York Daily News* gossip column said, "Jerry Seinfeld is working up a sweat with another beautiful brunette. The bachelor comic is keeping fit by having regular workouts alongside Jessica Sklar, a 26-year-old stunner with more than a passing resemblance to Seinfeld's last girlfriend, Shoshanna Lonstein . . . But there's one big difference between the two gals—Sklar is a newlywed."

The column went on to say that "the friendship may be taking its toll" on the marriage and noted that Jessica "denies chatter" that she was ready to leave Nederlander for Jerry. The column quoted Jessica as saying, "I'm friends with Jerry. There's no romantic interest whatsoever. He and I have, like, basically the same schedule right now. I'm looking for a job and we just, like, hang out and talk at the gym." She said her husband thought "the rumor of a romance is 'funny.' "

Oddly, Nederlander chose not to accuse his wife of what he knew was going on, but rather to deny it, at least publicly. He was quoted by columnists George Rush and Joanna Malloy as saying, "I know all about their friendship. I know they see each other at the gym . . . they talk to each other on the phone . . . I have no problem with it. Our relationship couldn't be more solid. Jerry Seinfeld is not going out with my wife, and if he were going out with my wife, then we'd have some real problems. That's not what's happening here." When the paper called Seinfeld for comment, his publicist Elizabeth Clark said the usual: "Jerry prefers to keep his private life private."

Meanwhile, Nederlander was devastated.

"I can't believe that Eric didn't end up in a mental institution," observed Drita Rosin. "After it all went down he spent some time with us in Aspen and he was really, really down in the dumps, and just very, very depressed. The only comfort that I could give him at the time was that she's going to end up with nothing, she's not going to have Jerry, and not have you, and have a horrible reputation. I told him, 'You'll find somebody better. It's good that you didn't have kids and find out later what type of person she is.' Those are really the only words of comfort I could find to give him."

In New York, Nederlander had support of friends who were once close to both him and Jessica, but who now had taken his side, shocked and revolted by the turn of events. All of them saw how their friend was emotionally wounded.

"Before Eric met Jessica, he was very low-key, the type of guy who would walk around in a T-shirt and a pair of sweatpants and watch baseball and football games," said James Meiskin. "He was an easygoing, regular sort of Joe. And she changed him into an angry and bitter person. Jessica was his life, the center of his universe. He really loved her. He was obsessing. He was very concerned about going out in public because every time he went out the press would take pictures of him. He was concerned about what it was doing to his reputation. And it totally destroyed him. I would go over to his apartment and he still had the dried wedding flowers. I said, 'Jesus, Eric, come on. You shouldn't be looking at those.' The thing is—he would have taken her back in a second."

Jessica wasn't the only married woman Jerry had pursued. Around the time he met Jessica, he was seeing Jennifer Crittenden, a one-time *Seinfeld* writer, who, typically, was much younger—in her case, fifteen years his junior. They were seen together in the summer at the 48th Annual Pebble Beach Concours d'Elegance car show, where they were photographed with his right arm around her shoulder and their hands entwined, both wearing sunglasses. She also accompanied him to the Emmy Awards in September, at least five months after he started seeing the engaged Sklar, and three months after the blushing bride left Nederlander for him.

Cliff Singer said Jerry and Jennifer "were attracted to each other when they were on the show, but they respected the work relationship. She was also married. It wasn't till that was done that they started dating. They waited."

Then he said, "Those boundaries are sacred to Jerry."

But according to Crittenden's husband, Jace Richdale, who had met his wife when they were writers on *The Simpsons*, Jerry had no scruples when it came to married women. He asserted that Jerry hit on his wife at a time when the couple was trying to resolve their problems. "It's almost like he gets a kick out of ending people's marriages," he observed. "It's like he has everything, and [married women] are a turn-on for him."

The furious husband said that after he and his wife separated "she was involved with him within a few weeks. We were still in marriage counseling when she started seeing him. I honestly couldn't say whether he was seeing her during our marriage, but I have wondered. I don't know Jerry Seinfeld. But this does seem to be a pattern with him."

—41—

And Baby Makes Three

If Jessica Sklar Nederlander wanted her fifteen minutes, she now got it delivered to her on a silver platter—and it continued virtually unabated for months, on page one, in the gossip columns, on tabloid TV, and on the Internet.

A paparazzi photo of her with Jerry appeared in the *National Enquirer* in supermarkets across the land, headlined "SEINFELD STEALS ANOTHER MAN'S WIFE—3 WEEKS AFTER WEDDING." As a sidebar, the king of the tabloids displayed the wedding *Playbill* photo of her and the touching caption regarding her love for the groom, which now seemed so ironic.

People magazine covered the scandal under a hilarious headline that asked, "MASTER OF WHOSE DOMAIN?" The celebrity-driven weekly with a circulation in the millions noted that it was a shame *Seinfeld* had gone off the air "because there's a great script idea floating around Manhattan," but wryly noted, "This isn't a sitcom; this is real life."

TV Guide declared, "Who says *Seinfeld* is over?" and went into depth about what it called Jerry's "own bizarre romantic storyline." The headline read: "THE SEINFELD AFFAIR: WORKING WITHOUT A SCRIPT, JERRY SEINFELD AD-LIBS HIS WAY INTO ROMANTIC ALTAR-CATION."

The magazine quoted Nederlander's mother, through a spokeswoman, as saying, "Eric is heartbroken." She said she had been delighted by the marriage and had welcomed Sklar into the family and thought Jessica was "elated." Regarding Jerry, she said, "Jessica, like Monica Lewinsky, became starstruck."

In a piece called "Brides Who Bed-Hop," *Cosmopolitan* led with Jerry and Jessica's affair and noted, ". . . whatever Jessica's reason for jilting Eric, she's not the first bride to get down and dirty with someone other than the groom." The magazine quoted psychotherapist Carolyn Bushong, author of *The Seven Dumbest Relationship Mistakes Smart People Make*, as saying, "Obviously, she wasn't in love [with her husband], and she's the type of woman who always looks for the better deal. There are a lot of women like that. Yes, she's probably in love with Jerry, but so is half the country."

The *New York Daily News* caught Jerry "stroking" Jessica's neck at a trendy downtown restaurant, but reported that they "hopped into separate cabs when photographers ambushed them outside." Later, the *New York Post* informed its straphanger readership that "the dallying duo had tongues wagging as they smooched publicly" while waiting for a table at an Upper West Side lox-and-eggs emporium, and the tabloid later served up the news that Jerry was helping Jessica look for a new apartment.

Always the lovable, laid-back pro, Jerry seemed to take it all in stride at first, and actually boasted, presumably with tongue in cheek, to the swarm of reporters tailing him daily, "I love scandal. It's fun. Being in show biz is like being in high school." He made the comment to the *News* while standing with his ever-present, longtime friend and traveling companion Mario Joyner.

To the *Post* Jerry laughed off *l'affaire Sklar Nederlander*, declaring outside the Reebok Sports Club while striking what was described as a "he-man" pose, "You know, *I'm* barely interested in my own life. I don't know how *you* could be interested in it."

Sklar supporters now began throwing bricks in the form of a published report that alleged Nederlander had "made overtures to another woman" during the short marriage. The cuckolded hubby's spokesman, Howard Rubenstein, scoffed at the charge. "Eric," he said, "is not going to get into a mudslinging contest with Jessica."

Unnamed friends who had stayed in Jessica's corner now described her to reporters as "the antithesis of a gold-digger. A gold-digger stays with someone for money, regardless of love," said one. "She opted to leave someone who could have given her a very comfortable life. She wants absolutely nothing from Eric or his family. She just wants her freedom. She wants to be on her own and get on with her life." But they never explained why she married and left him in record time.

When a reporter caught Jerry in the street and confronted him with whether he was serious about Jessica, he at first laughed aloud, doing one of his shticks. But then he gritted his teeth and snapped at the scribe, "You're a poor human," as if he had just confronted Newman at his apartment door. He went on to say that all the media interest in his life was "flattering," but quickly added, "Could you imagine asking people about their relationships, like it's high school? It's so sad."

By late October, though, faced with an onslaught of stories in papers, magazines, and on TV, Jerry—using unnamed friends as mouthpieces—denied he had done anything wrong. It was reminiscent of how he had handled the Shoshanna Lonstein affair initially, throwing the blame on the teenager by saying she led him to believe she was older than she was. Now, in this latest scandal, he kind of put the blame on Jessica by having one of his pals say that she "was already separating. She was not wearing a wedding band. Two days after they met she moved out," all of which was a curious response.

The unnamed source quoted by the *Daily News* under a headline reading "JERRY: I'M NO STAGE-MARRIAGE VILLAIN," said of his friend, "He doesn't feel he's done anything wrong. He's getting a little fed up with the way he's being portrayed."

The paper also quoted the ubiquitous Cliff Singer, Jerry's psychiatrist pal, pontificating about Jerry's upright morals and ethics. "Jerry holds himself up to a very high standard. I don't believe [home-wrecking] is something he would condone or accept. I'm sure he's holding up well because I doubt he believes he's done anything wrong. He obviously doesn't have to go into marriages to get women. If Jerry wanted sex, all he'd have to do is walk down into his apartment lobby. It's like a bear in salmon season. All you have to do is reach into the river." Continuing, he said, "He's very respectful of other men and people's relationships. He is resilient. But he's also very sensitive and self-critical. Jerry is not a child. He's very mature. He treats women respectfully."

Call him Saint Seinfeld.

When the *News* contacted Jerry's publicist, there was the usual Jerry-wants-to-keep-his-private-life-private no comment.

Jerry would remain publicly mum on the matter of how and why he and Jessica had hooked up, sparking the marital scandal. It wasn't until almost four years later, in the early spring of 2002, that he curiously raised the issue once again—then mostly long forgotten by the media and the public—in the most unlikely of forums, in a published letter in the Sunday Styles section of the *New York Times*.

By then, Jerry would be doing stand-up regularly again, and had become involved in a children's book project based on his routines about Halloween candy and costumes; with this new children's theme, he was following in the footsteps of his childhood idol Bill Cosby, who had also used humor to entertain little ones.

Jerry's hissy fit and rant in the *Times* had been sparked by a Styles section feature a week earlier that was headlined "A NIGHT OUT WITH ERIC NEDERLANDER," an update on what the one-time love of Jessica's life was doing. The story said Nederlander had moved downtown, opened a theater in Greenwich Village, produced an evening of songs by Janis Joplin, built a huge amphitheater in Connecticut, and was dating a woman eleven years his junior. All in all, a rather bland and uncontroversial piece.

But what ticked off Jerry was the writer's straightforward lead paragraph, which stated: "He surely doesn't want to hear anymore about the wife who left him, three months after their honeymoon, because she met Jerry Seinfeld at the gym and decided to become Mrs. Seinfeld. Except for that twist, Eric Nederlander, 36, is pretty much set."

Pointing to those words in probably the most trivial of mentions regarding his scarlet courtship of Jessica, Jerry went ballistic. He called the lead in the paper of record ". . . yet another example of slapdash celebrity journalism." He said, ". . . I am vaguely aware that your version of this utterly meaningless story is the general perception." He said that anyone reading those few words couldn't believe that such a "sequence of events" could take place. And then he went on to lay out, in stand-up fashion, how the published scenario of events sounded to him.

"So a woman marries a guy, then right after, meets another guy and tells him, 'Hang on while I get divorced, then I would like you to fall in love with me and ask me to marry you, a 45-year-old, famously single TV star, O.K.?"

He called all of the above, "Quite a scheme to plan and execute without a war-room-size map of Manhattan and one of those croupier sticks to push the various model characters around."

Was Jerry planning to bring back *Seinfeld*? The scene he painted sounded like the first draft of an episode for a *Seinfeld* reunion show, which, in fact, he was considering.

In any event, he then went on to say that the "true story" of what happened between him and Jessica had never been reported accurately because, he asserted, it wasn't as hot as the "gossip" that had been reported, which he never defined. Then he pointed out that while Jessica "doesn't have much interest in setting the tabloid record straight," he challenged the journalism community to "determine the actual events," but noted that the resultant story would be a "good deal less tantalizing" than the existing one.

So why didn't Jerry just tell the "real" story and set the record straight once and for all, instead of challenging the press, like Gary Hart who once famously said "follow me," and the world knows what happened to him. Setting the record straight, said Linda Lee, the author of the Nederlander piece, was exactly what she thought after reading Jerry's rambling missive, which she said was emailed to her by Jerry's publicist, Elizabeth Clark. Her other thought, Lee said, was, "I think he has too much time on his hands."

As the *New York Post* responded the next day, "Hey Jer, just give us a call and we'll clear the whole thing up. Heck, we don't want people thinking you fell in love on a treadmill when it was actually a stationary bicycle, do we?"

But all of that Sturm und Drang was still to come.

The *New York Post* was in tabloid heaven with the scandal. They contacted an "expert" named Sammi Sheen who had coauthored a 1998 book called *Is He Cheating: How to Recognize the Signs*, who said Jerry "has the makings of another Bill Clinton or Woody Allen." Another expert, a Long Island psychiatrist named Don-David Lusterman who wrote the 1998 *Infidelity: A Survivor's Guide*, saw in Jerry an element of "narcissism" and "a sense of entitlement," which he said was typical of entertainers and powerful men in general.

By the time autumn descended on New York, Jessica's newly minted husband had retained a divorce lawyer and set the legal wheels in motion to end his short, ill-fated marriage. He told the *Post*, "I was manipulated, misled, and completely caught off guard by Jessica's infidelity. Jerry and Jessica have no respect for decent values. They deserve each other."

There was speculation in the press that Nederlander might drag Jerry into court on charges of "alienation of affection," but he quickly let it be known he had no intention of suing because, as it turned out, the Empire State didn't have what are termed "heart balm" statutes—civil remedies for inappropriate behavior that causes a marriage to fall apart. But Nederlander made it clear that Jessica wouldn't see a red cent of his family money.

Around the time of the firefight, Jerry began the hunt for his biggest known real-estate investment, an oceanfront complex on Long Island, where he had started. Well, almost. This wasn't Massapequa, but rather the trendy Hamptons. Jerry looked at a number of places in the $10 million to $17 million range, some on East Hampton's storied Further Lane—one owned by Richard Avedon, the photographer; the other by businessman James Marsden. But Jerry rejected the places: he "didn't like the rocks on the beach" of Avedon's property, and felt Marsden's was too "Palm Beachy." On another house, the designer Helmut Lang outbid him by $1 million, snapping up a cottage that sold for $15.5 million. A subsequent deal on the East Dune Lane home of Lee Radziwill and Herb Ross for $19 million fell through after Jackie O's sister had a change of heart.

Finally, in 2000, Billy Joel took Jerry's money—a whopping $32 million, for his fourteen-acre estate with two houses, a little schlep from the beach. Joel later bought a house with more bedrooms for $8 million, pocketing millions from his sale to Jerry.

There were things about Chez Seinfeld that Jessica apparently didn't like. The *New York Post* quoted a friend of hers as saying, "Jessica is going to take one look at that nautical stuff and whoof. It's not her taste at all . . . she'll have it all pulled out faster than you can say Dramamine. She'll have adult trees brought in from Connecticut and replanted. She'll want a hard court. She'll want a solarium. She's a Be&Me gal—the Best and Most Expensive always gets her vote."

Joel revealed two years after the sale that he had taken Jerry for a ride on the outrageous price of his place. Doing what was described as "a wicked imitation" of Jerry, Joel said, "I said 'Blah, blah, blah.' He said, 'Well, okay!' I said, 'Okay?' He said, 'Okay!'" Jerry had given a thumbs-up to the $32 million number that Joel declared he had pulled "right out of my butt."

All of that, though, was still to come.

Back in the fall of '98, Jessica was in for a nasty surprise. Those Seinfeld feet, clad in signature virginal-white Nikes, suddenly became very cold, and by the turn of the year he told her he needed some space, needed to think things out, gave her the old "It's not you, it's me" line, and distanced himself from her.

The reason? Mr. Commitment Phobia had once again entered the picture. Jessica was not a happy camper, and told friends that Jerry was the one who wanted to meet her at the gym, not the other way around.

By late winter 1999, with the divorce about to be finalized and with Jerry cooling his heels, Jessica was on her own for a time.

"After she left Eric, Jessica was with Jerry all summer and fall and into the winter—and they even took trips to Hawaii and Sicily together, but then they broke up around February 1999," said a close friend of Jessica's. "Jerry just wasn't ready to get married. He was having commitment issues.

"I spent time with them and he's a strange character, fiercely private, controlling. I couldn't figure out what he saw in her, because intellectually they are not on the same level, but that may be a part of the attraction for him. They mostly just stayed in his apartment and hung out. He has a very close friend, George Wallace, and I thought they had a strange relationship. Jerry's a weird guy. I never knew if he was being sincere or sarcastic, and it can be taken either way. It was clear that he was the boss in their relationship. Jessica tries to be domineering and in most relationships she is, but with Jerry she really needed to play her cards right.

"He wasn't around all the time when they were together. He'd be in L.A. or somewhere and she would stay at his apartment and didn't seem to be concerned that he wasn't there, and I don't think she cared.

"They had a reconciliation and got back together for the rest of the winter. Then he broke off with her again in the summer of 1999 because she was putting a lot of pressure on him to get married, and he wasn't ready to get married. It was like a replay of her and Eric before they finally got engaged. So she had two breakups with Jerry in that first year—mainly because she was demanding marriage and he was refusing, and he didn't know if he was *ever* going to be ready. She said she loved him and when they broke up the second time she was really upset. She was devastated."

Meanwhile, Jessica also had to get on with her life. She rented an unfurnished one-bedroom apartment in an old elevator building near Jerry's, bought some inexpensive antique furnishings, and took a public relations job with the designer Tommy Hilfiger after briefly working for Tod's, the sporty shoe company, and Marvel, the comic book publisher.

Jessica's modus operandi was to befriend media people, and she had pals at the *New York Post*, *People*, CNN, and other news outlets. "That's the way she functions," the friend said. "She's very aware of press, getting her name in the press, always wants her face out there, her name out there—and in a positive light. She always tried to stay on Mitchell Fink's good side," a reference to the then *New York Daily News* gossip columnist. "She saved every clipping, every picture where she got a mention."

As part of her intense campaign to win back Jerry she used her media contacts to get items about herself in the papers in hopes of making him jealous. "She had a party and made sure the press knew about it and made sure they wrote about it—how she was out and about and having a good time," the pal said. "She always wanted Jerry to read that she was having a good time *without* him."

For instance, during her split with Jerry, Jessica went out with a producer at the *Today Show* and someone leaked the story. Mitchell Fink breathlessly reported that Jessica "is picking up the pieces following her recent breakup" with Jerry. "Word is, she's dating again, and the lucky guy this time is *Today Show* assistant producer David Friedman. Sources tell me the two have been inseparable . . . . As for Seinfeld, he's solo these days."

Meanwhile, Nederlander thought there might be another chance for him and Jessica. On a recommendation from James Meiskin, Nederlander sought counsel from an ultra-Orthodox rabbi who called himself the "Love Prophet."

Shmuley Boteach's schedule was almost as busy as Jerry's; the sprightly, personable, bearded rabbi even had a publicist and an assistant to handle his busy schedule: promoting *Dating Secrets of the Ten Commandments,* a copy of which Jay Leno once gave to zany basketball great Dennis Rodman when he and hottie Carmen Electra's odd marriage went down the tubes. Boteach also wrote a best-seller called *Kosher Sex*, which was translated into ten languages, and authored *The Jewish Guide to Adultery*, and bestowed his "blessing" on oral sex. He kibbitzed on TV with Matt Lauer and Charlie Gibson, and on radio with Howard Stern, and gave *Time* magazine dating tips for Monica Lewinsky. Along with all of that he was a close friend of Michael Jackson's.

Boteach met with Nederlander at an odd venue—a lunch for then Vice President Al Gore. "Eric told me about his wife, that he'd had a very difficult time with her, and about Jerry Seinfeld's involvement. I'm a great believer in trying to keep husbands and wives together, to keep the sparkle in relationships," the rabbi said later. "All my books are about the beauty of marriage, not just the sanctity, but its passion and its ecstasy, and to get husbands and wives to recommit themselves in faltering marriages. But in this case I told Eric to move on because from all indications the marriage was over, and I told him as much. I said, 'Look, it really sounds to me like your wife has moved on, and trying to salvage this painful relationship is probably a bad idea.' I felt very strongly about that, and he appreciated my advice and he agreed with me. I looked around the room and there were all these young, available women and I said, 'Look around you. You're an eligible bachelor, a handsome guy, you're a great catch. Move on.' "

With Nederlander and Jerry both out of the picture for now, Jessica's security blanket appeared frayed. Without a rich man to buy her Prada and Gucci, she was said to have kvetched to a friend, "Now I'm going to have to shop at Banana Republic again."

One of Jessica's obsessions was Shoshanna Lonstein, who had turned her relationship with Jerry, when it went south, into a publicity bonanza, a lucrative business, and celebrity. "When Jessica and Jerry were broken up," a confidant stated, "Jessica said, 'If we don't get back together I'm going to be like Shoshanna. I'm going to cash in and do my own thing. If things don't work out I'm going to do what she did. I'm going to turn this into something.' She *always* talked about Shoshanna."

Toward the end of the summer of '99, Jessica ran into Jerry a couple of times at the gym and they started seeing each other again, and during those dates she brought up the commitment issue, determined to win him back permanently. At the height of the scandal, alone and down, Jessica had become a believer in *The Rules*, a best-selling guide to getting a man. Part of the book's instructions was to put what you wanted on the line. Before a dinner in late October, Jessica told a confidant, "We're either going to decide to never speak again, or we're going to get married.'" "She called me," the friend remembers, "and said, 'We're getting married!' She laid it on the line and got what she wanted."

Jerry and Jessica picked out a ring together at Tiffany's on November 8. On November 20, at Balthazar, a trendy restaurant, forty-five-year-old Jerry popped the question to twenty-eight-year-old Jessica as they sat in a secluded booth.

Well, almost secluded.

Jessica had a contact at *People* magazine and dropped a dime about what was going down—and got a cover story in return. A *People* insider said, "We had it all set up. She told us where they were going for dinner. Jerry and Jessica knew the photographer. It was set up so that it would be easy for them."

People ran a photo of Jessica smiling and wearing sunglasses, trying to look her Jackie O best, while Jerry, also in dark glasses, was wearing a baseball cap with his lucky number nine on it.

"Jerry and Jessica are perfect together," declared Jerry's sister, Carolyn Leibling.

Curiously, even after the engagement, Jessica kept her apartment for a time. "It was Jerry's idea," said a friend. "It was a place for her to hang out with her friends and it gave him his own space. He didn't really like having her friends around, and she was very paranoid, when they were living together, about inviting people over."

In the minds of those in Jerry's circle the burning question was: why Jessica Sklar? Of all the women he knew, and of all the women who would have jumped at being Mrs. Jerry Seinfeld, why this one, who was tainted by scandal? None seemed to have an answer, and Jerry offered no explanations.

But James Meiskin considered himself an expert. With seven years of marriage to a Sklar woman under his belt, he was intimately familiar with their upside and downside, and was well versed in Jessica's ways and thought he knew what made her and women like her attractive to an inveterate bachelor like Jerry.

"The Sklar women are Jewish and that's a plus for guys like Jerry and Eric and myself," he observed. "They have strong personalities and they make you think they are bringing something to the table. In certain ways they are, and in other ways they're not. They're well-read, meaning they know what's 'in' and they know what's 'out.' They're very social and they like doing things and making plans. They're also able to show you how to spend money. We all have money, but we don't know what to do with it to have fun. They do. They have lots of energy and know how to do things that men don't. They become the executive assistant, the travel agent, the social director, the benefit consultant."

The wedding was set for Christmas Day 1999. Once again *People* got the inside word via the bride. Jerry had set the date himself—the year had three of his lucky numbers in it in succession, and Christmas Day two years earlier was when he had formally announced that there would be no more new *Seinfeld* episodes. The afternoon of his wedding he had a laugh-filled lunch of Cajun chicken at a luncheonette with the guy he selected to be his best man—George Wallace, who was after all those years an integral part of his life.

Meanwhile, Sklar was getting her hair done by Oscar Blandi, who owned the salon at the Plaza Hotel. He gave her braids called "love knots." Since his bride worked for Tommy Hilfiger now, Jerry wore a Tommy Hilfiger tuxedo, and she wore a Hilfiger sheath dress.

At 8:30 P.M., in a tony triplex party space called Sky Studio, on Lower Broadway, under a traditional *huppa,* Jerry Seinfeld gave up his much-vaunted bachelorhood—a running theme in his life and on his show—to

Jessica Sklar Nederlander, her second marriage in eighteen scandal-filled months, in a ceremony performed by Rabbi Charles Klein of the Merrick Jewish Community Center, in Merrick, on Long Island, where Jerry's sister was a congregant.

In the customary best man's toast, George Wallace cooed to Jessica, "If he gives you half the love he's given me, you'll be very fortunate."

Mario Joyner said, "They were very calm, very happy, very smiley."

Eighty-five-year-old Betty Seinfeld, flown up from Florida by Jerry, said, "Jessica's just right for my son . . . they are made for each other."

After the ceremony, Jerry, smoking a cigar—he never smokes in public—played pool with Wallace and some of his pals in a portion of the suite—a space that Hilfiger used for a fashion launch—that was designed like a pub.

That night the newlyweds disappeared into a suite at the Carlyle Hotel.

Eric Nederlander had heard from Jessica herself that the wedding was imminent, but got official word that she had gotten hitched while he was in Aspen attending a gala, celebrity-filled party at the home of pop-music composer and Democratic fund-raiser Denise Rich, who would soon be a central figure in the political scandal surrounding the last-minute pardons of Bill Clinton's administration.

There was no indication that Jerry and Jessica went on a formal honeymoon, and there was never a papparazi shot, or an authorized photo of the two romantically walking on the beach in St. Bart's, or strolling along the Seine. In fact, five days after the wedding Jerry was without his bride in the town of Cairo, in upstate New York, when he got into a late-morning fender-bender while driving in his $180,000 custom-made silver 1997 Mercedes. The collision involved a blue Ford Escort driven by a local elementary-school teacher, Donald Mosely, and was investigated by the New York State Police, but luckily there were no injuries, and no tickets were issued.

"When I got there, there was somebody sitting in the Benz wearing a baseball cap pulled down low, just staying away from everybody else," tow truck operator Dave Rogers recalled later. "Suddenly he was talking to me about taking the car to Manhattan, and I'm, like, I know that voice. So I turn around and I'm, like, eyeball to eyeball with him and I said,

'Jerry Seinfeld? I can't believe this!' I said, 'Were you driving?' and he said, 'Yeah," and I said, 'Oh, man, that stinks.' And he said, 'Yeah, a little bit.' He said he had to get the car back to New York. I knew he had just gotten married, but his wife wasn't with him. I never saw her, unless she was in the trunk."

Rogers offered Jerry a ride but he declined and left the scene with two friends, one of whom was Chris Misiano, who had a weekend house in the area. Later, Misiano's brother, Vincent, said Jerry was celebrating his wedding with longtime buddies.

In early April, someone tipped off Regis Philbin that Jessica was pregnant, and the big announcement that Jerry was going to be a father was made on the nationally televised *Live with Regis and Kathie Lee.*

"It feels fantastic," Jerry said. "It worked."

Sklar later said she had spent three months in bed after suffering difficult morning sickness. Jessica's competitive remarried sister, Rebecca Sklar Meiskin Shalam, jumped into the spotlight, announcing that she, too, was pregnant, and would have her baby first.

Later in April, the pregnant Jessica and three girlfriends vacationed at a posh spa called Rancho La Puerta, in Tecate, Mexico. But instead of being anonymous about who she was, she was very vocal. "At the pool, she talked loudly enough so that the hearing-impaired in Buenos Aires wouldn't miss a word," according to the *New York Post*'s gossip columnist, Cindy Adams. "Like about a magazine in the line of sight: 'Oh, look, there's the issue with the article about Jerry and me.' Or, in a strong, loud voice talking about their beach club putting them on notice for 'inappropriate behavior' because 'Jerry and I were sharing a beach chair.' Or, in a *forte vibrato*, talking about the baby and how excited Jerry is. Or, in a voice that definitely carried, talking to Him on a cell, then reading a fax out loud, 'From Jerry who says, *Honey I miss you.*' "

The headline over the column read: "MRS. SEINFELD GOES NOT QUITE INCOGNITO IN MEXICO."

Before Jessica went into labor at 7 P.M. on November 6, the expectant couple, along with Jerry's shadow, George Wallace, strolled through Central Park. On Tuesday, November 7, 2000—Election Day—at Weill Cornell Medical Center, with Jerry at her side, holding her hand, she gave birth to a happy and healthy seven-pound-one-ounce, twenty-inch girl whom the joyous couple named Sascha, a name they liked but one that didn't follow the Jewish practice of honoring a deceased family member.

Jerry was clearly overjoyed.

"The Upper West Side is simply baby-factory heaven," he quipped. "You look odd without a wife and a baby . . ."

One of the first to arrive at the hospital after the birth was Wallace, together with Jerry's mother and sister. "When I called him that morning," Wallace said, "[the baby] was crying. But when I got there and held the baby, she didn't cry at all."

A smiling Jerry and an unsmiling Jessica, as captured by a photographer, took Sascha home on November 8.

After the three Seinfelds settled in, Jerry did what he does best, and had been doing for all of his adult life. With new material, he went back on the road and told funny stories and made people laugh for lots of money.

Acknowledgments

The writing of a biography is a collaborative effort—the writer cannot succeed unless he gets the cooperation of scores of people from all facets of his subject's life. I'm fortunate that those who knew Jerry Seinfeld best from childhood, through his college years, and during his illustrious career—from his earliest days in stand-up, through the many seasons of his TV show—agreed to talk to me. Without their insightful observations, critical assessments, and candid anecdotes, this book could not have been written.

I'm especially grateful to certain key figures in Jerry's life who gave me hours of their valuable time, speaking publicly, most of them for the first time. I'm indebted to them for their candor, honesty, goodwill, and generosity of time and emotions. Among those I particularly wish to offer my thanks to are: Caryn Trager, Susan McNabb, Tawny Kitaen, Mike Costanza, Jesse Michnick, Joe Bacino, Larry Watson, Robert Williams, Lucien Hold, Richie Tienken, David and Vic Elkin, David Sayh, Glenn Hirsch, Ruth Goldman, Mike Wichter, John Egan, LuAnn Kondziela, Vincent Misiano, Tom Dreesen, Richard Belzer, Toby Seinfeld Rosenfeld, Harold Seinfeld, Lonnie Seiden Javurek, Joe Franklin, Lawrence Christon, and Fred DeCordova.

I'm grateful to scores of others who are quoted throughout the book—the list goes on and on—Seinfeld family friends, schoolmates of Jerry's, present and former friends, and show business colleagues. I also want to offer my appreciation to those very few who, for personal or professional reasons, agreed to be interviewed, but asked that they not be identified. Everyone tirelessly answered my difficult, probing questions.

A book of this magnitude cannot be written without the help of researchers, and I had two pros assisting in the process. A very special thanks to John Cummings, an award-winning former investigative reporter and author, and Judy Oppenheimer, also a veteran journalist, editor, and author. They know how to get the goods.

This is my second book with my editor, Larry Ashmead, a rarity in publishing, who got it from the start. He has won my everlasting respect. He is a gem. Krista Stroever, Larry's right hand, did a magnificent job overseeing the daily input of words and keeping me sane and on schedule—most of the time—during the entire project. She is tops. A special thanks to Beth Silfin and her fine-tooth comb. As a photo researcher, there is no one better or more insightful and creative than Vincent Virga. At HarperCollins, I'd also like to thank the best in the business—Jane Friedman and Cathy Hemming. Finally, in the world of publishing, Joni Evans, my agent at William Morris, is an icon. Without her creativity, honesty, and integrity this book would have never seen the light of day. I feel blessed. Thank you all.

Notes and Sources

Epigraph, page ix: "Stars can succeed": *Playboy* interview, 10/93.

Prologue

xi Details concerning Jerry's stay in Colorado on September 11, 2001, his involvement in the car rally, his immediate reaction to the terrorist attack, the fact that some participants in the event drove east, and Jerry's decision to wait until he could charter a private jet to return to New York were from author interviews with rally officials, and Charter at Beaver Creek employees. 3/14/02.

xi "That was the end": *Washington Post, 9/26/01.*

xiii "Which is more annoying": *Los Angeles Times, 4/2/01.*

xiii the servant had her own: *New York Daily News, 2/10/02.*

xiii "where somebody had just died": *Washington Post,* op. cit.

xiv "In a way it identifies": Ibid.

xiv "The world that I": Ibid.

Chapter 1. *Family Roots*

1–5 The early Seinfeld family history comes from various public records and author interviews with family members.

2 "He married my mother": Interview, Toby Seinfeld Rosenfeld, 11/13/00.

3 "After he left": Ibid.

3 But he did acknowledge: *US*, 9/92.

3 "My grandfather was not": Ibid.

3 "He used to come over": Rosenfeld, op. cit.

4 "Everyone was kind of": Interview, Harold Seinfeld, 10/26/00.

4 "I tried to get close": Rosenfeld, op. cit.

4 "He can keep it!": Ibid.

4 "He was a very nice": Ibid.

4 "All the kids left": *US*, op. cit.

4 "because of father problems": Seinfeld, op. cit.

5 "They were all very responsible": Ibid.

5 "He worked for the": Rosenfeld, op. cit.

5 "When he got older": Ibid.

5 "For most of Harold's": Ibid.

Chapter 2. *Tojo, Television, and Holy Water*

6–9 The facts, details, and color of Kalmen Seinfeld's early adult years are from his military records requested by the author under the Freedom of Information Act and supplied by the Department of Veterans Affairs, Department of the Army, and the National Personnel Records Center.

9 "Kal was basically very": Interview, Mike Wichter, 4/26/01.

Chapter 3. *And Betty Makes Two*

10–11 The details of Kal and Betty Seinfeld's meeting and their mutual interests, and Betty Seinfeld's background, are based on author interviews with Seinfeld family confidants and relatives.

11 The background on Syrian Jews was obtained from various online and print articles-including *The Jews of Syria*, by Michael Bard, Jewish Virtual Library, and the *New York Times*, May 6, 1992, "Delicate Path to U.S. for Jews From Syria."

11 "was ostracized from the family": *US*,. 9/92.

11 "There was no one there": Interview, Ruth Goldman, 6/20/01.

12 "developed a lot of spine": *US*. op. cit.

12 "He was very supportive": Ibid.

Chapter 4. *Near the Mall*

14 Jerry was three weeks shy: Nassau County, N.Y. real estate records.

14–15 History and color of Massapequa from *The Massapequas: From The Time of the Native Americans to the Year 2000*, by Lorraine Newman-Brooks, a local historian and journalist who wrote the booklet for placement in a time capsule, and author interviews with Brooks on 10/18/00 and 5/8/01. Additional history of the town came from author interview on 5/8/01 with Lillian Bryson, of the Massapequa Historical Society, and from the article "Past Meets Present in Massapequa," *Newsday*, 2/2/97.

15 "Most of us were struggling": Interview, Frances Katz, 5/15/01.

16 "I used to kid around": Interview, Norman Katz, 5/15/01.

16 "Kal was an outgoing": Katz, op. cit.

16 "He thought he was": Interview, Rabbi Leon Spielman, by researcher Judy Oppenheimer, 2/14/01.

17 "Betty was pretty much": Frances Katz, op. cit.

18 "Sometimes even Rabbi": Norman Katz, op. cit.

18 "He was out on the": Ibid.

19 "There was a sign": Interview, John Egan, 11/16/00.

19 "Kal started reading": Katz, op. cit.

19 "I heard there was": Interview, Kathy Greaney, 5/15/01.

20 "My children and Jerry": Frances Katz, op. cit.

20 "He was what I call": Spielman, op. cit.

20 "I've always liked": *Playboy*, 10/93.

21 "Very driven . . . If he wanted": *GQ*, 5/92.

Chapter 5. *Moving On Up*

23 The Seinfelds put their: Interview, Joel Halle, 11/29/00.

23 "It was Kal and I guess": Interview, Florence Jacks, 11/21/00.

24 "There was a simplicity": Interview, David Elkin, 1/5/01.

24 "*Everything* was a schlocky": Interview, source, 2/16/01.

24 One of the few interesting: Interview, Victor Elkin, (David Elkin's father), 11/21/00.
24 "That huge sign over the": Jacks, op. cit.
25 "I didn't care": *Rolling Stone*, 9/22/94.
25 "Betty wasn't a typical": Interview, Ruth Goldman, 6/20/01.
25 Jerry boasted in later: Interview, Caryn Trager, 2/12/01.
25 "There was a very": *US*, 9/92.

Chapter 6. *Tough Times*

26 "Basically, I was Jerry's": Interview, Lawrence McCue, by researcher Judy Oppenheimer, 1/5/01.
27 "I felt the milk": *Larry King Live*, 3/27/91.
27 "He knew he wanted": Interview, Scott Brod, by researcher Judy Oppenheimer, 1/24/01.
27 "I watched comedians": Betsy Borns, *Comic Lives* (New York: Fireside Book/Simon & Schuster, 1987), p. 101.
27 "I get tearful thinking": McCue, op. cit.
27 "When I first moved here": Interview, Angelo Roncallo, 11/10/00.
28 "Jerry was a tiny boy": *New York Daily News*, 11/23/97.
28 "You never knew the boy": Interview, Marlene Schuss, by researcher Judy Oppenheimer, 1/23/01.
28 "We're sitting in this": Ibid.
29 In his book *SeinLanguage*: Jerry Seinfeld, *SeinLanguage* (New York: Bantam Books, 1993).
29 "I really was into Flipper": *US*, 9/92.
29 "I led the swinging Stingray life": *Newsday*, 5/31/98.
29 "I had so many different:" Ibid.
29 "We used to run around": *New York Daily News*, op. cit.
29 Jerry liked to play: Interview, Caryn Trager, 3/8/01.
30 "I'd get on my bike": *New York Daily News*, op. cit.
30 "Kal was always trying": Interview, Dominic Totino, 1/19/01.
30 "They couldn't afford the": Interview, Victor Elkin, 11/21/00.
30 "Betty became the": Schuss, op. cit.
30 "Betty had a knack of": Elkin, op. cit.
31 "We were furious about": Interview, Bill Elio, 10/24/00.
31 "There was talk": Elkin, op. cit.
31 "I used to walk my": Interview, Harriet Harris, 4/25/01.
32 Background on the Gambino family from numerous news stories, and Nassau County, NY, real estate records.
32 "When we were building": Elkin, op. cit.
33 "beautiful finished basement": Interview, Ruth Hinden, 4/26/01.
33 "I used to say to Kal": Interview, Ruth Goldman, 6/20/01.
33 "He'd pinch his cheek": Interview, Joel Halle, 11/29/00.
34 "I'd hear Kal on": Interview, John Egan, 11/20/00.

Chapter 7. *Becoming a Man*

35 "belt buckles and knee caps": Interview, Lonnie Seiden Javurek, 11/28/00.
35 "Jerry was somewhere between": Interview, David Elkin, 1/14/01.
36 "I wasn't the happiest kid": *US*, 9/92.

36 "He was S-E-I-N": Javurek, op. cit.
37 "My mother owned": Interview, Javurek, 11/30/00.
38 "Another friend of mine": Ibid.
39 "We were in his room": Ibid.
39 "Some people's hobby": Interview, Scott Brod, by researcher Judy Oppenheimer, 1/24/01.
39 "I knew Jerry from": Interview, Cantor Max Klein, by researcher Judy Oppenheimer, 1/26/01.
40 "The whole town was": Interview, Marlene Schuss, by researcher Judy Oppenheimer, 1/23/01.
41 "There are a lot of people": Interview, Mrs. Leon Spielman, by researcher Judy Oppenheimer, 2/14/01.
41 "There should be something": Brod, op. cit.

Chapter 8. *High School Loner*

42 he pored obsessively: Interview, David Elkin, 1/5/01.
43 "There was a lot of": Interview, Spence Halperin, by researcher Judy Oppenheimer, 1/23/01.
43 "a Jewish Afro": Interview, Arnold Herman, by researcher Judy Oppenheimer, 1/8/01.
43 "Jerry was a listener": Ibid.
43 "Most of the kids": Interview, Don Nobile, 2/9/01.
44 "He bellied up to": Interview, John Egan, 11/20/00.
44 "Kal treated me like a": Interview, Jim Roncallo, 11/13/00.
44 "My son was a": Interview, Angelo Roncallo, 11/10/00.
45 "Jerry and I talked about": Interview, David Elkin, 1/14/01.
46 "I don't think there": *New York Daily News*, 11/23/97.
46 "I had it repainted blue": *Newsday*, 5/31/98.
47 "When I first listened": *US*, 9/92.
47 "There would be times": *New York Daily News*, 11/24/97.
47 "We did stuff about the": Ibid.
47 "Jerry would be at the house": Interview, Vincent Misiano, 4/5/01.
47 "Jerry was somebody who was": Ibid.
47 "I think what bonded": Interview, Tula Misiano, 4/5/01.
48 "I always wanted to talk": *Comic Lives*, p. 100.
48 "He had an intellectual": Interview, David Elkin, 1/5/01.
49 "Jerry was a gentle": Interview, Joseph McPartlin, by researcher Judy Oppenheimer, 1/23/01.
49 "I'm a comedian too": Interview, Al Bevilacqua, by researcher Judy Oppenheimer, 1/24/01.

Chapter 9. *Romeo and Juliet*

51 "He was a good student": Interview, Norma Rike, 12/5/00.
51 "My son saw that Jerry": Interview, Ruth Goldman, 6/20/01.
52 "In the crowd scene": Interview, Ann Pastorini McPartlin, by researcher Judy Oppenheimer, 1/23/01.
52 "The drama club": Interview, Spence Halperin, by researcher Judy Oppenheimer, 1/23/01.
53 "a lot of guys": McPartlin, op. cit.
53 Club member Grant King: Interview, Grant King, 3/29/01.
53 "a magnet for people": Ibid.
53 "Why Jerry left": Ibid.
53 "Jerry and Shepherd became": Goldman, op. cit.
54 "Let's put it this way": Ibid.

54 "There was this unspoken": Interview, Tracey Revenson, by researcher Judy Oppenheimer, 2/6/01.

54 "wasn't particularly happy": Interview, Marga Gomez, 3/22/01.

55 "My niece is an actress": Goldman, op cit.

Chapter 10. *To Israel, with Love*

56 "He whined and grumbled": Interview, source.

57 "We were all on the grass": Interview, Victor Elkin, 11/21/01.

57 When, in 1998: "Seinfeld's Kibbutz Days," *Yediot Aharonot*, 4/20/98.

57 "nineteen or twenty": *Playboy*, 10/93.

58 the self-described "fucked up": *Comic Lives*, p. 111.

58 "years of being fucked up": Ibid, p. 111.

58 "He was my first boyfriend": *E Celebrity Profile: Jerry Seinfeld.*

58 "We both confided": *New York Daily News*, 11/24/97.

58 "On your right": Ibid.

59 "I remember him sort of coming": Interview, David Elkin, 1/14/01.

59 "Jerry brought me back": Interview, Lonnie Seiden Javurek, 11/28/00.

59 "My memory holds": Interview, Caryn Trager, 2/2/01.

59 "We married our schedules" *New York Daily News*, op. cit.

59 "After that summer": Interview, Norma Rike, 12/5/00.

60 "Jerry was living with": Interview, Ruth Goldman, 6/20/01.

61 "When the Baldwin": Interview, Al Bevilacqua, by researcher Judy Oppenheimer, 1/24/01.

61 "The hall of fame is": Interview, Linda Roberts, 10/19/00.

62 "It was a great weekend": Interview, Michelle Cohen, 12/4/00.

62 "It was a last": Interview, Lonnie Seiden Javurek, 11/30/00.

62 where the Seinfelds, in partnership: Delaware County, N.Y. real estate records.

62 The Seinfelds and Blooms: Ibid.

62 "Being Cliff and I": Javurek, op. cit.

63 "was very subtle": Cohen, op. cit.

63 "We all kissed and hugged": Javurek, op. cit.

Chapter 11. *Sex, Drugs, and Suzukis*

64 "We came from different": Interview, Roseanne Mecca. 2/6/01.

65 "But I was hung up": Ibid.

65 "He said he had been": Ibid.

66 "kind of a hippie": Interview, Caryn Trager, 2/2/01.

66 "without the nose": Ibid.

66 "Doesn't she look like:" Interview, Garry Carbone, 2/8/01.

66 "Caryn was so much": Interview, LuAnn Kondziela, 4/12/01.

66 "I always thought": Interview, Michele DiCarlo Cerrone, 7/5/01.

67 "It was an unusually hot": Trager, op. cit.

67 "Jerry's sensitivity was": Trager, 2/5/01.

68 "LuAnn saw him sitting": Ibid.

68 "I never saw Jerry as": Cerrone, op. cit.

68 "It evolved very quickly": Trager, op. cit.

69 "When we began our": Trager, Ibid.

69 "I didn't want to talk to her": Trager, Ibid.
69 "He kept telling me": Trager, op. cit.
70 "Kal couldn't believe": Trager, Ibid.
70 "For most of the time": Trager, op. cit.
70 "He loved the freedom": Ibid.
70 "They had to be": Trager, op. cit.
70 "Jerry went white": Trager, Ibid.

Chapter 12. *Campus Odd Couple*

72 "jaw dropped and her eyes": Interview, Larry Watson, 12/1/00.
73 three black students had passed: Interview, Norma Rike, 12/5/00.
73 "what I could not": Watson, op. cit.
73 "the only soul sister": Ibid.
73 "really close relationship": Ibid.
74 "It was the end": Ibid.
74 "This was a big thing": Ibid.
75 "One of the things that": Ibid.
75 "measure out his shampoo": Interview, Caryn Trager, 2/5/01.
75 "In the length of time": Interview, Jesse Michnick, 1/9/01.
75 "Jerry brought absolutely": Watson, 12/3/00.
76 "Jerry didn't have": Trager, op. cit.
76 "That was my strategy": Watson, 12/1/00.
76 "I found Jerry": Ibid.
77 "Oh, Jerry was a big": Interview, Michele DiCarlo Cerrone, 7/5/01.
77 "when Jerry went off": Watson, 12/3/00.
77 "Larry was the quote": Interview, Carmel Reese Harris, 12/4/00.
79 "I'm certain the famous masturbation": Watson, op. cit.
79 "a brotherly, friendly, wholesome": Watson, 7/18/01.
79 "In that little circle": Trager, 2/7/01.
79 "was a really good": Watson, 12/3/00.

Chapter 13. *I Just Want to Be Funny*

82 "Jerry and I read the review": Interview, Garry Carbone, 2/16/01.
82 "There was never anything": Interview, Barton Ward, 2/12/01.
83 "He kind of was in that clique": Interview, Larry Watson, 12/3/00.
83 "The play was Joe's": Ward, op. cit.
83 "Everybody had been talking": Carbone, 2/8/01.
84 "Theater was a real community": Ward, op. cit.
84 "It was weird because": Carbone, 2/16/01.
84 spit up his spaghetti: Ibid.
84 "Jerry laughed about it": Ibid.
85 "I was in the middle": Ibid.
85 "There was a big": Ibid.
86 "very safe and very cushioned": From undated article in Oswego Alumni magazine, by Denise Owen Harrigan.

86 The marker event: Watson, 12/1/00.
86 "He was a guy": Harrigan article, op. cit.
86 "scared to death": Watson, op. cit.
86 "Why did you bring this": Ibid.
86 "I was lost": Harrigan article, op. cit.

Chapter 14. *Queens College Days*

87 "He had talked incessantly": Interview, Caryn Trager, 2/2/01.
87 "Jerry was impatient": From undated article in Oswego Alumni magazine, by Denise Owen Harrigan.
88 "He drove night and day": Trager, 2/16/01.
88 "It sounds like an exaggeration": Ibid.
88 "The downside of his": Interview, source.
89 "He thought they were": Interview, source.
89 "Jerry and everybody else": 2/16/01.
89 "She definitely didn't have": Interview, Ruth Goldman, 6/20/01.
89 "You should forget": Interview, John Egan, 11/16/00.
90 "the guy to see on": Mike Costanza with Greg Lawrence, *The Real Seinfeld* (New York: Wordwise Books, 1998), p. 12.
90 "buying special discount": Ibid.
90 "a full-fledged": Ibid.
90 "My life was like": Interview, Mike Costanza, 12/18/00.
90 "What does a bouncer": *The Real Seinfeld*, p. 16.
91 "Jesse had set me up": Costanza, op. cit.
91 "We convinced Jerry": Costanza, Ibid.
92 "Jerry was stiff": Interview, Joe Bacino, 12/11/00.
92 "The other characters are": Interview, Sue Giosa, by researcher Judy Oppenheimer, 12/20/00.
92 "He said laughs were the addiction": Ibid.
92 "We used to scam people": Costanza, op. cit.
93 "Hi, Mr. Cohen": Ibid.
93 "She was an absolutely gorgeous": Bacino, op. cit.
94 "Jerry actually competed": Ibid.

Chapter 15. *Girl Trouble*

95 "We didn't fight about": Interview, Caryn Trager, 2/5/01.
96 "I'm just preparing you": Ibid.
96 "They though he was": Ibid.
96 "Prestige mattered": Ibid.
96 "On the ride home": Ibid.
97 "I rang the bell": Ibid.
97 "With the other woman": Ibid.
98 "Back then I had this": Ibid.
98 "very strict . . . overbearing": *Comic Lives,* op. cit.
98 working various jobs: Ibid, p. 120.
99 She'd become so depressed: Ibid.

99 "You're one of the": Ibid, p. 121.
99 "My heart dropped": Trager, 2/7/01.
100 "I felt like he didn't": Ibid, 2/16/01.
100 "just to hear": Trager, op. cit.
100 "I don't remember anything": Giosa, op. cit.
100 "Your doctors advised": *The Real Seinfeld*, p. 31.
100 "The question is": Ibid.
100 "Once in a while": Costanza, 12/18/00.
101 "Even with me": Interview, Jesse Michnick, 1/9/01.

Chapter 16. *Catskills Forays*

102 "Jerry was a bright": Interview, Dr. Stuart Liebman, by researcher Judy Oppenheimer, 12/18/00.
102 "Knowing what he's become": Ibid.
103 "He was boasting": Ibid.
103 "Ed was very active": Interview, Dr. Raymond Gasper, by researcher Judy Oppenheimer, 12/18/00.
104 "It was fun": Ibid.
104 "There was no class": Interview, Sara Dillon, 9/25/01.
104 "When Jerry first started": Ibid.
104 When Greenberg died: *New York Times*, 11/27/95.
105 "I remember his": Gasper, op. cit.
105 "Jerry was funny": Costanza, op. cit.
105 "Jerry loved to taunt": *The Real Seinfeld*, p. 15.
105 "Pizza was a major": Ibid.
105 On occasion, Jerry: Ibid, p. 22.
106 "a guy in": Interview, Jesse Michnick, 1/9/01.
106 In the same media: Ibid.
106 "For the first time": Ibid.
107 "Too many buttons": Ibid.
107 "We've got a": Ibid.
107 In order for Jerry to get: *The Real Seinfeld*, p. 52.
108 "Jerry had great": Michnick, 1/7/01.
109 "The point of the film": Interview, Vincent Misiano, 4/23/01.
109 "He didn't work": Ibid.

Chapter 17. *I Want to Do Stand-up*

110 "Glenn was considered": Michnick. 1/9/01.
111 "Jerry pointed": Interview, Joe Bacino, 12/13/00.
111 "He felt very competitive": Costanza, 4/26/01.
111 "Jerry didn't stand out": Sue Giosa, op. cit.
112 "Listen, I gotta tell you": Costanza, 12/18/00.
113 "I remember doing our laundry": From undated article in Oswego Alumni magazine, by Denise Owen Harrigan.
113 "We laughed and said": Bacino, 12/11/00.
113 "It's the first time": Garry Carbone, 2/16/01.

114 "So it might kill him": *Newsday,* 5/31/94.
114 "It was a little more": *Entertainment Weekly*, 6/24/94.
114 "Jerry looked at the pass": *The Real Seinfeld*, p. 13.
114 "Everything we did": Costanza, 12/14/00.
114 "He really is a": *New York Post*, 6/3/94.
115 "When my parents were": Ibid.

Chapter 18. *Peanuts on the Bar*
117 "Mike, the cars": Costanza, 12/18/00.
117 "It's different getting on": Glenn Hirsch, 10/28/01.
118 "showed he had no": Costanza, 1/7/01.
118 "His nerves and his": *The Real Seinfeld*, p. 33.
118 "to get on and get off": Ibid.
119 "several long, silent": Ibid.
119 "I got up and": *St. Petersburg Times*, 5/19/89.
119 "That was Jerry": *The Real Seinfeld*, p. 34.
119 "After I was done": *People*, 9/5/88.
120 "was too anguished": *The Real Seinfeld*, op. cit.
120 "a seedy Times Square": Caryn Trager, 2/7/01.
120 "he was very nervous": Ibid.
120 "old and dyin' ": Costanza, op. cit.
120 "Sometimes it was me": Ibid.
120 "Comedians were nothing": Glenn Hirsch, op. cit.
121 "Franklin's pallor back": Jesse Michnick, 2/13/01.
121 "Jerry was very": Interview, Joe Franklin, 11/1/01.
121 "not so much nightclubs": *Comic Lives*, p. 171.
121 "They'd have a sixty": Ibid.
122 "I did shows": Ibid.

Chapter 19. *The Dianetics Kid*
123–25 Background on Scientology from Church of Scientology website.
125 "As far as being": Mike Costanza, 1/7/01.
125 "One day, out of the blue": Ibid.
126 "It was part of his": Costanza, 12/18/00.
126 "He always wanted": Costanza, op. cit.
126 "what meditating?": *The Real Costanza*, p. 23.
126 "Pasta fazool": Ibid.
127 "He got into TM": Caryn Trager, 2/7/01.
127 "Jerry told me his goal": Trager, 2/16/01.
128 "while the scientologists": Interview, Lucien Hold, by researcher Judy Oppenheimer, 12/28/00.
128 "He started out": Ibid.
128 "It all had to do": Costanza, 1/7/01.
129 "I remember thinking": LuAnn Kondziela, 4/12/01.
129 "When he told me": Interview, Susan McNabb, 12/14/00.
129 "I remember Chris": Ibid.

131 "I was never in the organization": *Washington Post*, 4/22/92.
131 "really poor journalism": Ibid.
131 "Look who's an": Ibid.
131 "I don't let that": Ibid.
131 "I'm very interested": Ibid.
131 "I've always had the skill": *Playboy*, 10/93.
132 "The most conspicuous example": *Washington Post*, 4/16/98.
132 "somewhere an anti-Semite": Ibid.
132 "took pains": Ibid.

Chapter 20. *Doctor Comedy*
133 "Those bastards give us nothing": Jesse Michnick. 1/9/01.
133 "I'd get home at": *US*, 9/92.
133 "He was a salesman": *Playboy,* 10/93.
134 "Kal used to come": Dominic Totino, 1/19/01.
134 "His nerves made": *The Real Seinfeld*, p. 35.
134 "The thing about": Vincent Misiano, 4/5/01.
134 "He really complained": Michnick, op. cit.
135 "Jerry came and saw": Mike Costanza, 12/18/00.
135 "Here I was running": *People*, 9/5/88, *GQ*, 5/92.
135 "He kept himself in the throes": Costanza, op. cit.
136 "Larry, could you please": Larry Watson, 12/1/00.
136 "The Comic Strip was": Lucien Hold, 12/20/00.
136 "There were some": Ibid.
137 "In those days": Ibid.
137 "The Catch was like": Interview, Richard Belzer, by researcher Judy Oppenheimer, 3/13/01.
138 "A little glitter": *The Real Seinfeld,* p. 38.
138 "Catch just always made": Interview, Rick Newman, by researcher Judy Oppenheimer, 3/13/01.
138 "He'd listen to it": Ibid.
138 "Jerry and Belzer": Ibid.
138 "We always treated": Lucien Hold, 12/28/00.
139 "We're good, paying": Ibid, 1/11/01.
140 "The power was very": Hold, op. cit.
140 "That was a big": Costanza, 1/7/01.
140 "So, the girl": Hold, op. cit.
140 "He knows funny": Ibid.
141 "If I hadn't": Hold, op. cit.

Chapter 21. *Major Changes*
142 "Lights. Camera. Action": Caryn Trager, 2/7/01.
142 "We had many": Mike Costanza, 12/14/00.
143 "I always thought": Trager, 1/12/01.
143 "Waiting on line": Ibid, 2/16/01.
144 "She and I": Costanza, 12/18/00.
144 "In the beginning": Trager, 2/5/01.

144 "There was a big": Costanza, 1/7/01.
146 "I want to be": *The Real Seinfeld*, p. 46.
146 "I was hesitant": Trager, op. cit.
146 "Slowly, he stopped": Ibid.
147 "He didn't feel": Ibid.
147 "The Seinfelds": Ibid.
147 "I've been happy": Trager, 3/8/01.

Chapter 22. *Strange Bedfellows*
148 "Jerry was really": *Fort Lauderdale Sun-Sentinal*, 5/10/98.
149 "Jerry and I had": Interview, Lucy Webb, by researcher Judy Oppenheimer, 2/5/01.
149 "I just want to": Susan McNabb, 12/27/00.
149 "He was working": *Dallas Morning News*, 12/27/97, and interview, 4/18/01.
150 "Wallace broke his": Larry Watson, 12/3/00.
150 "In advertising I": George Wallace, Richard De La Font Agency website, 1/3/01.
151 "George is the guy who": Glenn Hirsch, 10/28/01.
151 "There is no sex": George Wallace, op. cit.
151 "Jerry and George's": Hirsch, op. cit.
151 "Jerry and George spent": Interview, source.
152 "I know Jerry and": Interview, Nancy Parker, 2/7/01.
152 "Show business and": Interview, David Sayh, 5/2/01.
152 "She was after Jerry": 5/3/01.
153 "Suddenly I wasn't": McNabb, 12/17/00.
154 "There was conjecture": Interview, Robert Williams, 11/7/01.
154 "George was from": Jesse Michnick, 2/13/01.
155 "Everybody wants": *Excel* Magazine, Fall 1992.
155 "We try to keep": *Fort Lauderdale Sun-Sentinel,* 10/18/96.
155 "Seinfeld will open": *Washington Post*, 6/6/98.

Chapter 23. *Faster Than a Speeding Bullet*
157 "It was very unusual": Lucien Hold, 12/28/00.
157 "Everything has just": LuAnn Kondziela, 4/12/01.
157 "I still find time": Ibid.
157 "Sometimes I sit": Ibid.
157 "his energy just from": *The Real Seinfeld*, p. 43.
158 "Early on he": Glenn Hirsch, 10/28/01.
158 "Brenner is going": Kondziela, op. cit.
159 "Jerry made it clear": Hirsch, op. cit.
159 "There's a lot": Kondziela, op. cit.
159 "It was like one": Mike Costanza, 4/26/01.
159 "Ta tell ya": *The Real Seinfeld*, p. 42.
160 "How do you": Costanza, op. cit.
161 "He realized": Hold, op. cit.
161 "Jerry's eyes were": Costanza, op. cit.
162 "All's fair in": *The Real Seinfeld*, p. 41.
162 "Jerry just wasn't": Hirsch, op. cit.

163 "Jerry was always": Jesse Michnick, 2/13/01.
164 "I just did": David Sayh, 5/2/01.
164 "I was freaked": Ibid.
165 "It's a million": Ibid.
165 "He wasn't focused": Hold, op. cit.
165 "He talked about": Michnick, 1/9/01.
166 "Jerry took very": Costanza, op. cit.
166 "I don't think": Sayh, op. cit.
166 "were precipitated by": Ibid.
166 "going for it": Kondziela, op. cit.
167 "Jerry's idea was": Costanza, 1/7/01.
167 "For Jerry's going": Lucien Hold, 1/11/01.
168 "No trepidation": Costanza, op. cit.

Chapter 24. *Welcome to L.A.*

170 "pretty much the": Argus Hamilton, *E! True Hollywood Story: The Comedy Store.*
170 "I was given": Pauly Shore, Ibid.
171-76 Details about the L.A. comedy strike from interviews with Tom Dreesen, 4/21/01, 4/24/01.

Chapter 25. *The* Benson *Fiasco*

179 "He was hurting": *The Real Seinfeld*, p. 77.
180 "Wow. That Jerry": Interview, Fred Stoller, 4/26/01.
181 "His appearance was": Jesse Michnick, 1/9/01.
181 "It happened so fast": *TV Guide*, 5/23/92.
182 "When he came": Glenn Hirsch, 10/31/01.
182 "Paul said, 'I'm": Interview, Richie Tienken, by researcher Judy Oppenheimer, 1/4/01.
182 "I'm going to rededicate": Lucien Hold, 12/28/00.

Chapter 26. *Jerry Does Johnny*

184 "Getting George Shapiro": Mike Costanza, 1/7/01.
184 "Having George Shapiro": Glenn Hirsch, 10/31/01.
185 "We competed against": Bernie Brillstein with David Rensin, *You're No One in Hollywood Unless Someone Wants You Dead, Where Did I Go Right?* (New York: Little, Brown, 1999), p. 33.
185 "Everything is money": Bill Zehme, *Lost in the Funhouse: The Life and Mind of Andy Kaufman* (New York: Delacorte Press, 1999), p. 168.
185 "You're from New York": *You're No One*, op. cit., p. 91.
186 "I got Jerry": Robert Williams, 11/7/01.
187 "No matter where": Tom Dreesen, 4/24/01.
187 "I rehearsed the": Lucien Hold, 12/28/00.
188 "It was brave": *E! True Hollywood Story*.
188 "He went up": Interview, Rabbi Leon Spielman, by researcher Judy Oppenheimer, 2/14/01.
189 "From the Golden": *The Real Seinfeld*, p. 82.
190 "We stopped our mahjong": Interview, Fran Alter, 4/30/01.
190 "I was happy": Richie Tienken, 1/4/01.

191 "The White Rock": Jesse Michnick, 1/9/01.

Chapter 27. *Comedy Club Bonanza*

193 "Basically, there were": Robert Williams, 11/7/01.
194 "Four days is my": *People*, 9/5/88.
194 "That's when I decided": Ibid.
195 "All of a sudden": Glenn Hirsch, 10/31/01.
195 "The comedy clubs": Williams, op. cit.
196 "How did I": *The Real Seinfeld*, p. 83.
196 "You don't want": Ibid, p. 84.
197 "Can I ask": Joe Bacino, 12/19/00.
197 "So I arrive": Larry Watson, 12/1/00.
198 "If you're the": *The Real Seinfeld*, op. cit.

Chapter 28. *Love and Death*

199 "Jerry's life was": Robert Williams, 11/7/01.
199 "Turning thirty just": *People*, op. cit.
200 Biographical background on Jay Leno from Bill Carter, *The Late Shift* (New York: Hyperion, 1994), p. 40.
200 "If spending a": Ibid. p. 83.
201 "It was kind": Susan McNabb, 12/27/00.
201 "Jerry was like": Ibid.
201 "He was always": LuAnn Kondziela, 4/17/01.
201 Jerry was on: Mike Costanza, 12/18/00.
201 She was a pretty: Susan McNabb, 12/13/00.
201 "Basically, Jerry just": Costanza, op. cit.
202 "petrified that he": *The Real Seinfeld*, p. 85.
202 "She predicted the": McNabb, op. cit.
202 "get on with life": *GQ*, 5/92.
202 "He felt maybe": McNabb, 12/27/00.
203 "might be apocryphal": Lucien Hold, 1/11/01.
203 "The fact that": McNabb, 1/21/01.
203 "If you can't": Interview, Arthur Farb, 12/20/00.
204 "Kal was very": Ibid.
204 "You must not": Interview, Mrs. Arthur Farb, 10/31/00.
204 "His sense of": Arthur Farb, op. cit.
204 "They found a tumor": John Egan, 11/16/00.
204 "She called me": Farb, op. cit.
205 "Jerry's devastated": Larry Watson, 12/3/00.
205 "Jerry spoke at": Lucien Hold, 1/11/01.

Chapter 29. *Sex, Money, and Accolades*

207 "Hey, modeling's more fun": McNabb, 12/13/00.
207 "to go for it": Ibid.
207 "It's like I": Ibid.
207 "Oh, these are": Ibid.

208 "I was interested": Ibid.
208 "Jerry was not": Ibid.
208 "We talked": Ibid.
208 "I can't wait": Ibid.
209 "He was commitment": McNabb, 12/17/00.
209 "Unlike her": McNabb, 12/14/00.
210 "I'm kind of": Ibid.
210 "psychiatrists would have": Ibid.
210 "I trusted him": Ibid.
211 "I have lots": Ibid.
211 "Are you insane?": Ibid.
211 "Can you believe": Ibid.
212 "This was back": Robert Williams, 11/7/01.
212 "When I met": McNabb, 12/27/00.
213 "I was like": Ibid.
213 "We got to": Ibid.
213 "I remember at": Ibid.
215 "caught Jerry's nuances": Ernest Tucker, *Chicago Sun-Times*, 12/9/88.
215 "He's not just": *People*, 9/5/88.
215 "suburban-preppie cousin": *Time*, 8/25/87.
215 "I aspire to": *New York Daily News*, 9/3/87.
215 "human cartooning": Ibid.
215 "Just chill out": Ibid.
216 "He wouldn't seem": Ibid.
217 "he has, in": *People*, op. cit.

Chapter 30. *L.A. Story*
218 "In eight years": McNabb, 12/27/00.
218 "I don't think": Ibid.
218 "He was absolutely": Ibid.
219 "I'm very low": McNabb, 1/21/01.
219 "He used to love": McNabb, 12/14/00.
219 "never had a free": 12/17/00.
220 "my boyfriend's bringing": Ibid.
220 "I got pretty much": Ibid.
220 "She was so": 12/14/00.
220 "sort of past": Ibid.
221 "We all went": Interview, Barbara McNabb, 12/20/00.
221 "Susan talked about": Ibid.
222 "well-made, nice looking": LuAnn Kondziela, 4/12/01.
222 "He did a lot": Susan McNabb, 12/27/00.
223 "He took two": Ibid.
223 "The 944 was": Ibid.
224 "I was praying": Ibid, 12/17/00.
225 "While he was": Ibid.

225 "hospital with a": *People*, 8/5/88.
226 "It's all part of his": Mike Costanza, 12/18/00.
226 "I will never": McNabb, 12/14/00.
226 "I had no qualms": Barbara McNabb, op, cit.
227 "I remember thinking": Susan McNabb, op. cit.
227 "I was devastated": Ibid, 12/13/00.
228 "The whole article's": McNabb, 12/14/00.
229 "There was a time": Ibid, 1/21/01.
229 "In a lot of": *Calgary Sun*, 11/13/99 (from *People).*
229 "I was really": McNabb, op. cit.
229 "I remember Lyndi": Ibid.
230 "We were all": Ibid.
230 "I may have": Interview, source.
230 "I've done it": McNabb, 12/27/00.
230 "I just don't want": Ibid, 12/14/00.
230 "I said, 'Look' ": Ibid.
230 "We were talking": Ibid, 12/13/00.
231 "Why is commitment": *Redbook*, February 1991.

Chapter 31. *Birth of* Seinfeld

232 "Gee, you're really": Tom Dreesen, 4/24/01.
232 "So he comes": Ibid.
233 "Oh. Hey. Hi.": Ibid.
233 "What's with all": Ibid.
233 "If Nazi war": *The Onion*, interview with Larry David, by Stephen Thompson, 1998.
233 "We were walking": Glenn Hirsch, 10/31/01.
234 "a wonderful childhood": Larry David bio, www.castle-rock.warnerbros.com.
234 "He though, without": Mike Costanza, 4/26/01.
234 "I remember sitting": Interview, Bill Ervolino, 2/6/01.
234 "I'm really surprised": David Sayh, 5/4/01.
234 "I didn't like being': *The Onion*, op. cit.
235 "That's it!": Ibid.
235 "It was the only": Ibid.
235 "When you saw": Nancy Parker, 2/7/01.
236 "This is what the show": *New York Magazine*, 2/3/92.
237 "I would be the stand-up": Ibid.
237 "I was having": *The Onion*, op. cit.

Chapter 32. *Press War*

240 "Seinfeld was amusing": Interview, Lawrence Christon, by researcher Judy Oppenheimer, 2/27/01.
241 "on the cusp": *Playboy*, August 1990.
241 "That's an awful": Christon, op. cit.
241 "Here we are": *Playboy*, op. cit.
241 "Nothing shocking here": Christon review, *Los Angeles Times*.
241 "Jerry Seinfeld . . . bills": Ibid.

242 "It's one guy": *Playboy*, op. cit.
243 "a number of people": Interview, Paul Shefrin, 3/12/01.
244 "After it got back": Christon, op. cit.
244 "He called me": Interview, Stephen Randall, by researcher Judy Oppenheimer, 2/26/01.
245 "In ten years": Christon, op. cit.
245 "There are things": *Playboy*, op. cit.

Chapter 33. *Fame Is Free*

247 Bruns—who found: Interview, Phil Bruns, by researcher Caroline Howe, 3/20/01.
247 "My agent told": Ibid.
247 "No one ever": Ibid.
248 "Your job as": Ibid.
248 "Jerry never loses": *US*, 4/4/91.
248 "a wicked snideness": Ibid.
248 "the average person": Ibid.
248 "*Seinfeld* isn't": *Entertainment Weekly*, 6/1/90.
249 "gets good mileage": *People*, 6/4/90.
249 "I thought we'd": *New York Daily News*, 1/23/91.
249 "I love it": *GQ*, 11/91.
249 "It begins funny": *People*, 4/23/90.
250 Description of Jerry's New York apartment: *The Real Seinfeld*, p. 93.
250 "At his peak": Robert Williams, 11/7/01.
250 had to cancel: *Amusement Business*, 7/8/91.
251 Like a sports: Ibid.
251 Jerry's stand-up contract: Ibid.
251 "Jerry was, and": Williams, op. cit.
252 "I looked at the shelves": Jesse Michnick, 1/9/01.
252 "I think that": Ibid.
253 "He could have": Susan McNabb, 12/27/00.
253 "We bought all": LuAnn Kondziela, 4/12/01.
254 "*Seinfeld* is sort": *Entertainment Weekly*, 3/1/91.
254 "decidedly off-center": *New York Times*, 9/29/91.
254 "That ain't a": *Larry King Live*, 3/27/91.
255 "Fuck that shit": Lucien Hold, 12/28/00.
255 Jerry returned to: *US*, 4/4/91.
256 "It made me feel": *GQ* 11/91.
256 "sex symbol": *People*, 12/2/91.
256 "one of America's": *Ladies Home Journal*, 9/92.
256 "He laughed about": McNabb, 1/21/01.
256 "dozens, hundreds": Ibid, 12/14/00.
256 "Oh, cool": Ibid, 12/17/00.
257 "I'd say, 'No": Ibid.
257-58 Jerry seeking help from Jay Leno and Littlefield quote from *The Late Shift*, pp. 116-17.

Chapter 34. *Love in All the Wrong Places*

259 "painfully amusing": *Washington Post*, 4/22/92.

260 "He had kind": *Shoot*, 9/12/97.
261 "It was a gorgeous": Susan McNabb, 1/21/01.
262 "was like being": Mike Costanza, 4/26/01.
262-66 "Our sets, Jerry's": Interview, Tawny Kitaen, 2/15/01.
266 "At one time": Costanza, op. cit.
267 "There came a": Interview, source.
267 "Jerry's cruise director": Jesse Michnick, 2/13/01.
267 George Shapiro said his client: *People*, 5/10/93.
267 "We have fun": *Life*, 10/93.
267-68 Description of European trip and Seinfeld-Joyner quotes, Ibid.
269 "single, thin and neat": *Entertainment Weekly*, 4/9/93.

Chapter 35. L'Affaire Lonstein
272 "Jerry's coming down": Joe Bacino, 12/19/00.
273 "Jerry's a master": Robert Williams, 11/7/01.
275 "It was a prank": *Life*, 10/93.
275 "We went our": *Rolling Stone*, 9/22/94.
275 "I met this incredible": *The Real Seinfeld*, p. 111.
276 "you can take": Mike Costanza, 12/18/00.
276 "Jerry's one of": Jesse Michnick, 1/9/01.
277 "We all get along great": *People*, 12/27/93.
277 "So, you sit": *People*, 3/28/94.
277 The intrepid gossip: *New York Daily News*, 5/28/93.
278 "I didn't realize she was so young": *People*, 3/28/94.
278 "bulletproof—even in": *Playboy*, 10/93.
279 "Would that all": *Publishers Weekly*, 8/23/93.

Chapter 36. *Ups and Downs in TV Land*
281 "the hip sitcom": *Entertainment Weekly*, 4/9/93.
281 "The show has": *Time*, 11/8/93.
281 "Giddy": *Entertainment Weekly*, op. cit.
281 "It's a bit presumptuous": *Rolling Stone*, 9/22/94.
281 "If the script is not written": *Los Angeles Times*, 1/29/95.
282 "probably gave me more pleasure": *Entertainment Weekly*, 4/19/93.
282 "is probably my biggest": *Playboy*, 10/93.
283 "He risks overexposure": *People*. 12/27/93.
283 "a good idea": *Playboy*, op. cit.
283 "Watching Abbott and Costello": *Associated Press*, 1994.
283 "Crazy money": *Rolling Stone*, op. cit.
284 "That's what keeps" Ibid.
284 "You could kill": *Rolling Stone*, op. cit.
284 "the Seinfeld network": *Los Angeles Times*, op. cit.
284 "meddled with": Ibid.
285 "My antennae are": *TV Guide*, 2/4/95.
285 "still expands the": Ibid.
285 "We're on top of every" *Los Angeles Times*, op. cit.

285 "much in the tone of": *TV Guide*, op. cit.
285 "It doesn't matter what the people": *TV Guide*, Ibid.
286 "a Brink's truck": *Los Angeles Times*, op. cit.

Chapter 37. *Designing Woman*

287-88 Lonstein and family on road with Jerry, described in *Rolling Stone*, 9/22/94.
288 "Mario was always": Interview, source.
288 "There's not enough": *Rolling Stone*, op. cit.
288 "is a person, not an age": *People*, 3/28/94.
288 "I'm not an idiot": Ibid.
288 "It was like ugh": Susan McNabb, 12/13/00.
288 "I said, 'Oh": Ibid.
289 "Jerry was just": Lucien Hold, 1/11/01.
289 "the Shoshanna relationship": Robert Williams, 11/7/01.
290 "I figured he": 4/17/01.
290 "I used to tease": Interview, source.
290 "He was thrilled": Ibid.
291 "When he's with": *TV Guide*, 7/8/95.
291 "Jerry dangled the": source, op. cit.
291 "The most wonderful": *People*, op. cit.
291 "the age issue . . . forgotten": Ibid.
291 "it's serious between": Ibid.
291 "I've never seen": Ibid.
291 "I would like": Ibid.
292 Betty Seinfeld was: *People*, 1/24/94.
292 "She didn't think Shoshanna": Interview, Gloria Pell, 4/26/01.
292 SEINFELD TO WED: *National Enquirer*, 3/14/95.
292 One classic tabloid: Ibid, 10/10/95.
293 "He and Shoshanna": Interview, source.
293 "What size cups": *Entertainment Weekly*, 9/13/96.
293 "He's a fucking": Interview, source.
293 "I wanted to leave this show": *Entertainment Weekly*, 1997.
293 "Larry is sensitive": Ibid.
293 "It's a one-parent": *TV Guide*, 6/1/96.
294 "There's nothing in the contract": *MEDIAWEEK*, 1/22/96.
296 "Shoshanna was one": Jesse Michnick, 1/9/01.
296 "She was apparently": *The Real Seinfeld*, p. 122.
297 "It was right after": McNabb, 12/17/00.
297 "America first got": *Time*, 10/5/98.
297 "We did have": *TV Guide*, 10/16/98.
298 "a realistic romantic": *Cosmopolitan*, 6/99.
298 "JERRY SEINFELD, EAT": *New York Post*, 1/31/02.
299 "I don't know": *Dallas Morning News*, 12/27/97.
299 "not because we're": Ibid.
300 "Seinfeld just totally": Interview, Tod Gilman, 4/18/01.

Chapter 38. *The End of Life as We Know It*

301 "a thorny issue": *TV Guide*, 12/97.

302 "It was just": Interview, Larry Thomas, 5/2/01.

303 Description of meeting: *Time*, 1/12/98.

303 "I know the": *Washington Post*, 4/20/98.

304 "I felt the moment": *Time*, op. cit.

304 "They just started": Ibid.

304 "It was pretty": Ibid.

305 "work, real work": Ibid.

305 "A NATION OF": *Daily Mail*, 12/27/97.

305 "Millions of TV": 12/26/97.

305 "the success of *Seinfeld*": *National Review*, 2/9/98.

305 "a barometer of": *MEDIAWEEK*, 1/12/98.

306 "If you want him": Interview, source.

307 One involved her: James Ledbetter, *The Village Voice*, 4/21/98.

307 "I've got a feeling": Source, op. cit.

307 Another account had: *Village Voice*, op. cit.

308 "All hell broke": Source, op. cit.

308 "It's pretty good": Ibid.

308 "Don't bother coming": Ibid.

308 "had fallen in": Ibid.

309 "a very serious": *Village Voice*, op. cit.

309 "I did not": Ibid.

309 "I did not": *Brill's Content*, 5/2000.

309 Jerry had asked: Interview, source.

310 "The revealing personality": Interview, source.

310 "Seingate scandal masks": *Village Voice*, op. cit.

310 "underscored for me": Interview, James Ledbetter, by researcher Judy Oppenheimer, 3/12/01.

310 "I asked him": Mike Costanza, 12/18/00.

311 "there's nothing to": Ibid, 4/26/01.

311 "I always loved this guy": Costanza, op. cit.

311 "Look, when the": Ibid.

312 "We have to stop": Ibid.

312 "No, you're not": Costanza, op. cit.

312 "control freak and": Costanza, op. cit.

312 "It was just": Interview, Gregg Lawrence, 12/11/00.

313 "There was definite": Ibid.

313 "We had high": Costanza, 12/18/00.

313 "It's really amazing": Costanza, 1/7/01.

313 "the characters were at": *Washington Post*, 5/15/98.

314 "The New York": *Variety*, 5/18/98.

Chapter 39. *The Marrying Kind*

317 "I have always gone": from court records.

317 "We were very": Interview, source.

317 "My mother always": From January 2000 conversation between Rebecca Sklar and Debbie Salomon, reporter and columnist for the *Burlington Free Press*, as recounted by Salomon to author.

318 "Jessica definitely had": Interview, James Meiskin, 1/23/00.

318 "like a salon": From profile of Kathy Boudin, by Elizabeth Kolbert, *The New Yorker*, 4/16/01.

318 "They made no bones": Interview, source.

319 "Ellen talked to": Interview, Elizabeth Kolbert, 3/6/02.

320 "ordinary part": From notes of interview between Debbie Salomon and Ellen and Karl Sklar for a 1/2/00 story in *The New York Post*.

320 "when the family": James Meiskin, 1/29/01.

320 "Jason has something": From notes, op. cit.

320 "Aren't you proud": Ibid.

320 "I'm Rebecca's sister": *New York Post,* 1/2/2000.

321 "She was tremendously": From notes, op. cit.

321 "But we had": Ibid.

321 "Jessica didn't talk": Ibid.

321 "She was witty": Interview, David Lines, by researcher Judy Oppenheimer, 2/13/01.

322 "driven by ambition": From notes, op. cit.

322 "You had sixty-five": Interview, Brian Baliatico, 10/5/00.

323 "A lot of": Interview, source.

323 "It wasn't automatic": *New York Post*, op. cit.

323 "Even though they": Meiskin, 1/23/00.

323 "When Eric finally": Ibid.

323 "Her tiny dorm": Ibid.

324 "a little prissy": Interview, Brady Frost, by researcher Judy Oppenheimer, 2/16/01.

324 "Jessica had the": From notes, op. cit.

324 James Meiskin introduced: Meiskin, op. cit.

325 "We hung out": Interview, Rick Rosin, 2/13/01.

326 "Our parents would": Ibid.

326 "When Munson was": Ibid, 3/1/01.

326 "Eric would spend": Ibid.

326 "his father drove": Ibid.

327 "As kids, it": Rosin, op. cit.

327 "We definitely had": Rosin, op. cit.

327 "Bob, the father": Rosin, 2/12/01.

327 "He put all": Rosin, 2/13/01.

328 "Eric never had": Rosin, 3/1/01.

329 "Eric called me": Interview, Drita Rosin, 2/13/01.

329 "She was extremely": Rick Rosin, 2/12/01.

329 "Jessica became like": Drita Rosin, op. cit.

330 "Everything between them": op. cit.

Chapter 40. *Stop, Thief!*

331 "A year before": Drita Rosin, 2/13/01.

332 "It was like": Rick Rosin, 2/13/01.
332 "In fact, after": Interview, source.
332 "We would go": Interview, source.
333 "if he bought": Interview, source.
333 "It was a": Drita Rosin, 3/6/01.
333 "he wound up": James Meiskin, 1/23/00.
334 "her level of": Interview, source.
334 "If Jessica hadn't": Rick Rosin, 2/13/01.
335 "Look at Jessica": Drita Rosin, op. cit.
335 "I don't know": Interview, source.
336 "They knew each": *People*, 11/2/98.
336 Jerry's plan was: *New York Daily News*, 12/27/97.
337 "Jessica just didn't": Drita Rosin, op. cit.
337 "Drita, go talk": Drita Rosin, 3/6/01.
337 "They were having": Ibid.
338 "They were like": Interview, source.
338 "They weren't the kind": Drita Rosin, 3/6/01.
338 "Jessica didn't really": Ibid, 2/13/01.
338 "There were all": Ibid.
339 "They were just": Ibid, 3/6/01.
339 "Even Michael Kors": Meiskin, op. cit.
340 "Jessica was suddenly": Rosin, op. cit.
340 "I said, 'Wow": Rosin, 2/13/01.
341 "Yes, she did": Ibid.
341 "I said, 'Let's": Ibid.
342 "Rebecca told me": Meiskin, op. cit.
343 "Eric told me": Rosin, op. cit.
343 "all of her stuff": Meiskin, op. cit.
343 "arguably the most": *Variety*, 8/17/98.
344 "When was the": *Entertainment Weekly*, 8/21/98.
345 "I can't believe": Drita Rosin, 2/12/01.
345 "Before Eric met": Meiskin, op. cit.
345 48th Annual Pebble: *US*, 11/98.
345 Emmy Awards: *New York Post*, 10/21/98.
345 "were attracted to": *New York Daily News*, 10/21/98.
346 "It's almost like": *New York Daily News*, 11/11/98.

Chapter 41. *And Baby Makes Three*
347 "SEINFELD STEALS": *National Enquirer*, 9/15/98.
347 "MASTER OF WHOSE": *People*, 11/2/98.
348 "Who says *Seinfeld*: *TV Guide*, 11/24/98.
348 "*Brides Who Bed-Hop*": *Cosmopolitan,* 4/99.
348 "stroking" Jessica's: *New York Daily News*, 10/6/98.
348 "the dallying duo": *New York Post*, 10/26/98.
348 "I love scandal": *New York Daily News,* 9/20/98.

348 "You know, *I'm*": *New York Post*, 10/21/98.
349 "made overtures to": Ibid.
349 "the antithesis of": Ibid.
349 "You're a poor human": *New York Post*, Ibid.
349 "was already separating": *New York Daily News*, 10/21/98.
350 "A Night Out": *New York Times*, 3/17/02.
351 "yet another example": *New York Times*, 3/24/02.
351 *Seinfeld* reunion show: *Washington Post*, 1/7/02.
351 Setting the record: Interview, Linda Lee, 4/5/02.
351 "Hey, Jer": *New York Post*, 3/25/02.
352 "I was manipulated": *New York Post*, 11/9/98.
352 There was speculation: *New York Daily News*, 10/22/98.
353 Joel revealed: *New York Daily News*, 2/27/02.
353 "After she left": Interview, source.
354 "That's the way": Ibid.
355 "is picking up": New York daily News, 10/18/99.
355 "Eric told me": Interview, Rabbi Shmuley Boteach, 2/21/01.
356 "Now I'm going": Interview, source.
356 "When Jessica and Jerry": Interview, source.
356 "We're either going": Interview, source.
356 "We had it all": Interview, source.
356 "Jerry and Jessica are perfect": People, 11/22/99.
357 "It was Jerry's": Interview, source.
357 "The Sklar women": James Meiskin, 1/23/00.
358 "If he gives you": *People*, 1/1/00.
358 "They were very": Ibid.
358 "Jessica's just right": *New York Daily News*, 12/24/99.
359 "When I got there": Interview, Dave Rogers, 1/18/01.
359 "At the pool": *New York Post*, 4/20/00.
360 "The Upper West Side": *New York Post*, 11/8/00.
360 "When I called": *People*, 11/20/00.

Index